STRATEGIES FOR REDUCING DRUG AND CHEMICAL RESIDUES IN FOOD ANIMALS

STRATEGIES FOR REDUCING DRUG AND CHEMICAL RESIDUES IN FOOD ANIMALS

International Approaches to Residue Avoidance, Management, and Testing

Edited by

RONALD E. BAYNES
North Carolina State University
Raleigh, NC, USA

JIM E. RIVIERE
Kansas State University
Manhattan, KS, USA

WILEY

Published by John Wiley & Sons, Inc., Hoboken, New Jersey
Published simultaneously in Canada

Library of Congress Cataloging-in-Publication Data:

Strategies for reducing drug and chemical residues in food animals : international approaches to residue avoidance, management, and testing / edited by Ronald E. Baynes, North Carolina State University, Raleigh, North Carolina, USA, Jim E. Riviere, Kansas State University, Manhattan, Kansas, USA.
pages cm
Includes bibliographical references and index.
ISBN 978-0-470-24752-5 (cloth)
1. Food animals–Feeding and feeds–Contamination. 2. Food animals–Nutrition. 3. Veterinary drug residues. 4. Animal nutrition. I. Baynes, Ronald, editor of compilation. II. Riviere, J. Edmond (Jim Edmond) editor of compilation.
SF95.S88 2014
636.08′5–dc23

2014011452

Printed in the United States of America

10 9 8 7 6 5 4 3 2 1

CONTENTS

PREFACE

The focus of this book is to present strategies that are utilized to reduce drug and chemical residues in food from livestock production, and also to present some of the newer technologies and theories that will shape how drug residues will be managed in the future. One of the novel features of this book is that it will tie in the realities of veterinary clinical practice and the use of these drugs in food animals with regulatory standards and mitigation practices.

The first half of this book focuses on strategies that are part of public policy in national and international agencies and how these agencies assess the toxicology of veterinary drugs and contaminants. This involves some discussion of how to compute safe levels (tolerances and maximum residue levels, MRLs) of these drugs and chemicals in meat and milk so that human health is not adversely affected. This section highlights the efforts at harmonization as well as differences across such jurisdictions as United States, European Union, Canada, Australia, South America, China, and Asia, where this issue has a significant impact on the trade of livestock products. This ***section also*** focuses on novel computational strategies that incorporate more statistical and mathematical approaches that are now possible with the advent of modern computers to derive safe withdrawal times. These chapters provide the reader with a general introduction to basic pharmacokinetic principles, especially those principles that are applicable in subsequent chapters in this section as it pertains to estimating a safe withdrawal time for veterinary drugs and contaminants. PK parameters and their derivation are defined in the Chapter 1. These chapters also focus on how the WDT is established in US vs. EU.

The second half of this book focuses on the use of major drug classes in livestock food animal production systems and the drugs most likely targeted for regulatory policy, pharmacokinetic modeling, and chemical residue monitoring. Each chapter in this section will be focused on subtherapeutic (feed) and therapeutic use of drugs in major livestock species such as dairy and beef cattle, swine, poultry, fish aquaculture, and small ruminant production systems. Each production system requires species-specific management practices of drug residues. Quality assurance programs are discussed for each major species with regards to species-specific management practices for controlling drug residues as well as subtherapeutic versus therapeutic drug use in livestock, and how these practices are related to the emergence of antimicrobial resistance.

CONTRIBUTORS

Glen Almond, DVM, PhD
Department of Population Health and Pathobiology
College of Veterinary Medicine
North Carolina State University
Raleigh, NC, USA

Kevin Anderson, DVM, PhD, Dipl. ABVP
Department of Population Health and Pathobiology
College of Veterinary Medicine
North Carolina State University
Raleigh, NC, USA

Reha Azizoglu, PhD
Department of Population Health and Pathobiology
College of Veterinary Medicine
North Carolina State University
Raleigh, NC, USA

Ronald E. Baynes, DVM, PhD
Department of Population Health and Pathobiology
College of Veterinary Medicine
North Carolina State University
Raleigh, NC, USA

Jennifer Buur, DVM, PhD, Dipl. ACVCP
College of Veterinary Medicine
Western University of Health Sciences
Pomona, CA, USA

Isaura Duarte, DVM
European Medicines Agency
London, UK

Virginia Fajt, DVM, PhD, Dipl. ACVCP
Department of Veterinary Physiology and Pharmacology
College of Veterinary Medicine and Biomedical Sciences
Texas A&M University
College Station, TX, USA

Kornelia Grein, PhD
European Medicines Agency
London, UK

Dee Griffin, DVM, PhD
Great Plains Veterinary Education Center
University of Nebraska
Lincoln, NE, USA

Hui Li, PhD
Division of Residue Chemistry, Office of Research
Center for Veterinary Medicine, FDA
Laurel, MD, USA

Sharon E. Mason, DVM, PhD
Department of Biological Sciences
Campbell University
Buies Creek, NC, USA

Sanja Modric, D.V.M., PhD
Center for Veterinary Medicine
Food and Drug Administration
Rockville, MD, USA

Renate Reimschuessel, V.M.D., PhD
Veterinary Laboratory Investigation and Response Network
Center for Veterinary Medicine, FDA
Laurel, MD, USA

Jim E. Riviere, DVM, PhD, DSc(hon), ATS
Department of Anatomy and Physiology
College of Veterinary Medicine
Kansas State University
Manhattan, KS, USA

G. Johan Schefferlie, BSc, MSc
Veterinary Medicines Unit
Medicines Evaluation Board
Utrecht, the Netherlands

Stefan Scheid, PhD
Federal Office of Consumer Protection and Food Safety (BVL)
Berlin, Germany

Geof Smith, DVM, PhD, Dipl. ACVIM
Department of Population Health and Pathobiology
College of Veterinary Medicine
North Carolina State University
Raleigh, NC, USA

Lisa Tell, DVM, Dipl. ABVP, Dipl. ACZV
Department of Medicine and Epidemiology
University of California
Davis, CA, USA

Thomas W. Vickroy, PhD
Department of Physiological Sciences, College of Veterinary Medicine
University of Florida
Gainesville, FL, USA

Dong Yan, PhD
Division of Human Food Safety, Office of New Animal Drug Evaluation
Center for Veterinary Medicine, United States Food and Drug Administration
Rockville, MD, USA

1

IMPORTANCE OF VETERINARY DRUG RESIDUES

Ronald E. Baynes[1] and Jim E. Riviere[2]

[1] *Department of Population Health and Pathobiology, College of Veterinary Medicine, North Carolina State University, Raleigh, NC, USA*
[2] *Department of Anatomy and Physiology, College of Veterinary Medicine, Kansas State University, Manhattan, KS, USA*

1.1 INTRODUCTION

Food animal production over the last 50–60 years has significantly increased with the implementation of modern genetics, breeding, husbandry, and nutrition. During this same time period, livestock producers have relied on the use of veterinary drugs as one of several strategies to ensure economic viability of the industry. This need for increased use of veterinary drugs, and especially antimicrobial drugs, has been linked to changes in standard livestock practices where the objective is to increase feed and space efficiency and to a need to generate greater quantities of meat, milk, and egg products in an ever increasing competitive global market. While the consumer appreciates the need to increase livestock production and generate reliable and affordable animal-derived products, this is tempered by the consumers' requirement that the food items be "free" of drugs or chemicals introduced in the production system. The wide availability of related information via the Internet has exposed the consumer to useful facts but all too often to controversial

Strategies for Reducing Drug and Chemical Residues in Food Animals: International Approaches to Residue Avoidance, Management, and Testing, First Edition.
Edited by Ronald E. Baynes and Jim E. Riviere.

statements and hypotheses with very little factual support from the scientific literature regarding the prevalence of drug residues in our food, how veterinary drugs are used, and what safeguards are implemented to reduce these residues. This introductory chapter will briefly review the role of drugs in modern livestock production, quality assurance programs, adverse human health effects of drug residues, and economic impact of these residues to the livestock industry.

1.2 VETERINARY DRUG USE IN LIVESTOCK

Modern livestock production can be described as involving intensive animal production practices that often use veterinary drugs at subtherapeutic level in feed and water in order to improve feed efficiency for growth and production and maintain animal health. In such close animal–animal contact practices, prevention of disease is more important than treating for disease that would require therapeutic levels (higher doses) of the drug. The United States defines subtherapeutic use of an antimicrobial as a feed additive less than 200 g of drug per ton of feed.

Subtherapeutic drug use may take the form of (i) *antimicrobials* delivered to the animal as a feed or water additive and (ii) *hormones* delivered via ear implants or feed additives.

The antimicrobials approved in the United States and EU to be used in this legal manner often belong to the tetracycline, sulfonamide, or macrolide class of antimicrobials. Several EU countries and others banned or limited the use of these drugs as growth promoters as there are concerns that their use promotes the emergence of antimicrobial resistance. This cause-and-effect relationship is continually being debated across various jurisdictions; although epidemiological evidence continues to accumulate, definitive conclusions from rigorous research in livestock production systems has not been forthcoming. This issue will be further explored in this and other chapters of this book.

The use of hormone growth promoters in livestock has also been a controversial debate as various regulatory authorities in different jurisdictions regulate these drugs in a different manner. The U.S. FDA has approved the legal use of 17β-estradiol, testosterone, progesterone, trenbolone, and zeranol as solid ear implants and melengestrol acetate (feedlot heifers) and ractopamine (swine) as feed additives. Compared to the United States, the EU in 1988 issued a total ban of all hormonal active growth promoters in livestock production. Prior to 1988, in the Netherlands (1961) and Belgium (1962–1969), there was a total ban on anabolic agents for growth promotion purposes in slaughter animals in order to protect consumers and for the benefit of international trade (Stephany, 2010). It should be noted that the

United States challenged the EU's ban, and in 1998, the WTO found that the EU's ban was not supported by science and inconsistent with WTO obligations (USTR, 2009).

Therapeutic drug use in veterinary livestock involves administration of veterinary drugs according to label to treat an individual animal or herd or flock of animals by various approved routes of administration. The use of water additives is recognized in all countries as a form of therapeutic drug use and not subtherapeutic drug use or for growth promotion purposes. It has however been our experience (Mason et al., 2012) that treatment of large herds via water medication does not always result in each animal in the herd receiving therapeutic drug levels. This has often been associated with competition between animals in the herd and/or malfunctioning medicators. The approved use of the many therapeutic drugs will be outlined in the species-specific chapters of this book.

The passage of the Animal Medicinal Drug Use Clarification Act (AMDUCA) 1994 in the United States allows food animal veterinarians to administer drugs in an extralabel manner within certain guidelines as outlined in the following text. Veterinarians often have to resort to using these drugs in an extralabel manner for a number of reasons. *New generics of old drugs are approved based on bioequivalence to pioneer formulation*, which allows the same dosage and milk discard/meat withdrawal times. The problem with this approach is that new bacteria being treated have much higher MICs than bacteria and microorganism many years ago, and thus, higher dose must now be used. Veterinarians often consult with FARAD to find out new withdrawal times, and this is described in more detail in Chapter 14. The scientific issue is that most antimicrobials used in dairy practice today are old drugs (or generic copies of old drugs) that are now not effective unless given at higher doses, necessitating extended milk discard times. Risk of exposure to low-level residues of most other drugs out there is "theoretical," but low label dosages of antimicrobials, used to insure adequate withdrawal times, will promote resistance, which is the major public health issue. There are more modern approaches that would allow dosage adjustments with new withdrawal times, but we are stuck in the science of the 1970s. *Legal precedence and business issues tend to hand tie the FDA* (in approving all generics just like the first one that was approved even if science has advanced in 30 years). Production use of antibiotics as growth promoters may very well be banned, and therapeutic use at higher doses by licensed vets maintained.

Phytoceuticals are increasingly being used on organic farms with varying degrees of success. These drugs are not regulated by the FDA-CVM as they are often described as "generally recognized as safe" (GRAS). There are however several guidance documents and requirements that organic livestock farms are required to follow and are discussed elsewhere in this book.

1.3 QUALITY ASSURANCE PROGRAMS

Consumers are very aware of drug and chemical use in the livestock industry, and oftentimes, there is general misinformation about how these drugs are used in the industry. The infrequent catastrophic drug residue violations are often a direct result of careless farm management. *The subsequent economic cost to the livestock industry is not ignored by the many stakeholders involved in livestock production and distribution and sales of meat, milk, fish, and egg products.* This will be discussed in more detail in a later section of this chapter.

In lieu of these scenarios, the livestock industry has been aggressively policing itself to make sure that producers are educated and trained to prevent drug residue violations on their farms. Many if not most livestock producers follow and adhere to their respective quality assurance programs for their commodity group that attempt to minimize drug residue violations and promote judicious use of veterinary drugs. A summary of the steps producers are encouraged to follow whether it is the beef, dairy, pig, goat, or poultry industry is as follows:

1. Improve husbandry practices by maintaining appropriate husbandry, hygiene, examinations, and vaccinations.
2. Consult with a veterinarian prior to use of drugs or medicated feed or water as therapeutic alternatives may be more appropriate.
3. Use drug according to veterinary label and only resort to using veterinary drugs as a last resort. This is especially important for antimicrobial drug use.
4. Antimicrobial drug use is inappropriate for viral infections without bacterial complication.
5. Optimize antimicrobial drug regimen using current pharmacological information and principles.
6. Mitigate veterinary drug spillage into the environment.
7. Keep good records of drug use on each farm.
8. Extralabel drug use in the United States must follow the FDA regulations: prescriptions, including extralabel use of medications must meet the AMDUCA amendments to the Food, Drug, and Cosmetic Act and its regulations. This includes having a valid veterinary–client relationship.

The passage of the AMDUCA in the United States in 1994 allows food animal veterinarians in the United States to administer drugs in an extralabel manner within certain guidelines. Several chapters in this book will focus on

PK principles that can be used to extrapolate across and within species, across routes of administration, and across doses. To date, legislation similar to AMDUCA does not exist in other major livestock-producing countries.

1.4 ADVERSE HUMAN HEALTH EFFECTS OF DRUG RESIDUES

Inappropriate use of several of veterinary and human drugs in livestock production can result in significant residue levels in meat, dairy, and poultry products that can cause adverse health effects in consumers. Although approximately 80% of all food animals are given drugs during their lifetime, residue violations are often less than 1% thanks to rigorous surveillance and testing in major livestock-producing countries and increasing so in smaller developing states. However, many consumers in developed and developing states rely on livestock products as their major source of protein. The average American consumes 200 pounds of meat and fish, 67 pounds of poultry, 30 pounds of eggs, and 600 pounds of dairy products annually. In spite of the low level of drug residue contamination, this high level of consumption of livestock products increases the possibility that any one violative incident can result in adverse health effects affecting more than one individual or community following acute or chronic exposure.

A very good example of the aforementioned case was associated with clenbuterol residues. In one 6-month period in 1993, more than 1200 hospitalizations and 3 deaths in France and Spain were reported to have resulted from eating beef livers contaminated with clenbuterol. One study documented in Portugal four cases of acute food poisoning, involving a total of 50 people, due to the ingestion of lamb and bovine meat containing residues of clenbuterol (Barbosa et al., 2005). An outbreak with hospitalization was described in Italy in 1997 involving 15 people within 0.5–3.0 h after the consumption of veal and not livers (Brambilla et al., 2000). No deaths were reported but clinical signs and symptoms disappeared within 3–5 days. More recently, 286 villagers in Changsha, capital of Hunan province in China, were hospitalized and suspected to have been made sick from consuming clenbuterol-tainted pork (UPI, 2011). Symptoms of clenbuterol intoxication can be described as predominantly gross tremors of the extremities, tachycardia, nausea, headaches, and dizziness. This drug is a beta-agonist, acts as a bronchodilator, and can have anabolic effects such as increase lean body mass and weight gain. It is not approved for use in humans or in food animals by the U.S. FDA, and extralabel use in food animals is strongly prohibited. However, there is approval for use in horses with recurrent airway obstruction (heaves), and there are no studies to support meat withdrawal times for this drug given to horses intended for food.

1.5 WITHDRAWAL TIME DETERMINATIONS

Several chapters in this book will describe in brief several of the methods used by the U.S. FDA (2006) and the European Medicinal Agency (EMA, 1996, 2000) to derive regulatory withdrawal times that ensures the consumer is protected from exposure to drug concentration that will cause adverse health effects. The guidance documents for these calculations from each of these regulatory authorities are always changing with new revisions, and they may vary slightly, but there are some common features that the reader should appreciate.

For example, in assigning a milk withdrawal time, the U.S. FDA uses an algorithm that calculates the upper 99th percentile of the population and 95th percent confidence limit. As with the tissue withdrawal period, this assures that when the drug product is used according to its approved label, there is only a 5% chance that one animal in 100 will have milk residues above the milk tolerance concentration. In the EU, the recommended method is also a statistical method based on a linear regression model in which the upper 95% tolerance limit of the 95% percentile of the residue depletion curve is used to determine the withdrawal period. As per the U.S. FDA, the minimum number of animals in a milk residue study is 20, based on the statistical requirements for the calculation of the withdrawal time. In the EU, milk withdrawal periods are established for individual animals and not for tank milk as per the U.S. FDA-CVM. The reader is encouraged to consult with updated guidance documents in the respective jurisdictions with regard to recommended regulatory methods to calculate the meat and milk withdrawal times. There are several chapters in this book that describe alternative and more flexible pharmacometric methods that utilizes the current advances in mathematical modeling and well-accepted software that considers a larger population of animals and other variables such as production and disease status that are often overlooked in the current regulatory methods in many jurisdictions. These novel methods are not currently accepted by regulatory agencies in the establishment of meat and milk withdrawal periods for veterinary drugs. However, several of them such as physiologically based pharmacokinetic (PBPK) modeling have been adapted with success by the U.S. EPA in their guidance for conducting a human health risk assessment of environmental contaminants.

1.6 ANTIMICROBIAL RESISTANCE

The U.S. FDA in 2010 provided guidance on the judicious use of antimicrobial drugs in livestock and recognized that failure of antimicrobial therapies in humans can be related to human and animal use of antimicrobials among other factors. The FDA believes that "the use of medically important antimicrobial drugs in food-producing animals for production purposes (e.g., to promote growth or improve feed efficiency) represents an injudicious use of these important drugs. **Production uses are not directed at any specifically identified disease, but rather are expressly indicated and used for the purpose of enhancing the production of animal-derived products**. In contrast, FDA considers uses that are associated with the treatment, control, or prevention of specific diseases, including administration through feed and water, to be uses that are necessary for assuring the health of food-producing animals." This topic is discussed in more detail in subsequent chapters of this book that describe the prudent drug use of antimicrobials in ruminant and pig production systems.

1.7 ECONOMIC IMPACT OF DRUG RESIDUES

There is a significant economic impact associated with drug residues in meat, milk, or egg products. Besides loss in sales of product, public perception can have the greatest impact on consumers already weary about drug and chemical use in food production systems in developing and developing countries. Oftentimes, the consumer is exposed to misinformation from media sources whose understanding are limited with regard to how these drugs are used on livestock farms and the many stages between the farm and table where residue violations are prevented. The remainder of this book will highlight many of the established practices that are effective in the mitigation of drug residues and scenarios where residue violations are likely to occur and warrant future research and attention by regulatory authorities.

REFERENCES

Barbosa J, Cruz C, Martins J, Silva JM, Neves C, Alves C, Ramos F, Da Silveira MI (2005 Jun). Food poisoning by clenbuterol in Portugal. Food Addit Contam 22(6):563–566.

Brambilla G, Cenci T, Franconi F, Galarini R, Macrì A, Rondoni F, Strozzi M, Loizzo A (2000 Apr). Clinical and pharmacological profile in a clenbuterol epidemic poisoning of contaminated beef meat in Italy. Toxicol Lett 114(1–3):47–53.

EMA (1996). Note for Guidance: Approach towards harmonisation of withdrawal periods (EMEA/CVMP/036/95). Available at http://www.ema.europa.eu/docs/en_GB/document_library/Scientific_guideline/2009/10/WC500004428.pdf. Accessed April 6, 2014.

EMA (2000). Note for guidance for the determination of withdrawal periods for milk (EMEA/CVMP/473/98-FINAL). Available at http://www.ema.europa.eu/docs/en_GB/document_library/Scientific_guideline/2009/10/WC500004496.pdf. Accessed April 6, 2014.

Mason SE, Baynes RE, Almond GW, Riviere JE (2012). Variables important to the population pharmacokinetic modeling of tetracycline water medication in swine. Am J Vet Res 73(10):1641–1649.

UPI (2011). Food poisoning sends dozens to hospitals. Available at http://www.upi.com/Top_News/World-News/2011/04/25/Food-poisoning-sends-dozens-to-hospitals/UPI-92881303704598/#ixzz2RDoZPiuf. Accessed April 6, 2014.

US FDA (2006). FDA Center for Veterinary Medicine's Guidance for Industry #3: General principles for evaluating the safety of compounds used in food producing animals. Rockville: US FDA.

USTR (2009). USTR Announces Delay of Trade Action in Beef Hormones Dispute. Available at http://www.ustr.gov/about-us/press-office/press-releases/2009/march/ustr-announces-delay-trade-action-beef-hormones-disp. Accessed March 20, 2013.

2

PHARMACOKINETIC PRINCIPLES FOR UNDERSTANDING DRUG DEPLETION AS A BASIS FOR DETERMINATION OF WITHDRAWAL PERIODS FOR ANIMAL DRUGS

Sanja Modric

Center for Veterinary Medicine, Food and Drug Administration, Rockville, MD, USA

2.1 INTRODUCTION

According to the U.S. Food and Drug Administration's (FDA) Center for Veterinary Medicine (CVM), an approved animal drug is considered to be safe and effective, if it is used according to its label instruction—safe for use in the intended species as well as for human consumption of the edible products derived from animals treated with the drug. An evaluation of drug safety for human consumption includes an assessment of toxicology and residue chemistry—as described in "FDA CVM's Guidance for Industry (GFI) #3: General Principles for Evaluating the Safety of Compounds Used in Food Producing Animals (FDA GFI #3, 2006)," and all the toxicology-related GFIs. In addition, the human food safety evaluation for active pharmaceutical ingredients (API) possessing antimicrobial activity also includes an assessment of the effect of the transmission of food-borne bacteria of human health

Strategies for Reducing Drug and Chemical Residues in Food Animals: International Approaches to Residue Avoidance, Management, and Testing, First Edition.
Edited by Ronald E. Baynes and Jim E. Riviere.

concern through the consumption of animal-derived food products (FDA GFI #152, 2003) and an evaluation of the safety of drug residues with respect to the human intestinal flora for establishing a microbiological acceptable daily intake (FDA GFI #159, 2011).

The use of approved drugs in food-producing animals can lead to the presence of unsafe drug residues in the edible products above the established tolerances (21 CFR 556) if drugs are not used according to their label directions (i.e., if animals are sent to slaughter before the established withdrawal period has been observed). Brynes (2005) provided a more recent review on the history of tolerances for residues of new animal drugs in food. In addition to providing updated definitions and concepts of establishing and promulgating regulations on tolerances, Brynes provides a historical perspective on how the establishment of residues changed over time (e.g., the earliest tolerances generally referred to the parent drug, which was later changed to establish tolerances that would reflect the total residue). The presence of violative drug residues in food may result in potential risks to humans consuming residues, including acute and chronic toxicities, such as allergic reactions, various adverse reproductive and developmental effects, carcinogenicity, as well as a risk of the development of antimicrobial resistance (Horrigan et al., 2002). The edible products considered in the human food safety evaluation include muscle, liver, kidney, skin with or without fat, and milk and eggs (when appropriate). Residues of human food safety concern may include the APIs and excipient(s) of a drug product, drug metabolites, and any substance formed in or on the edible food products as a result of drug treatment.

Risk assessment principles based on the standard of reasonable certainty of no harm for human consumption are applied in the human food safety evaluation of animal drug residues in food animals. The assessment comprises an evaluation of the traditional toxicological effects of drug residues on human health, the amount of residues human consumers are exposed to, the risk of developing antimicrobial drug–resistant bacteria due to the use of antimicrobial drugs in animals, and the effects of drug residues on the human intestinal flora (Friedlander et al., 1999). It should be emphasized that multiple and robust layers of safety are factored in during the risk assessment process for the evaluation of human food safety of new animal drugs to accommodate various intrinsic (i.e., related to the animal physiology) and extrinsic factors (i.e., factors influencing the drug's characteristics, food, environment, concomitant medications, etc.) that can influence a drug's behavior in the body. In this chapter, a brief summary of basic pharmacokinetic (PK) principles is provided to help readers understand the pharmacologic principles underlying the human food safety evaluation of new animal drugs.

The depletion of residues of a compound may vary considerably due to the impact of various intrinsic and extrinsic factors, as described below.

Total residue evaluation consists of evaluation of the parent compound, free metabolites, and metabolites that are covalently bound to endogenous molecules. The levels of residues depend on the types of tissues, the amount of the drug administered, and the time following the last drug administration to the animal (FDA GFI #3, 2006). Therefore, FDA typically requires that residue chemistry studies be conducted in each species/class of animal for which the sponsor is seeking approval. Studies also may be needed in specific classes of animals (e.g., veal calves, lactating dairy cattle for milk, laying hens for eggs). The studies are typically conducted in a limited number of animals in order to minimize the economic and ethical impacts on new animal drug sponsors (Martinez et al., 2000). However, it is critical that the human food safety evaluation is conducted in the appropriate population of animals for which the drug is intended. The human food safety studies use the highest intended treatment dose, the longest intended treatment duration (or a duration that ensures that the drug concentrations have reached steady state), and the intended administration route (and therefore represent the worst-case scenario in terms of drug residue exposure to humans). Because of these conditions, the human food safety studies for new animal drug approvals are typically conducted once the dose and dosing regimen of the drug have been firmly established and once when the sponsor has identified the final formulation for their new animal drug product. Depending on the dosing regimen, the design of required studies may differ considerably. For example, if a drug is intended to be administered once for a specific therapeutic effect, then a single dose of drug in the target animals will capture appropriate exposure; on the other hand, if a drug is intended for prolonged treatment, it is critical to evaluate residue depletion after the drug concentrations have reached the steady state.

Before discussing specific study designs for various kinds of residue depletion studies (which will be covered in Chapter 3), it is important to understand the pharmacologic basis for recommending those study designs. This chapter reviews the impact of various internal (endogenous) factors on *in vivo* drug behavior, which includes both blood and tissue levels. The withdrawal time, a critical factor for ensuring human food safety, is in essence a PK parameter based on the legal target tissue tolerance and reflecting the drug's rate of depletion from that target tissue (Riviere, 1999).

The amount of drug substances in edible animal products is a complex function of the rate and extent of absorption of the parent compound, the formation of metabolites (free and covalently bound to endogenous molecules), and the distribution and clearance of the parent compound and its metabolites. Drug distribution depends on the physicochemical properties of the drug, the concentration gradient between the blood and tissue, the ratio of the blood flow to tissue mass, and the affinity of the drug for tissue (Riviere, 1999).

Tissue depletion reflects the drug's partitioning characteristics between blood and tissue, the blood flow to that tissue, and the rate at which the drug is depleted from the systemic circulation. In some instances, the tissue itself (especially the liver and kidney) may also be involved in drug metabolism, which then further contributes to the overall tissue rate of depletion. Therefore, drug residues in the various edible products will deplete at different rates, and their respective tissue elimination half-lives have to be determined for the establishment of the withdrawal time. The final withdrawal period assignment is based upon the time it takes for the marker residue to deplete from the slowest depleting tissue (the target tissue).

2.2 BASIC PHARMACOKINETIC PRINCIPLES UNDERLYING DRUG DEPLETION

For any desired level of drug exposure (typically expressed in terms of an area under the concentration versus time curve, AUC), critical points to consider are the dose administered (D), total body clearance (Cl), and bioavailability (F). Drug exposure is determined by the following equation:

$$\mathrm{AUC}_{0\text{-inf}} = \frac{\mathrm{D} \times \mathrm{F}}{\mathrm{Cl}} \tag{2.1}$$

It is important to note that the targeted dose is equally influenced by the Cl and F. Bioavailability (or fraction of administered dose that is absorbed) is the proportion of the administered dose that reaches the systemic circulation. It is a function of animal physiology, route of administration, and the physicochemical characteristics of the API and the formulation (Martinez and Amidon, 2002). Clearance represents the volume of whole blood, serum, or plasma completely cleared of drug per unit of time. Unlike the F, Cl is solely a function of the physicochemical properties of the API and the host physiology (unless a specific ingredient interacts with the elimination process).

All aspects of the PK response (absorption, distribution, metabolism, and elimination (ADME)) are important in understanding the human food safety effects, as they can ultimately influence drug depletion profiles. Although not a critical factor for immediate-release drugs, absorption can significantly affect depletion times of modified release dosage forms, due to the presence of flip-flop kinetics (where the rate of drug absorption rather than elimination is the rate-limiting factor determining the slope of the terminal phase of the concentration versus time profile).

Distribution of a drug to peripheral tissues is affected by the binding of the drug to blood and tissue macromolecules, blood flow, partition coefficient of the drug between the blood and the organs into which it distributes, and the physicochemical properties of the drug. Tissue binding, which tends to increase drug distribution, is an important underlying consideration in the evaluation of the human food safety of edible products derived from animals treated with a new animal drug. The tissue drug concentrations determine the time needed for drug-related residues to deplete to legally established tissue tolerances, which are, in turn, based upon the safety of the residues to humans consuming edible products of animals treated with a new animal drug, extrapolated from studies in toxicological model species.

The drug distribution between plasma and tissues is described by the PK parameter, the volume of distribution (Vd). Vd is not a physiologic value, but rather a reflection of how a drug gets distributed throughout the body, the latter depending on its physicochemical properties, such as solubility, charge, and size. Drugs that remain in the circulation tend to have a low Vd, whereas drugs that are highly bound to tissue tend to have a very high Vd. Vd relates the mass of drug in a compartment to the volume into which it is diluted and is described by the following equation:

$$\mathrm{Vd} = \frac{\mathrm{Dose}}{\mathrm{Drug\ plasma\ concentration}} \tag{2.2}$$

The term Vd may be expressed as either Vd_c, Vd_{area}, Vd_{ss}, or Vd_β. Volume of distribution of the central compartment, Vd_c, reflects the volume of the central compartment, before any distribution has taken place, and relates the dose to the drug concentration at time 0.

The apparent volume of distribution, Vd_{area}, is based on the total AUC. It relates plasma concentration to the amount of drug in the body at all times after distribution equilibrium is reached after a single dose or multiple discrete doses. It is calculated as follows:

$$\mathrm{Vd}_{area} = \frac{\mathrm{Dose} \times \mathrm{F}}{\mathrm{AUC} \times \beta}, \tag{2.3}$$

where β is the slope of the terminal portion of the plasma concentration–time curve (plotted as a natural logarithm of concentration versus time).

The volume of distribution at steady-state, Vd_{ss}, provides an estimate of drug distribution independent of elimination processes, which is most useful for predicting the plasma concentrations at steady state. Steady state is reached when the free concentration of drug in the plasma equals the free concentration in the tissue. It is a correct measure for continuous intravenous infusion or at a single instant in time (when the rate of elimination equals that of distribution) and is calculated as:

$$\mathrm{Vd}_{\mathrm{ss}} = \frac{\mathrm{Dose} \times \mathrm{F} \times \mathrm{AUMC}}{\mathrm{AUC}^2}, \tag{2.4}$$

where AUMC is the area under the moment curve, which is the integral of the curve plotting the product of concentration and time by the time the concentration was observed.

The apparent volume of distribution in the postdistribution (or terminal) phase, Vd_β, neglects the distribution phase of drug disposition and is calculated as follows:

$$\mathrm{Vd}_\beta = \frac{\mathrm{Dose} \times \mathrm{F}}{B}, \tag{2.5}$$

where B is a value obtained from extrapolating the linear terminal portion of the plasma concentration–time curve to its intercept on the y axis (plasma drug concentration).

Because the Vd_β ignores the distribution phase, it is valid only for drugs that fit a one-compartment model (it generally overestimates the true volume of distribution of multi-compartmental drugs). The only measure of volume that is independent of the rate of chemical elimination is the $\mathrm{Vd}_{\mathrm{ss}}$.

V_d may also be used to determine how readily a drug will displace into the body tissue compartments relative to the blood using the following equation:

$$V_\mathrm{d} = V_\mathrm{p} + V_\mathrm{t}\left(\frac{f_\mathrm{u}}{f_\mathrm{ut}}\right), \tag{2.6}$$

where V_p is the plasma volume, V_t is the apparent tissue volume, f_u is the fraction unbound (free) in plasma, and f_ut is the fraction unbound (free) in tissue.

Understanding the drug distribution and the presence of peripheral compartments is important when evaluating drug depletion from the body. A drug that selectively binds to tissues or sequestered into a deep compartment may have several different half-lives, and it is critically important to understand drug depletion for the determination of the withdrawal time. In addition, the analytical method has to be sufficiently robust and sensitive to address the tissue distribution of a drug and potential presence of deep peripheral compartments, which could result in "spikes" in residue concentrations above the tolerance.

As the drug is absorbed and distributed throughout the body, drug elimination becomes the most predominant process. Mechanisms of drug elimination include biotransformation (metabolism) and excretion. In general, both mechanisms are involved in drug elimination, although one mechanism is usually dominant over the other. Of the physicochemical properties that determine the mechanism of elimination, lipid solubility and degree of ionization seem to play the most critical role. For example, lipid-soluble drugs undergo biotransformation by hepatic microsomal enzymes, while many polar drugs and metabolites are excreted by the kidney (Brown, 2001).

Most commonly, a constant proportion of the dose is cleared over time, which is termed first-order elimination. By definition, in first-order or linear processes, the elimination rate (ke) is constant, while the actual rate of the process varies in direct proportion to the dose. The concentration (C) at any time (t) after a single intravenous dose administered can be calculated as:

$$Ct = \left[\frac{\text{Dose}}{V_{\text{d}}}\right] \times e^{-\text{ke} \times t}, \tag{2.7}$$

where e is the base of the natural logarithm ($e = 2.713$), and the elimination constant (ke) is represented by the ratio of clearance to volume of distribution and is usually expressed in units of 1/h.

$$\text{ke} = \frac{\text{Cl}}{V_{\text{d}}} \tag{2.8}$$

Most equations in this chapter describe a one-compartment body model with no absorption. This is an oversimplified description of the PK processes, because most drugs are not adequately described by a one-compartment body model: the body does not behave as a single homogeneous compartment and there is usually no instantaneous distribution through this one compartment. These one-compartment equations are included as illustrations of the principles used in PK, but the reader should keep in mind that for most drugs the body does not behave as a single compartment and that the understanding of multiple compartments is critical for understanding drug depletion and the risks for violative residues in tissues.

In a multi-compartmental model, different body compartments are characterized by different rates of drug distribution. Most typically, there are two major body compartments, although there can be more than two, depending on the rates of drug distribution among the compartments. A two-compartment body model is schematically represented in Figure 2.1. It consists of a central compartment, which comprises blood plasma and the extracellular fluid of

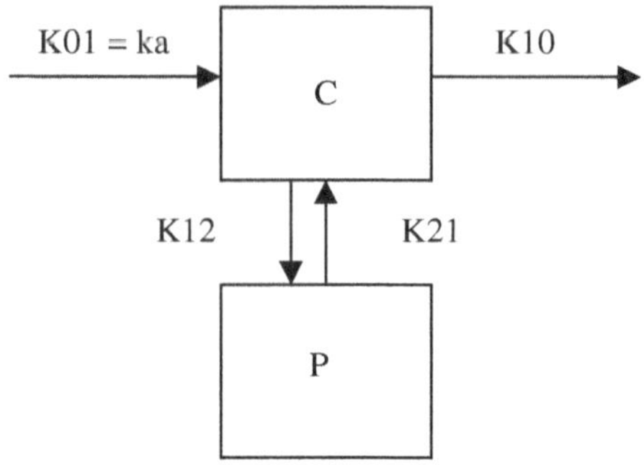

FIGURE 2.1 A two-compartment body model with first-order absorption.

highly perfused organs (such as the heart, lungs, kidney, and liver), and a peripheral compartment, in which the distribution occurs more slowly (such as in muscle and fat). In addition to the presence of multiple compartments, many drugs are administered extravascularly, so there is an absorption phase that needs to be taken into consideration when modeling the drug PK response (illustrated by the $k01 = \text{ka}$ arrow).

Figure 2.1 shows a two-compartment model with a first-order absorption, where C is the central compartment (1), P is a peripheral compartment (2), $k01$ is the absorption rate constant, $k10$ is the elimination rate constant, and $k12$ and $k21$ are the inter-compartmental constants reflecting distribution.

For a drug administered extravascularly and assuming a non-instantaneous distribution (e.g., a two-compartment model), the plasma concentration at any time can be calculated as:

$$Ct = A \times e^{-\alpha \times t} + B \times e^{-\lambda z \times t} - C \times e^{-\text{ka} \times t}, \tag{2.9}$$

where A, B, and C are the y-axis intercepts for the slopes α (rapid redistribution phase), λz (elimination phase), and ka (absorption phase), respectively.

Elimination rate for a two-compartment model is calculated as follows:

$$\text{ke} = \frac{k10 \times k21}{(k21 + k12)} \tag{2.10}$$

Absorption rate of a drug is determined by the slope of the relationship between the logarithm of the amount of the drug absorbed and time. In first-order absorption, a constant fraction of the drug is absorbed per unit of time and the absorption process is thus linear. In contrast, if the saturation of the absorption mechanism occurs, the process may become nonlinear due to capacity limitations (with a lower percent of the dose absorbed at higher doses).

Saturable absorption processes involve active efflux and carrier-mediated transport mechanisms. To further complicate the estimation of absorption, many drugs undergo a first-pass biotransformation, in which the concentration of a drug is reduced in the liver prior to reaching the systemic circulation, thereby affecting its bioavailability. Besides the hepatic enzymes, other minor sources of the first-pass metabolism include the enzymes in the intestinal wall and the bacterial enzymes.

The elimination half-life ($t_{1/2}$), the time it takes for the plasma concentration of the drug to reduce by 50%, depends on the elimination rate constant and on both Cl and Vd. It can be calculated as:

$$t_{1/2} = \frac{0.693}{\mathrm{Cl} / \mathrm{Vd}} \tag{2.11}$$

Therefore, the observed half-life is dependent upon both the extent of a drug's distribution in the body and its rate of clearance.

Because various factors can influence the Cl and Vd, they may also affect the half-life of a drug, and as a result, have an impact on the drug depletion. For example, a disease process or a dose increase can both prolong tissue half-lives, potentially resulting in violative residues. As a rule of thumb, if the dose is doubled, the depletion time should be increased by one half-life, and if the half-life doubles, then the depletion time needs to be doubled as well.

Hepatic disposition plays an important role in drug distribution, metabolism, and elimination of xenobiotics and can affect bioavailability. Drug metabolism alters the chemical structure of the parent molecule, leading to a change in drug lipophilicity/hydrophilicity, distribution kinetics, clearance, and physiological effects (Riviere, 1999). Hepatic metabolic processes are traditionally divided into Phase I and Phase II reactions; however, Phase III reactions have been introduced more recently (Coleman, 2007). Phase I metabolism mainly describes oxidative cytochrome P450 (CYP450) reactions, but non-CYP oxidations, such as reduction and hydrolyses, are also included in the Phase I metabolism. In general, Phase I reactions involve the introduction of functional groups to drug molecules, which may lead to the formation of active and/or toxic metabolites. In Phase II processes (e.g., glucoronidation/glucosidation, sulfation, methylation, and acetylation), hepatic enzymes conjugate water-soluble endogenous sugars, salts, or amino acids with xenobiotics or endogenous chemicals. Conjugates can also be secreted back into the intestines, where bacteria deconjugate the drug, allowing the reabsorption of the parent molecule. In many cases, Phase I processes occur first, followed by conjugative reactions. However, it should be noted that although this sequence happens often, conjugation can also occur directly without a prior Phase-I "preparation" (Coleman, 2007). Similarly, it is often believed that

Phase-II reactions always lead to detoxification and formation of highly water-soluble products. Although this is often the case, there are situations in which conjugative processes lead to formation of toxic species or less water-soluble molecules. Finally, Phase III metabolism involves further processing of conjugates, including the system of efflux pumps that exclude water-soluble products of metabolism from the cells, blood, and ultimately, the organism.

Hepatic clearance ($\mathrm{Cl_h}$), the ability of the liver to remove the drug from the blood, is related to the intrinsic hepatic clearance ($\mathrm{Cl_{int}}$) and hepatic blood flow rate (Q_h):

$$\mathrm{Cl_h} = Q_\mathrm{h}\left[\frac{f_\mathrm{u} \times \mathrm{Cl_{int}}}{Q_\mathrm{h} + f_\mathrm{u} \times \mathrm{Cl_{int}}}\right] = Q_\mathrm{h} E_\mathrm{h}, \tag{2.12}$$

where $\mathrm{Cl_{int}}/(Q_\mathrm{h} + \mathrm{Cl_{int}})$ is the hepatic extraction ratio or E_h and f_u is free fraction (unbound) of a drug.

As described earlier, the primary factors that influence hepatic clearance are the blood flow (Q_h), extent of protein binding (f_u), and inherent capacity of the hepatocytes to metabolize a drug ($\mathrm{Cl_{int}}$). When metabolic capacity is low (low $\mathrm{Cl_{int}}$), the drug is described as a low-extraction drug. The clearance of low-extraction drugs is primarily determined by the intrinsic metabolizing capacity of the liver and by the free drug fraction, while it is relatively independent of hepatic blood flow. The extraction is said to be restrictive or capacity limited. When this is the case, $f_\mathrm{u} \times \mathrm{Cl_{int}} \ll Q_\mathrm{h}$, Equation 2.12 can be simplified to

$$\mathrm{Cl_h} = f_\mathrm{u} \times \mathrm{Cl_{int}} \tag{2.13}$$

On the other hand, when drugs are rapidly and extensively cleared from the blood by the liver (e.g., in a single pass), they are described as high-extraction drugs. Their clearance depends primarily on hepatic blood flow, and binding to blood components is not an obstacle for extraction; the extraction is said to be nonrestrictive or blood flow dependent. When this is the case, $f \times \mathrm{Cl_{int}} \gg Q$, and the equation can be simplified to

$$\mathrm{Cl_h} = Q \tag{2.14}$$

There is another group of drugs, the so-called intermediate-extraction drugs, for which the hepatic clearance is dependent on hepatic blood flow, intrinsic metabolizing capacity of the liver, and the free drug fraction. In general, it is considered that high extraction drugs have the extraction ratio above 0.7, intermediate extraction drugs between 0.3 and 0.7, and low extraction drugs below 0.3.

This discussion illustrates the importance of considering the drug's physicochemical properties, hepatic metabolism enzyme activity, and the hepatic blood flow rate when estimating hepatic drug clearance. Because of its nonlinear response, the intrinsic hepatic clearance can be described by the Michaelis–Menten process, in which

$$\frac{dC}{dt} = \frac{V_{\mathrm{max}} \times C}{K_{\mathrm{m}} + C}, \tag{2.15}$$

where dC is the rate of change in concentration, dt is the rate of change in time, V_{max} is the maximum rate (velocity) of reaction and K_{m} is the Michaelis–Menten constant, which represents the drug concentration at which half-maximal reaction velocity occurs.

Because of a rapid and extensive clearance by the liver, the situation described by Equation 2.15 is really a concern only for the high-extraction drugs.

While most of the discussion so far focused only on the hepatic clearance (because of its complexity and a critical role in determining human food safety), other clearance mechanisms should also be considered. Of nonhepatic clearance mechanisms, renal clearance represents the most common way of drug elimination from the body. Renal clearance depends on the glomerular filtration rate (GFR), tubular reabsorption, and tubular secretion. For a drug that is excreted unchanged in the urine, its renal clearance is the most predominant component of body clearance. Glomerular filtration is a unidirectional process by which a drug is removed from the blood by bulk flow and is restricted to small nonprotein-bound molecules. Unlike glomerular filtration, active tubular secretion and absorption are both energy-dependent processes described by the Michaelis–Menten enzyme kinetics principles (Eq. 2.15). Besides from being saturable (due to the limited capacity of carrier-mediated processes), these renal mechanisms are characterized by competitive inhibition by similar drugs. As a result, when two or more drugs from the same ionic class are coadministered, their rate and extent of renal excretion will be affected.

Although the kidney is the ultimate route for drug elimination from the body, renal clearance is less critical than hepatic clearance in terms of determining the depletion times, as drugs are typically excreted by kidneys rapidly and efficiently. In addition to those two major eliminatory pathways (renal and hepatic), there are other routes of elimination, such as biliary, pulmonary, salivary, and mammary excretion. Each of these pathways contributes to the total body clearance, $\mathrm{Cl_B}$, which is the sum of all elimination clearances:

$$\mathrm{Cl_B} = \mathrm{Cl_{renal}} + \mathrm{Cl_{hepatic}} + \mathrm{Cl_{other}}. \tag{2.16}$$

For most therapeutic drugs, however, these additional routes are quantifiably negligible and therefore largely unimportant for estimating the total elimination of a drug from the body.

2.3 THE IMPACT OF PK ON DRUG DEPLETION

The basic PK principles described earlier are important in terms of understanding the possible impact of intrinsic and extrinsic variability on drug depletion and on the overall evaluation of human food safety. As already discussed, the drug depletion is a function of the initial concentration of the drug in the edible tissues and the rate at which the drug and its active metabolites are eliminated from those tissues (Gehring et al., 2004). The initial concentration is dependent on the administered dose, drug bioavailability, drug formulation, and route of administration. The rate of elimination varies as a function of Cl and Vd. In cases where there is flip-flop kinetics, the depletion rate can also be influenced by drug formulation and route of administration. All ADME processes that affect the drug depletion can be influenced by host physiology and pathophysiology. As the formulation, route, species, and dose are already established by the time pivotal human food safety studies are conducted, only the potential effects of various physiologic and pathophysiologic conditions that adversely affect tissue elimination and drug depletion are considered in this chapter.

The drug depletion is closely related to the drug's rate of elimination and to the tissue half-life. For example, if the target tissue half-life of a drug is increased due to a systemic disease (or any other factors described later), then the time for the residues in the target tissue to deplete to the tolerance level may increase. The withdrawal time is based on the depletion of the marker residue, which is selected to reflect the total concentration of the drug and metabolites in a tissue. Disease processes that alter the parent-drug-to-metabolite ratio may not predict the necessary changes in withdrawal time in order to prevent residue violations (Riviere, 1999). Factors that can impact the expression of genes that encode drug-metabolizing isoenzymes can have a similar effect on the drug depletion time.

Finally, it should be noted that even relatively minor changes in the ADME processes that would not necessarily result in a different therapeutic outcome might potentially affect the residue kinetics in edible products and cause violative residues. However, multiple robust safety factors are applied in the process of establishing the tissue tolerances so that these relatively minor changes in ADME are accounted for in the overall evaluation of human food safety.

2.4 FACTORS INFLUENCING ADME

Drug exposure is influenced by both intrinsic and extrinsic factors and understanding the possible impact of those factors on drug depletion will help explain study requirements for the evaluation of human food safety under the most appropriate conditions of use (i.e., using the worst case scenario). Some of the factors that can alter a pharmacological response are easily identified, such as breed, gender, and age. These factors are typically considered when designing the drug residue depletion studies. Others, such as disease, body composition, heritable traits, and environmental factors, are either less obvious and/or are difficult to predict. In any case, failing to acknowledge the possible impact of these factors on drug behavior in the body can lead to substantial error in predicting the dose exposure–response relationship.

Within the field of veterinary medicine, there is limited information on the impact of physiological variables and the impact of population variability on the dose exposure–response relationship. Two recent manuscripts (Martinez and Modric, 2010; Modric and Martinez, 2011) provide a comprehensive review of the role of patient variation in veterinary pharmacology: Part I of the series focuses on the potential influence of disease processes, stress, pregnancy, and lactation on drug PK and pharmacodynamics (PD), while Part II considers other covariates, such as gender, heritable traits, age, body composition, and circadian rhythm, and their effect on PK and PD. These covariates are just briefly summarized in this chapter in order to help readers understand the PK principles underlying the human food safety evaluation of new animal drugs.

2.4.1 Gender

Gender disparity in PK has been identified for numerous drugs, and mechanistically, it can be attributed to either molecular or physiological factors. Among veterinary species, gender-related differences in PK have been established in cats (Erichsen et al., 1980; Lainesse et al., 2007), cattle (Dacasto et al., 2005; Janus and Antoszek, 1999), dogs (Bruss et al., 2004; Hay Kraus et al., 2000), and fish (Vega-Lopez et al., 2007). For example, Janus and Antoszek (1999) reported marked sex-linked differences in plasma antipyrine clearance and urinary excretion of the main metabolites of antipyrine in cattle over 12 months of age, with females being the more active metabolizers. However, many domestic animals are castrated prior to reaching full maturity, which can dampen the magnitude of expected gender effects in the PK response. Interestingly, Hutson et al. (2008) showed that castration does not completely eliminate gender effects, as shown in humans undergoing medical castration.

The authors hypothesized that therapeutic reduction in testosterone concentrations would affect the metabolism of other drugs, as testosterone is a substrate of the CYP3A4 drug-metabolizing enzyme. However, their study results showed that the decrease in testosterone concentrations did not lead to a significant change in the activity of the CYP3A4 enzyme.

The major gender-related differences in drug disposition are related to the hepatic CYP450 enzymes, whose function is correlated to sex hormones. Some drug-metabolizing enzymes have overlapping substrate specificities, and they can also metabolize endogenous substances (Riviere, 1999). Therefore, some of the gender differences may be the result of hormones being metabolized by the same enzymatic systems. Different levels of circulating hormones in females at various stages of the reproductive cycle may influence the PK behavior of some drugs.

Many domestic animals are seasonal breeders, with one or more estrous cycles occurring during certain periods of the year. Although seasonal breeders have dormant phases in the reproductive cycle, which could suggest potential differences in the PK response due to the different levels of sex hormones, no studies have specifically compared the impact of gender on drug PK between seasonal breeders and nonseasonal breeders.

As discussed in the "FDA's Guidance for Industry #3: General Principles for Evaluating the Safety of Compounds Used in Food Producing Animals (FDA GFI #3, 2006)," if a drug is intended for use in both male and female animals, then the sponsor should use animals of both sexes in the residue depletion study because males and females of a species can metabolize chemicals at a different rate. By including the animals of both genders, the variability is increased, which will result in a more conservative estimation of the withdrawal time (because the tolerance is calculated based on the 99th percentile of the population and the 95% confidence level). In addition, as emphasized above, animals selected for the study are generally market-weight animals of the proposed target species and production class, such that the residue depletion data would be representative of the residues to which humans would be exposed when consuming animal-derived food products.

Finally, it should be noted that extent and direction of gender differences vary among the veterinary species. Therefore, interspecies extrapolations should be done with great caution (Witkamp et al., 1991). Rodents in general (and rats in particular) tend to exhibit marked gender differences in PK (Niwa et al., 1995; Reinoso et al., 2001), which has been attributed to the different daily rhythm of rat hepatic enzymes and a secretion of the growth hormone between male and female rats (Czerniak, 2001; Furukawa et al., 1999). However, the gender difference observed in rats rarely extends to other species. This is an important point in evaluation of human food safety as toxicology studies are routinely conducted on rodent species (including both genders).

2.4.2 Age

The impact of age on drug PK should be considered for two separate life stages: maturity and senescence versus infancy, adolescence, and immaturity. Maturity and senescence bring about major changes on drug PK response, which are physiologically based on decreased plasma protein binding, declining liver function, and impaired kidney function. However, advanced age has no major importance on the evaluation of human food safety, as food animals rarely reach the level of maturity in which the physiologic processes associated with aging would become affected.

Unlike senescent organisms, neonates have a considerably different physiologic makeup than adults, with higher skin surface area per body weight, greater percentage of body water, less body lipid, insufficient renal blood flow, and functionally immature hepatic and renal function (Bartelink et al., 2006). In addition, variable gastric emptying rates, irregular peristalsis, and increased permeability of intestinal mucosa further influence drug bioavailability in neonates. Although the same physiologic and pathologic processes occur as described for aging humans, information on the impact of age on drug PK is limited in veterinary species. However, based on the same physiologic processes associated with aging, similar impact on drug response (as already well established in humans) is to be expected in veterinary patients (Martinez and Modric, 2010; Modric and Martinez, 2011).

In terms of the role of age on blood and tissue PK, the largest impact of immaturity is observed in the rate of maturation of various enzyme systems associated with drug elimination: some systems take days or weeks, and others months to reach their full capacity. An exception to the rule of general immaturity of metabolizing enzymes includes conjugative reactions which may have considerable activity in the fetal and early neonatal period (Short and Davis, 1970). Patterns of development of hepatic enzyme activities differ in relation to age, species, substrate, and sex (Shoaf et al., 1987). In the pig, adult levels of hepatic enzyme activity and CYP450 contents are reached by about 6 weeks of age (Short and Davis, 1970), compared to 8–10 weeks of age in goats (Burley and Bray, 1983), and 7–42 days in cattle (Shoaf et al., 1987). In sheep, the total CYP450 reached the maximum activity at 7 months (Galtier and Alvinerie, 1996), with CYP2B reaching adult levels at 4 weeks, and CYP3A peaking at 1 week. Different monooxygenases reach maximum activity between 4 weeks and 11 months, and transferases between 1 week and 11 months. In chickens, the rate of microsomal cytochrome P450 reactions progressively increased during the first 9 weeks and decreased thereafter (Coulet et al., 1996). Hepatic monooxygenases were characterized by different developmental patterns. The demethylase activities increased progressively up to 9 weeks, then they declined, in 12 weeks reaching the activity level observed in 3-week-old chickens. Glutathione S-transferase was found to be less active in chickens aged from 3 to 9 weeks compared to 12-week-old ones.

The extent of difference in drug disposition between neonates and adults also depends on the drug involved. For example, low oxidative capacity has been demonstrated in cattle during the first days of life for sulfatroxazole, sulfamethazine, sulfadiazine, and sulfamerazine with the maximum capacity reached at several weeks of life (Nouws, 1992). On the other hand, phenylbutazone[1] is also oxidized, but it takes 2–3 months to reach a maturation point. Therefore, depending on the age of young animals treated with the drug, the elimination may be impaired, possibly (or likely) resulting in violative residues if the residue studies were not conducted in that subpopulation.

Because of different rates and extent of maturation of enzyme systems involved in xenobiotic metabolism, the percent marker to total residue can vary as a function of age. If the marker residue was determined in adults (and it is a metabolite whose metabolic pathways are not fully developed in young animals), when the drug is used in an extralabel manner in neonates, there is a risk that the marker residue will not be relevant for that subpopulation.[2] Therefore, both qualitative and quantitative changes in drug metabolism are important in understanding the possible impact of age on drug metabolism. For example, metabolism is the main contributor to the elimination of trimethoprim by adults, with approximately 90% of the administered dose being o-demethylated and glucuronidated in adult calves and pigs. However, in newborn pigs this percentage is approximately 11% and the glucuronide metabolite is totally absent. As a result, Friis et al. (1984) reported that the average trimethoprim elimination half-life in pigs was 8.1 h at birth, 3.7 h at 9 days, and 2 h at 2 months of life. Similarly, in calves, the elimination half-life at birth was three to eight times longer than in adults (Shoaf et al., 1989). Another example of the effect of age on drug metabolism is the metabolism of sulfonamides. Acetylation of sulfonamides is part of the acetylation–deacetylation equilibrium. While the acetylation reaction in calves and pigs is fully developed at birth, deacetylation is immature, resulting in the higher ratio of N4-acetyl metabolites in newborns (Nouws et al., 1983, 1989). However, in terms of the residue depletion, the relative concentrations are not

[1] Examples of drugs used in this chapter are from literature and their listing does not imply that a drug is approved in the United States for that species and/or indication. These literature examples are used to illustrate a point of how the PK may differ under various conditions of use and do not reflect the official policy of the FDA on use of these drugs. No official support or endorsement by the FDA is intended or should be inferred.

[2] Extralabel use of drugs in the discussion throughout this chapter is used to illustrate possible pharmacokinetic/human food safety concerns if drugs are used outside of their approved label condition. No official support or endorsement of such extra-label use by the FDA is intended or should be inferred.

important; it is the residue in tissue that is the slowest to deplete that becomes the most critical component for determination of drug depletion times.

In addition to the immaturity of excretory systems, there is a substantial difference in body composition between young and mature animals. Volume of distribution is typically greater in neonates due to the greater body water content than in adults (in particular the extracellular fluid volume), while there is a decreased plasma protein binding (due to lower plasma albumin concentrations). With aging, the ratio of lean to fat tissue changes, with more fat tissue present, which will particularly affect the PK behavior of lipophilic drugs.

2.4.3 Body Composition

Tissue distribution of drugs is affected by body composition, regional blood flow, and the affinity of the drug for plasma proteins and/or tissue components. Body composition can affect various physiologic processes involved in the distribution, metabolism, and elimination of drugs, and may, therefore, affect the clearance and half-life. In general, the PK behavior of drugs characterized by low lipophilicity is rather predictable, as they are mostly distributed in lean tissues, whereas the PK behavior of lipophilic drugs is highly affected by body composition changes (Cheymol, 2000).

Body condition has been found to influence the kinetics of various veterinary drugs. For highly lipophilic drugs, such as macrocyclic lactones, studies have shown that the distribution in fat has significant importance in defining their PK behavior. There have been reports of adverse drug reactions in cattle with lower body condition scores, suggesting rapid release of drug from fat storage under weight loss conditions or an inability of drug to accumulate in adipose tissue, resulting in increased plasma levels. Several PK studies indicate different rates of absorption and exposure to various macrocyclic lactones in cattle (Dupuy et al., 2007) and sheep (Echeverría et al., 2002) with different body condition scores and body fat composition, which may be explained by the rapid release of drug from fat storage under weight loss conditions or a lack of adipose tissue into which the drug distributes. The evidence of the body condition effect was further confirmed in thin pigs, in which the plasma ivermectin concentration peaked more rapidly and levels were less persistent, compared to fat pigs, while the concentration in backfat was significantly lower in thin animals slaughtered 3 weeks after treatment (Craven et al., 2002a, b). This raises target animal safety concerns in malnourished animals but may also pose human food safety risks because large fat depots may increase the drug's volume of distribution in overconditioned animals, which in turn can slow drug elimination from fat tissues, possibly resulting in violative residue levels.

2.4.4 Pregnancy and Lactation

Physiological changes accompanying various stages of pregnancy and lactation have been well described in literature for most veterinary species and different classes of compounds. A variety of physiologic processes in the gestational and lactation periods can affect all aspects of ADME, including endogenous hormonal changes, changes in plasma volume, altered body fat proportion, changes in weight or muscle mass, delayed gastric emptying and prolonged gastrointestinal transit time; increased cardiac output, stroke volume, and heart rate; decreased albumin concentration with reduced protein binding; increased blood flow to the various organs; increased GFR; and altered hepatic enzyme activity (Cono et al., 2006).

The most common PK-related effects of pregnancy and lactation are associated with changes in the plasma clearance and volume of distribution, both of which may have an effect on residue depletion in edible tissues of food animals. Because of the continual changes in physiologic parameters during pregnancy and lactation, different stages of pregnancy and lactation will affect drug PK in a different way, therefore requiring that the PK information be interpreted with regard to the gestational and lactational stage. For drugs metabolized by the liver, pregnancy and lactation might further influence the ADME processes. Oukessou and Toutain (1992) studied the influence of the stage of pregnancy on the kinetic disposition of gentamicin in the ewe and showed that the steady state volume of distribution was significantly increased from mid to end of pregnancy (from 0.09 to 0.194 l/kg). Similarly, plasma clearance was increased by about 150% at the end of pregnancy. The authors concluded that these modifications must be taken into account in order to adapt the dosage regimen and determine an appropriate depletion time for gentamicin in ewes. Similar to the effect of the stage of pregnancy on drug PK, it has also been shown that the stage of lactation alters the PK response. Bengtsson et al. (1997) showed that serum concentrations of beta-lactam antibiotics in ewes and cows were markedly lower in early than in late lactation, with significantly higher weight-corrected values of clearance and volume of distribution, and markedly shorter mean residence time and half-life.

As already emphasized for other factors, residue depletion studies are conducted in animals that are representative of the target population for which a drug is being developed. Therefore, if a drug is intended to be used in pregnant or lactating animals, studies may have to be conducted in those classes of animals, unless there is a scientific justification for not conducting the study in those classes. For example, if a product is intended for use in both beef cattle and lactating dairy cows, a withdrawal period determined based on a tissue residue depletion study conducted in beef cattle would likely be used for lactating dairy cows as well, as it represents the worst case scenario. A separate milk residue depletion study would be conducted to establish the milk discard time in lactating dairy cows.

2.4.4.1 Disease/Stress All aspects of the PK response can be affected by the presence of concomitant diseases, with hepatic, renal, and cardiovascular diseases having the most impact. The most profound changes in drug disposition usually accompany renal failure, but various hepatic diseases are also known to markedly impact various aspects of the PK response. Other physiologic conditions, such as inflammation, endotoxemia, and stress, can also significantly alter a drug response and therefore impact drug depletion times. If the drug dose is not modified as a result of hepatic or renal disease, there may be a risk of violative residues in edible products of animals that were treated if they are sent to slaughter following the established withdrawal period. The most common reason for residue violations under clinical conditions of use is the failure to observe the correct withdrawal period (Riviere and Sundlof, 2001). Other common causes include extensive usage and/or excessive dosage, failure to identify treated animals, and prolonged drug clearance, which may be due to the presence of various underlying diseases (Radostits et al., 2007).

The most obvious effect of renal impairment on drug therapy is a decrease in renal excretion, but other processes, such as renal metabolism, protein binding, volume of distribution, and metabolism, can also be affected. Drugs that are normally excreted in urine unchanged will accumulate in the body in case of renal insufficiency if no dose adjustments are made. As a result, there could be a possible risk of violative residues in edible products (depending on the stage of renal impairment), as renal disease can prolong tissue concentrations of drugs and therefore affect the drug depletion time. Besides a decreased renal excretion, the other major impact of renal disease is an alteration in drug distribution patterns, which may also result in violative residues in food animals. In addition, a marked reduction in protein binding that occurs in uremia may affect drug disposition, as a result of higher levels of free drug. However, it should be noted that these situations of significantly impaired renal function are relatively rare in food-producing animals, as production animals generally do not reach levels of maturity typically associated with degenerative renal processes.

Liver disease can have complex effects on intestinal absorption, plasma protein binding, hepatic extraction ratio, liver blood flow and portal–systemic shunting, biliary excretion, enterohepatic circulation, and renal clearance. If the drug dose is not modified as a result of a severe hepatic disease, similar to what was discussed with the presence of renal disease, there is a risk of violative residues in edible products of animals that were treated if they are sent to slaughter following the established withdrawal period. Despite a decreased activity of both Phase I and Phase II enzymes associated with liver disease, liver metabolism is usually not significantly affected until there is a major loss (>80%) of liver function (Novotny, 2001), which again is not a common situation for food-producing animals.

Drug elimination from both the central and peripheral compartments may be impaired in the presence of various diseases, which may become a concern in terms of presence of violative residues in edible products of animals treated with drugs. Because withdrawal times are established in healthy market weight (but typically young) animals, any effect that significantly slows the depletion of drug from tissues, such as with an underlying renal or hepatic disease, could risk the presence of violative residues in edible products. However, the statistical analyses incorporated into the determination of withdrawal times by the Center for Veterinary Medicine (Martinez et al., 2000) ensure that the impact of disease will not cause violative residues *per se*, unless there are some other contributing extrinsic factors (such as not observing withdrawal times or using the drug in an extralabel manner).

2.4.5 Heritable Traits/Breeds

Breed-related differences in pharmacological response to xenobiotics have been reported in various veterinary species, including cattle (Dacasto et al., 2005; Sallovitz et al., 2002), sheep (Ammoun et al., 2006), chickens (Opdycke and Menzer, 1984), and pigs (Sutherland et al., 2005). Genetic factors that are known to cause variability in the PK/PD response include genetic variability in both drug-metabolizing enzymes and in drug transporters. The CYP450 enzymes that are involved in the metabolism of >80% of all clinically used drugs are known to be highly polymorphic. Although not nearly as well recognized and researched as in companion animal species, breed differences in pharmacological response have also been identified in food-producing animals. Sallovitz et al. (2002) reported a significantly slower absorption and lower systemic availability of moxidectin in Aberdeen Angus compared to Holstein calves. Depelchin et al. (1988) reported that the Friesian calves eliminated antipyrine twice as fast as the Blue White Belgian breed, suggesting a breed-related difference in the hepatic microsomal oxidative function. Ripoli et al. (2006) reported on significant polymorphism of the microsomal enzyme diacylglycerol *O*-acyltransferase (DGAT1) in 14 populations of cattle from Argentina, Bolivia, and Uruguay.

Breed differences in the hepatic CYP3A apoprotein and CYP3A-dependent catalytic activities have been reported among Limousine and Piedmontese cattle (Dacasto et al., 2005). Giantin et al. (2008) recently demonstrated the effect of breed upon various Phase I and Phase II drug metabolizing enzymes between three meat cattle breeds: Charolais, Blonde d'Aquitaine, and Piedmontese.

Therefore, similar to what was described for processes that alter the expression of genes modulating drug-metabolizing isoenzymes, the different PK responses confirmed in various cattle breeds could also affect the establishment of marker residue among these breeds. However, these breed differences are not expected to significantly affect the drug depletion time and assignment of withdrawal time, as they are "accounted for" in the statistical analysis of typical residue depletion studies.

2.5 CONCLUSION

In summary, when evaluating factors that may influence drug depletion times, it is important to consider the impact of individual and population differences on drug behavior in the body when animal drugs are used in food-producing animals. Although the importance of these factors was discussed in terms of their potential to influence drug depletion times generally, it should be emphasized that the residue depletion studies that are used to assign regulatory withdrawal times are designed to provide a "worst case" scenario. These residue depletion studies, however, evaluate the maximum dose and treatment duration at which the new animal drug is intended to be used, assess residue depletion in market weight animals of the intended target species and in appropriate gender(s), use the intended (labeled) routes of administration, and incorporate a 99th percentile upper tolerance limit with a 95% confidence interval. We believe that intrinsic and extrinsic factors influencing ADME are factored into the assignment of withdrawal times by the Center for Veterinary Medicine, as described in subsequent sections of this book.

REFERENCES

Ammoun I, Encinas T, Veiga-Lopez A, Ros JM, Contreras I, Gonzalez-Añover P, Cocero MJ, McNeilly AS, Gonzalez-Bulnes A (2006). Effects of breed on kinetics of ovine FSH and ovarian response in superovulated sheep. Theriogenology 66:896–905.

Bartelink IH, Rademaker CM, Schobben AF, van den Anker JN (2006). Guidelines on paediatric dosing on the basis of developmental physiology and pharmacokinetic considerations. Clin Pharmacokinet 45:1077–1097.

Bengtsson B, Jacobsson SO, Luthman J, Franklin A (1997). Pharmacokinetics of penicillin-G in ewes and cows in late pregnancy and in early lactation. J Vet Pharmacol Ther 20:258–261.

Brown SA (2001). Pharmacokinetics: Disposition and fate of drugs in the body. In: Adams R, editor. Veterinary Pharmacology and Therapeutics. 8th ed. Ames: Iowa State University Press; p 15–56.

Bruss ML, Margolin SB, Giri SN (2004). Pharmacokinetics of orally administered pirfenidone in male and female beagles. J Vet Pharmacol Ther 27:361–367.

Brynes Steven D (2005). Demystifying 21 CFR Part 556: Tolerances for residues of new animal drugs in food. Regul Toxicol Pharmacol 42:324–327.

Burley FE, Bray TM (1983). Effect of rumen development and preexposure to chemicals on the activity of the mixed function oxidase system in goats. Comp Biochem Physiol 75C:137.

Cheymol G (2000). Effects of obesity on pharmacokinetics implications for drug therapy. Clin Pharmacokinet 39:215–231.

Coleman MD (2007). Human Drug Metabolism. An Introduction. Chichester: John Wiley & Sons, Ltd.

Cono J, Cragan JD, Jamieson DJ, Rasmussen SA (2006). Prophylaxis and treatment of pregnant women for emerging infections and bioterrorism emergencies. Emerg Infect Dis 12:1631–1637.

Coulet M, Eeckhoutte C, Galtier P (1996). Ontogenic development of drug-metabolizing enzymes in male chicken liver. Can J Physiol Pharmacol 74:32–37.

Craven J, Bjorn H, Hennessy DR, Friis C (2002a). The effects of body composition on the pharmacokinetics of subcutaneously injected ivermectin and moxidectin in pigs. J Vet Pharmacol Ther 25:227–232.

Craven J, Hennessy DR, Friis C (2002b). Does the rate of fat deposition influence the pharmacokinetic disposition of subcutaneously administered moxidectin and ivermectin in pigs? J Vet Pharmacol Ther 25:351–357.

Czerniak R (2001). Gender-based differences in pharmacokinetics in laboratory animal models. Int J Toxicol 20:161–163.

Dacasto M, Eeckhoutte C, Capolongoa F, Dupuy J, Carletti M, Calleja C, Nebbia C, Alvinerie M, Galtier P (2005). Effect of breed and gender on bovine liver cytochrome P450 3A (CYP3A) expression and inter-species comparison with other domestic ruminants. Vet Res 36:179–190.

Depelchin BO, Bloden S, Michaux C, Ansay M (1988). Effects of age, sex and breed on antipyrine disposition in calves. Res Vet Sci 44:135–139.

Dupuy J, Sutra JF, Alvinerie M (2007). Pharmacokinetics assessment of moxidectin long-acting formulation in cattle. Vet Parasitol 147:252–257.

Echeverría J, Mestorino N, Errecalde JO (2002). Comparative pharmacokinetics of ivermectin after its subcutaneous administration in healthy sheep and sheep infected with mange. J Vet Pharmacol Ther 25:159–160.

Erichsen DF, Harris SG, Upson DW (1980). Therapeutic and toxic plasma concentrations of digoxin in the cat. Am J Vet Res 41:2049–2058.

FDA (October 23, 2003). Center for Veterinary Medicine's Guidance for Industry #152: Evaluating the Safety of Antimicrobial New Animal Drugs with Regard to Their Microbiological Effects on Bacteria of Human Health Concern. Rockville: U.S. Department of Health and Human Services.

FDA (June 25, 2006). Center for Veterinary Medicine's Guidance for Industry #3: General Principles for Evaluating the Safety of Compounds Used in Food Producing Animals. Rockville: U.S. Department of Health and Human Services.

FDA (June 3, 2011). Center for Veterinary Medicine's Guidance for Industry #159: Studies to Evaluate the Safety of Residues of Veterinary Drugs in Human Food: General Approach to Establish a Microbiological ADI, VICH GL-36). Rockville: U.S. Department of Health and Human Services.

Friedlander Lynn G, Brynes Steven D, Fernandez Ana H (1999). The human food safety evaluation of new animal drugs. Vet Clin North Am Food Anim Pract 15:1–11.

Friis C, Gyrd-Hansen N, Nielsen P, Olsen CE, Rasmussen F (1984). Pharmacokinetics and metabolism of sulphadiazine in neonatal and young pigs. Acta Pharmacol Toxicol 54:321–326.

Furukawa T, Manabe S, Watanabe T, Sharyo S, Mori Y (1999). Sex difference in the daily rhythm of hepatic P450 monooxygenase activities in rats is regulated by growth hormone release. Toxicol Appl Pharmacol 161:219–224.

Galtier P, Alvinerie M (1996). Pharmacological basis for hepatic drug metabolism in sheep. Vet Res 27:363–372.

Gehring R, Baynes RE, Craigmill AL, Riviere JE (2004). Feasibility of using half-life multipliers to estimate extended withdrawal intervals following the extralabel use of drugs in food-producing animals. J Food Prot 67:555–560.

Giantin M, Carletti M, Capolongo F, Pegolo S, Lopparelli RM, Gusson F, Nebbia C, Cantiello M, Martin P, Pineau T, Dacasto M (2008). Effect of breed upon cytochromes P450 and phase II enzyme expression in cattle liver. Drug Metab Dispos 36(5):885–893.

Hay Kraus BL, Greenblatt DJ, Venkatakrishnan K, Court MH (2000). Evidence for propofol hydroxylation by cytochrome P4502B11 in canine liver microsomes: Breed and gender differences. Xenobiotica 30:575–588.

Horrigan L, Lawrence RS, Walker P (2002). How sustainable agriculture can address the environmental and human health harms of industrial agriculture. Environ Health Perspect 110:445–456.

Hutson PR, Oettel K, Douglas J, Ritter M, Messing E, Staab MJ, Alberti D, Horvath D, Wilding G (2008). Effect of medical castration on CYP3A4 enzyme activity using the erythromycin breath test. Cancer Chemother Pharmacol 62:373–377.

Janus K, Antoszek J (1999). The effect of sex on antipyrine metabolism in cattle at different ages. J Vet Pharmacol Ther 22:163–169.

Lainesse C, Frank D, Beaudry F, Doucet M (2007). Effects of physiological covariables on pharmacokinetic parameters of clomipramine in a large population of cats after a single oral administration. J Vet Pharmacol Ther 30:116–126.

Martinez M, Modric S (2010). Patient variation in veterinary medicine: Part I. Influence of altered physiological states. J Vet Pharmacol Ther 33:213–226.

Martinez M, Friedlander L, Condon R, Meneses J, O'Rangers J, Weber N, Miller M (2000). Response to criticisms of the US FDA parametric approach for withdrawal time estimation: Rebuttal and comparison to the nonparametric method proposed by Concordet and Toutain. J Vet Pharmacol Ther 23:21–35.

Martinez MN, Amidon GL (2002). A mechanistic approach to understanding the factors affecting drug absorption: A review of fundamentals. J Clin Pharmacol 42:620–643.

Modric S, Martinez M (2011). Patient variation in veterinary medicine: Part II. Influence of physiological variables. J Vet Pharmacol Ther 34:209–223.

Niwa T, Kaneko H, Naritomi Y, Togawa A, Shiraga T, Iwasaki K, Tozuka Z, Hata T (1995). Species and sex differences of testosterone and nifedipine oxidation in liver microsomes of rat, dog and monkey. Xenobiotica 25:1041–1049.

Nouws JF (1992). Pharmacokinetics in immature animals: A review. J Anim Sci 70:3627–3634.

Nouws JF, van Ginneken CA, Ziv G (1983). Age-dependent pharmacokinetics of oxytetracycline in ruminants. J Vet Pharmacol Ther 6:59–66.

Nouws JF, Vree TB, Mevius D, Degen M (1989). Pharmacokinetics, metabolism and renal clearance of sulphatroxazole in calves and cows. J Vet Pharmacol Ther 12:50–57.

Novotny MJ (2001). Clinical pharmacology: Principles of therapeutics. In: Adams R, editor. Veterinary Pharmacology and Therapeutics. 8th ed. Ames: Iowa State University Press; p 57–68.

Opdycke JC, Menzer RE (1984). Pharmacokinetics of diflubenzuron in two types of chickens. J Toxicol Environ Health 13:721–733.

Oukessou M, Toutain PL (1992). Influence of the stage of pregnancy on gentamicin disposition in the ewe. Ann Rech Vet 23:145–150.

Radostits OM, Gay CC, Blood DC, Hinchcliff KW (2007). A Textbook of the Diseases of Cattle, Sheep, Pigs. 10th ed. Philadelphia: Saunders Elsevier.

Reinoso RF, Farrán R, Moragón T, García-Soret A, Martínez L (2001). Pharmacokinetics of E-6087, a new anti-inflammatory agent, in rats and dogs. Biopharm Drug Dispos 22:231–242.

Ripoli MV, Corva P, Giovambattista G (2006). Analysis of a polymorphism in the DGAT1 gene in 14 cattle breeds through PCR-SSCP methods. Res Vet Sci 80:287–290.

Riviere JE (1999). Comparative Pharmacokinetic: Principles, Techniques, and Applications. Ames: Iowa State University Press.

Riviere JE, Sundlof SF (2001). Chemical residues in tissues of food animals. In: Adams R, editor. Veterinary Pharmacology and Therapeutics. 8th ed. Ames: Iowa State University Press; p 1166–1174.

Sallovitz J, Lifschitz A, Imperiale F, Pis A, Virkel G, Lanusse C (2002). Breed differences on the plasma availability of moxidectin administered pour-on to calves. Vet J 164:47–53.

Shoaf SE, Schwark WS, Guard CL, Babish JG (1987). The development of hepatic drug-metabolizing enzyme activity in the neonatal calf and its effect on drug disposition. Drug Metab Dispos 15:676–681.

Shoaf SE, Schwark WS, Guard CL (1989). Pharmacokinetics of sulfadiazine/trimethoprim in neonatal male calves: Effect of age and penetration into cerebrospinal fluid. Am J Vet Res 50:396–403.

Short CR, Davis LE (1970). Perinatal development of drug-metabolizing enzyme activity in swine. J Pharmacol Exp Ther 174:185–196.

Sutherland MA, Rodriguez-Zas SL, Ellis M, Salak-Johnson JL (2005). Breed and age affect baseline immune traits, cortisol, and performance in growing pigs. J Anim Sci 83:2087–2095.

Vega-Lopez A, Galar-Martinez M, Jimenez-Orozco FA, Garcia-Latorre E, Dominguez-Lopez ML (2007). Gender related differences in the oxidative stress response to PCB exposure in an endangered goodeid fish (*Girardinichthys viviparus*). Comp Biochem Physiol A Mol Integr Physiol 146:672–678.

Witkamp RF, Lohuis JA, Nijmeijer SM, Kolker HJ, Noordhoek J, van Miert AS (1991). Species- and sex-related differences in the plasma clearance and metabolite formation of antipyrine. A comparative study in four animal species: Cattle, goat, rat and rabbit. Xenobiotica 21:1483–1492.

3

EVALUATION OF DRUG RESIDUE DEPLETION IN THE EDIBLE PRODUCTS OF FOOD-PRODUCING ANIMALS FOR ESTABLISHING WITHDRAWAL PERIODS AND MILK DISCARD TIMES

Dong Yan

Division of Human Food Safety, Office of New Animal Drug Evaluation, Center for Veterinary Medicine, United States Food and Drug Administration, Rockville, MD, USA

3.1 INTRODUCTION

Therefore, as part of the approval process for new animal drug products intended for use in food-producing animals, the U.S. Food and Drug. Administration's (FDA's) Center for Veterinary Medicine (CVM) requires that pharmaceutical companies demonstrate that the edible products of treated animals are safe for human consumption. The edible products considered in the safety evaluation for human food derived from treated animals, referred to as "human food safety evaluation" for the purpose of this document, include muscle, liver, kidney, fat (or skin with fat) (collectively referred to as edible tissues in the document), and milk and eggs (when appropriate).

Strategies for Reducing Drug and Chemical Residues in Food Animals: International Approaches to Residue Avoidance, Management, and Testing, First Edition.
Edited by Ronald E. Baynes and Jim E. Riviere.

The drug residues considered in human food safety evaluation include not only the active pharmaceutical ingredient(s) and excipient(s) of a drug product but also their metabolites and any substance formed in or on the edible food products as a result of drug treatment. For antimicrobial products, drug residues may include antimicrobial-resistant bacteria. Animal drug products may be xenobiotics, containing compounds foreign to the target (food-producing) animals (including modified compounds of endogenous origin), or endogenous compounds naturally present in the target animals and humans. There are differences in the approaches used to evaluate the human food safety of residues of xenobiotics and endogenous compounds used in food-producing animals.

Food safety evaluation of animal drug residues in food-producing animals follows risk assessment principles and is based on the standard of reasonable certainty of no harm to human consumers. The evaluation takes into consideration the traditional toxicological effects of the drug residues on human health, the risk of developing antimicrobial-resistant bacteria due to the use of antimicrobial drug products in animals, the effects of drug residues on human intestinal flora, human exposure to the drug residues, and approaches needed to mitigate the exposure. Excellent comprehensive discussions on these topics have been published previously (Friedlander Lynn et al., 1999).

3.1.1 Residue Safety Standards for Total Residues

The evaluation of traditional toxicology information for residues of a new animal drug establishes the acceptable daily intake (ADI), which is a food safety standard for all drug-related residues (total residues) of toxicological concern. In some cases, information is available that identifies the specific components of the total residues that cause the toxicological concerns and those components that do not. For some antimicrobial drugs, an ADI can also be determined using a microbiological end point based on the evaluation of the effect of the drug residues on human intestinal flora. Therefore, the final ADI for residues of an antimicrobial drug product could be a toxicological ADI or a microbiological ADI, whichever is lower.

The residue chemistry evaluation of drug residues is based on one of the following three residue safety standards:

1. ADI value. An ADI is usually assigned for drug residues derived from xenobiotics.
2. Allowable incremental increase limits. An allowable incremental increase limit means that no residues resulting from the drug use are permitted in excess of the increment above the concentration of the endogenous compound naturally present in untreated animals.

3. No human food safety concerns for the drug residues. In that case, neither an ADI nor an allowable incremental increase limit is assigned. When this conclusion is reached, there are usually no tolerance, no target tissue, and no analytical methods required.

3.1.1.1 Partition of an ADI and Calculation for Safe Concentrations The ADI is used to calculate the safe concentrations for total residues in each edible product. The safe concentration represents the maximum concentration of the total residues that is allowed in a specific animal-derived edible product at the time the animal is slaughtered or when milk or eggs are collected for human consumption. The procedure for calculating safe concentrations ensures that when the total residues in edible products derived from animals treated with the drug product are below their assigned safe concentrations, the daily exposure of consumers to the drug residues does not exceed the ADI.

The calculation for safe concentrations assumes that, on a daily basis, consumers of a full portion of one edible tissue from any meat-producing species (poultry, pork, beef, etc.) will not consume another edible tissue from either the same animal species or another species, but may consume milk and eggs on the same days they eat meat (US FDA, 2006). It also takes into consideration the average body weight of human adults, 60 kg, and the anticipated maximum amount of edible products consumed per day (food factors). These factors are applicable for the calculation of safe concentrations for all the species for which the drug product is approved. Currently, an average adult consumer is assumed to consume up to 300 g muscle, 100 g liver, 50 g kidney, or 50 g fat, plus up to 1.5 l milk and 100 g eggs (US FDA, 2006). Therefore, if a drug product is proposed for use in dairy cattle and laying hens, a fraction of the ADI would be reserved for the calculation of safe concentrations in milk and eggs. In such cases, only the remainder of the ADI is available for calculating safe concentrations for the edible tissues (meats). This reduction of the percentage of the ADI assignable to tissues affects the safe concentration calculations for tissues from all food-producing animal species. For example, if 30% of the ADI is reserved for cattle milk, only 70% of the ADI would be available for the calculation of safe concentrations for the edible tissues of cattle. The same 70% of the ADI would be used for calculating the safe concentrations for the edible tissues of swine or poultry, even though there are no milk products for human consumption derived from swine or poultry. It should be noted that the portion of the ADI allocated to the edible tissues is not further partitioned among individual tissues; thus the whole portion of the ADI allocated to the edible tissues is used for the calculation of safe concentrations for muscle, liver, kidney, and fat.

$$\text{Safe concentration} = \frac{\text{ADI} \times \text{human weight}}{\text{Food factor}}$$

3.1.1.2 Calculation for Allowable Incremental Increases for Residues of Endogenous Substance Some drug products consist of endogenous substances. For residue chemistry evaluation of these products, it is not possible to differentiate the substances that are endogenously produced from those that are exogenously administered. There is usually no ADI assigned for residues of endogenous substances. Instead, allowable incremental increase limits may be assigned for residues of endogenous substances without the need for additional animal studies. This is because when the substances are endogenous in people, an individual is exposed to large quantities of these substances by *de novo* synthesis relative to the much smaller quantities from edible products derived from food-producing animals that are treated with these substances (US FDA, 2006). For example, the FDA has concluded that the allowable incremental increase for progesterone should be based on the daily production of progesterone by prepubertal boys. This subset of the human population is considered to provide lowest daily production of 150 μg progesterone from *de novo* synthesis (US FDA, 2006). Using this as the limit, the permitted increased exposure from foods derived from treated animals is 1.5 μg (1% of 150 μg) per person per day (US FDA, 2006). The allowable incremental increase is calculated for each edible tissue by dividing the permitted increased exposure value by the individual food factor in a way that parallels the safe concentration calculations using the ADI. The allowable incremental increases, although different from safe concentration and tolerances, essentially serve the same function as food safety standards. However, when the allowable incremental increase is used as the regulatory paradigm, the analytical method requirements to support a drug product approval are likely to be more flexible. This is because the safety assessment compares residues in untreated control animals with those in treated animals.

3.1.2 Residue Exposure Evaluation and Mitigation

The residue chemistry evaluation of new animal drug products ensures human food safety by characterizing the quantity and composition of drug residues in the edible products derived from treated animals and mitigating human consumer exposure to the residues through the assignment of a withdrawal period or a milk discard time as a condition of approval for new animal drug products. A withdrawal period is the interval between the time of the last drug treatment and the time at which the treated animals can be slaughtered for human consumption. A milk discard time is the interval between the time of the last drug treatment and the time before which the milk from the treated animals is not safe for human consumption. For a new animal drug product intended for use in laying hens, the product needs to qualify for a zero withdrawal time for the eggs.

3.2 INFORMATION NEEDED FOR DETERMINATION OF WITHDRAWAL PERIODS OR MILK DISCARD TIMES

The residue chemistry evaluation for the assignments of withdrawal periods and milk discard times makes use of the target tissue, marker residue, and tolerance concepts (Brynes, 2005; US FDA, 2006). When completed, the evaluation assigns withdrawal periods and milk discard times, ensuring that the edible products from treated animals are safe for human consumption.

A discussion of these concepts is provided below:

1. Target tissue—The target tissue is the edible tissue that is selected for monitoring drug residues in the target animals. It is usually the tissue from which the total drug residues deplete most slowly to the safe concentration. Target tissue assigned is often liver or kidney but can be fat or muscle, as appropriate.
2. Marker residue—The marker residue is the residue (analyte) selected for use in the postmarket monitoring program. A marker residue may be the parent drug, a metabolite, or, in rare cases, a combination of residues, if a common analytical method can be established for their measurement. The concentration of the marker residue has a known relationship with the concentration of the total residues. As such, the amount of marker residue is indicative of the amount of the total residues in edible products. A finding that the marker residue is at or below the tolerance in the target tissue from a treated animal indicates that all edible tissues from that animal are safe for human consumption.
3. Tolerance—The tolerance is the maximum concentration of a marker residue for a new animal drug that can legally remain in a specific edible tissue or milk or eggs of a treated animal and raises no concerns for human food safety. The tolerance is the concentration of the marker residue in the tissue at the time when the total drug residues deplete to the safe concentration. The tolerance values are determined using the data generated in the specific species and, in some cases, the production classes of animals for which the drug product will be approved. Therefore, the tolerance value pertains to only the species and production classes of animals for which a drug product has been approved. If a compound is approved for different uses in a single species (e.g., different doses, or different routes of administration), the codified

tolerance is the lowest of all the tolerances that may be calculated based on the available data for supporting the drug product approvals in that species.

The target tissue, marker residue, and tolerance assignments usually rely on the residue chemistry information obtained from the total residue depletion and drug metabolism studies.

3.2.1 Total Residue Depletion Study

Generally, the first residue chemistry study a sponsor conducts is the study that monitors drug metabolism and depletion of total residues in the tissues and excreta of the target animals. The study establishes a correlation between the total residues and marker residue in the edible products as a function of time. The study enables the calculation for the marker residue to total residue ratio. If the drug product is intended for use in lactating dairy cows and laying hens, milk and egg samples are also collected for the total residue analyses. The study results provide essential information on the pattern of total residue distribution, the rate of total residue depletion, and the metabolite profile of the drug.

To enable tracking of total residues in the target animal species, a common approach is to radiolabel the drug used in the study and dose animals at 1 or 1.5 times the proposed label dose. The total residues in tissues and excreta of the treated animals is then measured by measuring the radioactivity without separating or identifying the residue components. Because the safe concentrations, which are calculated from the ADI, reflect the maximum concentrations of total residues that are allowed in the edible products of treated animals, the total residue concentrations determined in the total residue depletion study are directly comparable to the safe concentrations for determining safety of the edible products.

However, in most cases, the total residue depletion study is not used for the determination of a withdrawal period. This is because:

1. The study is conducted with a small number of animals and is not suitable for addressing the issue of residue variation in individual animals.
2. The study is not conducted under field use conditions due to the constraints associated with the use of radioactive material. For example, a product proposed for use in animal feed or drinking water may be administered *via* oral gavage in the total residue depletion study, which ensures complete animal dosing and limits environmental contamination.

3. The study monitors total radioactivity in the edible products by combustion and liquid scintillation counting, which are not methods suitable for post approval monitoring.

The total residue depletion study results, however, are useful for the determination of whether a drug product could qualify for a zero-day withdrawal period. If the total residue concentrations in the edible tissues of the animals that received the proposed label dose are below half the respective safe concentrations for the drug residues at the practical zero withdrawal (between 8 and 12 hours after the last administration of the drug product for large animals or 4 and 6 hours after the last administration of the drug product for poultry species), or if the total residue concentrations in the edible tissues of the animals that received 1.5 times the proposed label dose are below the respective safe concentrations, then the drug product would qualify for a zero-day withdrawal.

When a compound is to be used in milk- or egg-producing animals, a tolerance may need to be established for the marker residue in these food commodities. In some cases, it may be necessary to select a marker residue for milk or eggs that is different from the marker residue selected for the target tissue representing the edible tissues. The approach to setting a tolerance for milk and eggs parallels that for edible tissues. To provide the information needed to establish the marker to total residue ratio and tolerance in eggs, a total residue depletion study is conducted in laying hen dosed for 12 days, the time necessary to encompass complete egg yolk development.

3.2.2 Metabolism and Comparative Metabolism Studies

Drug metabolism in the target animals is evaluated using edible products, urine, and feces obtained from the radiolabeled total residue depletion studies. High-performance liquid chromatography and mass spectrometric detection are often used for residue identification and quantitation. The residue components, especially major residue components of the total residues, are characterized and quantified in the metabolism study. Major metabolites are those residue components that account for at least 10% of the total residues or whose concentration is at least 0.1 parts per million (ppm) (US FDA, 2006). The metabolite profiles in the target animal species are compared with the profiles in the toxicology model species treated orally with the same drug substance. The purpose of the comparison is to confirm that the metabolites present in the target animal species are also present in the toxicological model species used for establishing the residue safety standard for human food (ADI and safe concentrations). It is helpful to use the same analytical method for analyzing samples from the target species and toxicology model species as this would help minimize difficulties associated with comparing the metabolites (US FDA, 2006).

3.2.3 Target Tissue, Marker Residue and Tolerance Determination

As discussed earlier, the total residue depletion and metabolite profile data in the target animals are used to support the assignments of the target tissue, marker residue, and tolerance for the marker residue in the target tissue. These residue chemistry parameters are needed for determining the withdrawal period and facilitating the postmarket monitoring of drug residues. By establishing the relationship between the total residues and the marker residue in the target tissue, we now have a way to monitor total residues indirectly by measuring the marker residue concentration in the target tissue using a typical analytical method without the need to radiolabel the drug. Because tolerance is the concentration of a marker residue in the target tissue at the time when the total residues in the target tissue deplete to the safe concentration, when the marker residue in the target tissue has depleted to the tolerance, the total residues in all the edible tissues from the treated animal are below their respective safe concentrations and are safe for human consumption. Likewise, where applicable, when the concentrations of the marker residue in milk and eggs are at or below their respective tolerances, the milk and eggs are safe for human consumption.

3.2.4 Marker Residue Depletion Study

If drug residues in edible products at zero-day withdrawal are not determined to be safe in the total residue study, a withdrawal is assigned as part of the conditions of use for drug product approval. The determination of the final withdrawal period is based on the data from studies that monitor the depletion of the marker residue from the target tissue over time. The studies are conducted in accordance with good laboratory practice regulations, under field use conditions, using the commercial (final) formulation of the drug product. Samples of tissue are collected at different times following cessation of drug treatment, and marker residue concentrations are analyzed.

To facilitate proper design of the depletion studies, it is important to have knowledge of the presumptive target tissue, marker residue, and tolerance assignments before the study is initiated. This allows for selection of appropriate sampling times and collection of relevant tissue samples. Because the tolerances established by the FDA are linked to the analytical methods by which they were determined, it is essential that the final residue depletion studies that establish the withdrawal period or milk discard time are based on the same analytical methodologies.

If a drug product is labeled for a variety of treatment modalities (e.g., multiple doses, a dose range, or a variety of treatment durations), the residue depletion study used to set the withdrawal period should be conducted with the highest label dose for the longest treatment duration on the label. This will provide depletion data that represent the worst-case residue scenario and support the assignment of a withdrawal period that ensures residue safety of all the treatment modalities included in the label. Where appropriate, and if it can be followed in practice, a drug product label can list multiple withdrawal times consistent with specific treatments.

If a drug product is intended for use in lactating dairy cattle, a milk discard period may be needed as a condition for approval. A milk discard period, if followed, would mean that milk collected after the milk discard period is safe for human consumption. As with tissue residue depletion studies, milk residue depletion studies are conducted in accordance with good laboratory practice regulations, under field use conditions, using the commercial (final) formulation of the drug product. Milk samples are collected at different times following cessation of drug treatment, and drug residue concentrations are analyzed. When the concentration of the marker residue in milk depletes to or below the milk tolerance, the milk from the treated animals is safe for human consumption.

Reflective of the way eggs are produced in the United States, a drug product intended for use in laying hens needs to qualify for a zero withdrawal for eggs.

3.3 FACTORS FOR CONSIDERATION IN CONDUCTING A MARKER RESIDUE DEPLETION STUDY

3.3.1 Animals Species, Class, Gender, and Maturity

The animals selected for the study should be healthy animals and representative of the proposed target species, production class, and gender, such that the residue depletion data obtained from the study are representative of the residues to which humans would be exposed when consuming animal-derived food products. Animals are generally of market weight, even if the drug product is intended to treat conditions associated with or limited to younger animals, because food for human consumption is derived from market weight animals. For a drug product intended for use in animals of more than one subclass, a selected subpopulation of animals may be used to establish the withdrawal period for other relevant subclasses and maturities, if there are scientific justifications for doing so. For example, if a product is intended for use in both beef cattle and lactating dairy cows, a withdrawal period determined based on a tissue residue depletion study conducted in beef cattle will likely be appropriate for lactating dairy cows as well. However, if different subclasses of animals are expected to be significantly different in drug pharmacokinetics and metabolism (e.g., preruminating calves typically marketed as veal), separate studies should be conducted. Of course, a separate milk residue depletion study would be conducted to establish the discard time for milk for human consumption derived from animal species (i.e., lactating dairy cows, sheep, and goats).

3.3.2 Animal Husbandry

The normal husbandry for the animals should be maintained according to industry standards. The animal health records, including feed and water intake and immunizations administered, should be maintained as part of the study records. A particularly important aspect of record keeping is the animal history, including any prior or concurrent treatment of the study animals with other veterinary drugs or feed additives. Concomitant administration of drugs or feeds can potentially affect residue depletion in these animals or the performance of the analytical method for monitoring the marker residue depletion in the incurred samples of the treated animals and are generally discouraged in depletion study design.

3.3.3 Number of Animals

The number of animals assigned to each sampling point should be sufficient to allow a meaningful assessment of the residue concentrations in treated animals to provide sufficient residue data. Readers are referred to VICH GL 48 (MRK—Metabolism and Residue Kinetics, February 2011, for Implementation at Step 7—Final) for recommendations regarding the number of animals needed for marker residue depletion studies.

3.3.4 Dose and Administration

The study should be conducted using the final drug product formulation, because excipients in the drug formulation could affect residue depletion in the treated animals. In general, the study should be conducted based on the dose, route of administration, treatment frequency, treatment duration, and restrictions that will appear on the product label.

3.3.5 Sampling Time Intervals

Tissue samples should be collected from animals after the last treatment with the product. The choice of tissue sampling time points should include those closest to the time when residues are anticipated to be just at or below the tolerance, on the linear portion of the marker residue depletion curve of the target tissue. The total drug residue depletion study data in the target animals often provide preliminary information for determining the number of sampling time points needed in the marker residue depletion study for establishing a withdrawal period.

3.3.6 Tissue Sample Collection

Samples of the target tissue and other edible tissues, as needed, should be collected at the intended sampling times. For injectable drug formulations, collection and analysis of the injection site residue information may be necessary. If milk discard time needs to be established, milk samples should be collected to determine the residue depletion pattern as a function of time in milk.

3.3.7 Tissue Sample Analysis and Data Report

To establish a withdrawal period or milk discard time, a reliable analytical method must be available to assay quantitatively the amount of marker residue in the target tissue or milk. In most cases, the samples from the residue depletion study are analyzed using the determinative analytical method developed by the drug sponsor. Because the tolerance assigned for the marker residue is linked to the analytical method by which it was determined, the analytical method used for the marker residue depletion study is usually based on the same analytical methodology. Deviations from the tolerance-determining method require bridging data to determine how the residue concentrations determined with the two methods are related.

Prior knowledge of storage stability of the analytes in the tissue matrix may be helpful in planning the in-life phase and analytical phase of the marker residue depletion study. The analysis of the samples from the depletion studies must be carried out within the time for which sample storage stability has been properly demonstrated.

3.3.8 Withdrawal Time Calculation

A withdrawal period is determined based on the time needed for the upper tolerance limit of the marker residue concentration in the target tissue to deplete to the established tolerance. The FDA uses the 99th percentile of the population and the 95% confidence level for the statistical withdrawal period calculation.

In some cases, a single time point statistical calculation of residue data for the upper 99% tolerance limit at 95% confidence level may be appropriate. This approach is commonly used to confirm the appropriateness of a proposed withdrawal period, based on the residue data at the single time point selected for the calculation, and may be applicable when there are slight changes in drug product formulations or for proposed generic copies of pioneer products where a withdrawal time has already been assigned for the pioneer product.

3.3.9 Special Case Considerations Related to Setting a Withdrawal Period

3.3.9.1 Drug Products that Cause Persistent Residues at the Injection Site Some injectable products administered at high doses or in long-acting formulations cause drug residues at the injection site that deplete more slowly to the injection site safe concentration than those in the target tissue to the target tissue safe concentration. However, because injection sites may be difficult to identify during post-mortem sampling at the slaughter facility and because injection site residue concentrations are highly variable, injection sites are usually not suitable as the target tissue for postmarket residue monitoring. In these cases, a more traditional target tissue (e.g., liver or kidney) is generally chosen for residue monitoring of these types of products. Where persistent injection site residues are likely, the residue chemistry evaluation may consider the marker residue depletion from both the target tissue and the injection site.

A possible first step for establishing the withdrawal period for such injectable products is to determine an injection site research tolerance. The injection site research tolerance is calculated using the safe concentration for the injection site muscle, which is usually 10 times the safe concentration for remote muscle calculated from an ADI. In cases where an acceptable single-dose intake (ASDI) is established for acute exposure to injection site residues, the injection site safe concentration is calculated from the ASDI values directly by multiplying the ASDI by 60 kg (average body weight of an adult) and dividing it by 300 g (muscle consumption value). The calculation for the injection site research tolerance is similar to the calculation for a tolerance in the marker residue of the target tissue, except that the injection site research tolerance calculation uses the injection site marker to total residue ratio. The injection site research tolerance is called "research tolerance" because it serves as the basis for determining the withdrawal period for the product to ensure that injection site residues are within the safe limit, but it is usually not codified nor is it used as a residue safety standard for postmarket residue safety monitoring.

The tolerance assigned for the target tissue (e.g., liver or kidney) of these injectable products should conform to the withdrawal period needed to ensure the safety of residues at the injection site. For that reason, an iterative approach can be used to arrive at a target tissue tolerance for these drug products that ensures the safety of all of the edible tissues, including the injection site. The tolerance assignment for the target tissue through this process is not based on the marker residue to total residue ratio at the time when the total residues in the target tissue deplete to the safe concentration, but is adjusted to reflect the marker residue concentration in the target tissue at the withdrawal period (i.e., at the time when the injection site residue is safe for human consumption). This approach acknowledges that the injection site is rate-limiting for determining the withdrawal period. However, by setting the tolerance for the target tissue to conform to the withdrawal period needed for the injection site residues to deplete to the injection site research tolerance, the target tissue can be retained as a practical and effective tissue for postmarket monitoring.

3.3.9.2 Withdrawal Period Assignment for Large Molecule Products Large molecule drug products, such as peptides, proteins, or nucleic acids, are often subject to enzymatic digestions in the biological systems of food animal species treated with such drugs. The intrinsic instability of these products makes it difficult to determine a target tissue, a marker residue, or a tolerance, or to develop regulatory methods for quantitating and confirming residues of these drug products in biological matrices. In such cases, it is often beneficial for drug sponsors to contact the FDA for guidance on the path forward for addressing human food safety concerns with regard to the large molecule drugs of their interest.

ACKNOWLEDGMENTS

The author wishes to thank Dr. Lynn G. Friedlander of CVM, FDA, for providing many constructive comments on the manuscript.

REFERENCES

Brynes SD (2005) Demystifying 21 CFR part 556: Tolerances for residues of new animal drugs in food. Regul Toxicol Pharmacol 42:324–327.

Friedlander LG, Brynes SD, Fernandez AH (1999). The human food safety evaluation of new animal drugs. Vet Clin North Am Food Anim Pract 15:1–11.

US FDA (2006). Center for Veterinary Medicine's Guidance for Industry #3: General principles for evaluating the safety of compounds used in food producing animals. Rockville: US FDA.

4

ESTABLISHING MAXIMUM RESIDUE LIMITS IN EUROPE*

Kornelia Grein and Isaura Duarte
European Medicines Agency, London, UK

4.1 INTRODUCTION

One of the fundamental principles for the protection of public health in relation to veterinary medicines in the European Union (EU) is that no veterinary medicinal product intended for use in food-producing animals can be authorized unless an evaluation for the establishment of maximum residue limits (MRLs) has been undertaken for the pharmacologically active substances it contains and specific MRLs have been established, or it has been concluded that no MRL is necessary for the protection of human health. This requirement applies not only to the active substances but also to excipients, if these are capable of having a pharmacological effect.

An MRL is defined as follows: "the maximum concentration of a residue of a pharmacologically active substance which may be permitted in food of animal origin," with the residue definition given as follows: "'residues of pharmacologically active substances' means all pharmacologically active substances, expressed in mg/kg or μg/kg on a fresh weight basis, whether

*All views expressed in this contribution are strictly personal and should not be understood or quoted as being made on behalf of the European Medicines Agency.

Strategies for Reducing Drug and Chemical Residues in Food Animals: International Approaches to Residue Avoidance, Management, and Testing, First Edition.
Edited by Ronald E. Baynes and Jim E. Riviere.

active substances, excipients or degradation products, and their metabolites which remain in food obtained from animals."

MRLs are the points of reference for the establishment of withdrawal periods for veterinary medicinal products to be used in food-producing animals and for the control of residues in food of animal origin in the EU and at border inspection controls.

The evaluation for the establishment of MRLs is carried out in a separate procedure from the evaluation of the application for a marketing authorization of the product concerned and precedes the marketing authorization application. The establishment of the withdrawal period is part of the marketing authorization procedure as it is dependent on the formulation of the final product and not only on the active substance.

4.2 PROCEDURE FOR THE ESTABLISHMENT OF MRLs

The principles for the establishment of MRLs are laid down in Regulation (EC) No 470/2009 (European Parliament and Council, 2009) and those for a marketing authorization in Directive 2001/82/EC (European Parliament and Council, 2004a) and Regulation (EC) No 726/2004 (European Parliament and Council, 2004b). Regulation 470/2009 also applies for the establishment of MRLs for substances included in biocidal products for use in animal husbandry. Any specific procedures regarding this group of substances are not addressed here.

The Committee for Medicinal Products for Veterinary Use (CVMP) of the European Medicines Agency carries out the scientific evaluation of the data submitted with the request for the establishment of MRLs. Following its evaluation, the CVMP delivers a scientific opinion on the MRL application, which includes a recommendation with regard to the establishment of MRLs.

The CVMP opinion is transmitted to the European Commission. On the basis of the scientific recommendation, the Commission prepares a draft Commission Regulation for the MRL classification of the substance for approval by EU member states prior to adoption by the European Commission. Once adopted, the Commission publishes the MRL classification for the pharmacologically active substance concerned in the *Official Journal of the European Communities* (European Commission, 2009), which is directly applicable in all countries of the EU.

The possible outcomes of the evaluation and the subsequent classification for the pharmacologically active substances are as follows:

a. An MRL
b. A provisional MRL

c. No MRL required
d. Prohibited substance

The specific animal species and target tissues or foodstuffs are indicated for the classifications (a), (b), and (c), as appropriate. In addition, with regard to classification (c), specific conditions may apply such as route of administration (e.g., topical use only).

If an MRL has been established for a substance (a), this means that in the opinion of the CVMP, sufficient data were available to allow a complete risk assessment to be undertaken, and in light of the risk assessment the European Commission agreed that final MRLs could be established.

If a provisional MRL has been established (b), this means that there are no grounds for supposing that residues of the substance at the provisional MRL will present a hazard for the health of the consumer but that some scientific data that are not crucial for the risk assessment are lacking. Normally, the outstanding issues for the establishment of final MRLs relate to aspects of the validation of the analytical method proposed for monitoring of residues. Provisional MRLs apply for a defined period of time not exceeding 5 years in the first instance. That period may be extended once for a maximum of 2 years to allow for the completion of scientific studies in progress. The period granted for addressing the outstanding issues takes into account the nature of the issues and the time anticipated necessary for its resolution.

If it is concluded that no MRL is required (c), this means that at the time when the evaluation is completed, residues of the substance concerned are not considered to present a public health hazard at any level or at the levels observed following the intended use of the substance. The classification of a substance as "no MRL required" may only be granted after the evaluation of the safety of residues of the substance concerned. Thus the designation of this classification has the same effect as the allocation of MRL values to a substance, meaning that the substance may be used in veterinary medicinal products intended for food-producing species.

If a prohibition on the administration of a substance is established (d), this means that the pharmacologically active substance cannot be used in veterinary medicinal products intended for food-producing animals—and biocidal products for use in animal husbandry—either because residues of the substance concerned, at whatever concentration, in foodstuffs of animal origin would constitute a hazard to the health of the consumer or because no final conclusion concerning the effect on human health of a substance can be drawn.

The procedure for extending existing MRLs to further species or to include additional target tissues or food commodities (e.g., milk or eggs) or for modifying MRLs is in principle the same as for a full application. However, the data to be submitted differ, and only new data relevant for the extension or modification of MRLs are required. Typically, the data to be submitted would be residue depletion studies in the new target species. Safety data would normally be required only if the request concerns the modification of the acceptable daily intake (ADI).

The requirement for the establishment of maximum residues limits for all pharmacologically active substances in medicines for food-producing animals was introduced in the EU in 1992. The legislation made the establishment of MRLs compulsory prior to the granting of new marketing authorizations but also for all substances included in veterinary medicinal products already authorized at that time for use in food-producing species. A transition period was granted for the so-called old substances, that is, substances already in use in veterinary medicines in the EU on the day of entry into force of the MRL legislation to allow for the evaluation of these substances. Following the end of the transition period (December 31, 1999) all marketing authorizations for medicines containing substances that could not be classified under (a), (b), or (c) had to be withdrawn.

In total, over 800 substances have been evaluated by the CVMP and appropriate classifications regarding MRLs have been made. The list of all substances and their classifications are published in the Annex to Regulation (EU) No 37/2010 (European Commission, 2009).

Since 2009, with the entry into force of Regulation (EC) No 470/2009, MRLs adopted by the Codex Alimentarius will subsequently be adopted as EU MRLs without requiring an additional MRL application and evaluation in the EU, provided that the EU delegation at the Codex Alimentarius Commission did not object to the MRLs and the scientific data have been made available to the EU delegation prior to the decision at Codex.

4.3 SCIENTIFIC EVALUATION

Details on the data requirements and the scientific approach for setting MRLs are described in Volume 8 of the *Rules Governing Medicinal Products in the European Union* (European Commission, 2005). The scientific principles for setting MRLs are largely similar to the approach taken by the Joint FAO/WHO Expert Committee on Food Additives (JECFA). The approach used by the CVMP is based on the determination of the ADI, on which, in turn, considering pharmacokinetic and residue depletion studies in the target species, MRLs are based.

The data requirements for the safety part for an MRL application have been largely harmonized at the international level under the International Cooperation on Harmonization of Technical Requirements for Registration of Veterinary Medicinal Products (VICH). The VICH has developed harmonized guidelines on data requirements, criteria, and standards for the registration of veterinary medicinal products, which includes a set of guidelines regarding the safety of residues. These guidelines have been implemented in the countries and regions participating in VICH, which are the VICH members (the EU, Japan, and the United States) as well as the observer countries in VICH (Australia, Canada, and New Zealand). VICH guidelines have been agreed on with regard to data requirements for the standard battery of safety studies to be provided, which comprise repeated dose toxicity (subchronic and chronic), reproductive toxicity, developmental toxicity, genotoxicity, carcinogenicity, and are specific for antimicrobial substances and the effects on the human gut flora (VICH, n.d.). Where relevant for the substance under consideration, other effects, for example, neurological or immunological effects, may be considered in addition. Pharmacological studies are also considered within the safety assessment.

The ADI for a specific substance may be set on the basis of toxicological, pharmacological, or microbiological data—whichever level is the lowest. In order to determine the toxicological ADI, the lowest no-observable-adverse-effect-level (NOAEL) with respect to the most sensitive parameter in the most sensitive test species is identified from the battery of toxicological studies, or in some cases, where such data are available, from observations in humans. A standard uncertainty factor of 100 (10×10, to correct for intraspecies variability and for interspecies extrapolation) is considered to extrapolate from the NOAEL to the ADI. However, in case adequate human data relevant to the establishment of the ADI are available an uncertainty factor of only 10 is acceptable. On the other hand, depending on the relevance and the quality of the available toxicological data, higher uncertainty factors from 100 to 1000 can be applied. Pharmacological effects are also studied as the intended pharmacological effect in the target animal is considered an adverse effect for consumers at (involuntary) exposure to residues in food. The approach for establishing a pharmacological ADI is similar to that for a toxicological ADI, with the lowest NOAEL determined in specific studies, which are tailored to the type of substance under consideration, being the basis for the ADI by applying an uncertainty factor, as appropriate.

For most antimicrobial substances, a microbiological ADI is established on the basis of sensitivity testing in relevant microorganisms of the human gut flora by investigating on the basis of *in vitro* or *in vivo* studies the disruption of the colonization barrier and the increase of the population(s) of resistant bacteria. The microbiological ADI is then calculated using the approach as detailed in the VICH Topic GL 36: Studies to evaluate the safety of residues of veterinary drugs in human food: General approach to establish a microbiological ADI (VICH, 2012).

A qualitative comparison of the metabolic profiles in the laboratory animals used for establishment of the toxicity profile and ADI and the food-producing species under consideration is undertaken in order to ensure that laboratory animals were exposed to the same residues to which humans may be exposed when they consume products from treated animals. Studies that demonstrate the total drug-derived residues, which are usually studies using the radiolabeled substance, are required to be conducted considering that the residue definition includes all residues, active substance(s), as well as metabolites and degradation products, and all drug-derived residues are assumed to have the same hazard potential as the parent compound. Based on these studies the marker residue that will enable residue control is selected. The marker residue is normally the residue (i.e., the parent compound or one of its metabolites) present at the highest concentration at the expected withdrawal time. In most cases it is the parent compound.

The ratio of marker to total residues needs to be known in order to determine the MRL value and to calculate the theoretical intake of residues. It is also essential that the analytical method available is capable of analyzing the marker residue. In addition, a non-radiolabeled ("cold") marker residue study in the target animal is required to establish the residue distribution in the target tissues, which is used as a basis for establishing MRLs. The approach applied considering the kinetics of the residues for each tissue means that MRLs are allocated following the actual residue distribution observed in the different target tissues so that tissues with highest residues get the highest MRL, and vice versa (see Fig. 4.1). Thus, each target tissue MRL is a monitor for depletion of residues from all other edible tissues to below the ADI and

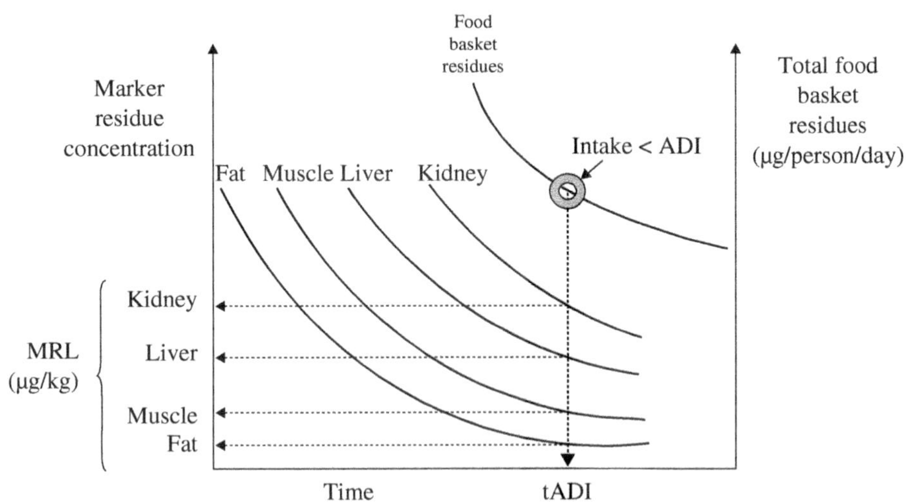

FIGURE 4.1 How the ADI is divided between the target tissues.

can serve as tissue for control of residues. In addition, with this approach the withdrawal period for each target tissue is approximately the same.

The MRLs are established for each target tissue of each target species in which the pharmacologically active substance will be used. For mammals, the target tissues are muscle, liver, kidney, and fat (or for pigs, fat + skin in natural proportions). For poultry, the target tissues are muscle, liver, kidney, and fat + skin in natural proportions.[1] For fish, the target tissues are muscle + skin in natural proportions. As a general rule, where residue concentrations can be quantified in the target tissues, MRLs are established for all target tissues based on the tissue residue distribution pattern of the pharmacologically active substance investigated.

In case residues are not measurable in some target tissues, MRLs must be established in at least one target tissue of the carcass (muscle or fat) and one target tissue of the offal (liver or kidney). An arbitrary value of twice the limit of quantification of the analytical method is set as the MRL.

Normally, an MRL established for muscle will also be applicable to muscle at the injection site for those substances administered by the intramuscular or subcutaneous routes. The CVMP has published specific guidance on this subject (European Medicines Agency, 2004). In certain cases where the depletion of residues from the injection site is significantly slower than from "normal" muscle (particularly relevant for long acting formulations) a so-called injection site residue reference value (ISRRV) is recommended in addition to the MRL for muscle. This value is intended to provide a reference point for use by competent authorities when setting withdrawal periods based on residues at the injection site. However, the ISRRV is not used for routine residue surveillance, for which the established MRL for muscle applies. The ISRRV is derived in a manner that allows for residues in 300 g of muscle to correspond to the unused portion of the ADI while the intake from a food basket containing the injection site should not exceed the ADI. The CVMP has published detailed considerations on the rationale for the determination of the ISRRV and the principles applied to its calculation (European Medicines Agency, 2013b).

The MRLs are set in such a way so as to ensure that the ADI is not exceeded after considering intake from all sources. To consider the potential consumer intake of residues, the levels of consumption of residues in foods of animal origin are calculated based on arbitrarily high fixed consumption values, to ensure the protection of the majority of consumers (different populations with different types of diets, e.g., children) and by calculating the theoretical maximum daily intake (TMDI).

[1] For mammals, 500 g comprises 300 g of muscle, 100 g of liver, 50 g of kidney, and 50 g of fat (for pigs, fat and skin in natural proportions); and for poultry, 500 g comprises 300 g of muscle, 100 g of liver, 10 g of kidney, and 90 g of fat and skin in natural proportions.

The standard daily food basket used for such TMDI calculation comprises 500 g of meat or 300 g of fish, plus 1500 g of milk, 100 g of eggs, and 20 g of honey.

If the veterinary medicinal product is intended to be used in lactating cattle, laying hens, or bees, MRLs also need to be established in milk, eggs, or honey, as appropriate. Usually, a certain portion of the ADI is reserved for these commodities, depending on the substance and anticipated uses. For setting an MRL in milk or eggs, the calculation is done like the MRL calculation for edible tissues considering the residue kinetics, based on the residue depletion study in milk or eggs. The calculation for honey can be made directly on the basis of the ADI equivalent available, and a radiolabeled study is not required (European Medicines Agency, 2006).

For substances that also have other uses that may lead to residues in other food products, for example, substances that are used as pesticides or biocides or that might occur in drinking water, combined exposure is considered. The combined exposure from these sources is subtracted from the ADI, thus lowering the portion of the ADI available for veterinary use.

Effects on food processing, in particular on dairy starter cultures, are also considered when establishing MRLs as risk management considerations, where appropriate. This applies mainly to microbiological substances.

In summary, the determination of MRLs is based on the ADI, the identified marker residue and total residues, the tissue distribution, and the EU food basket. MRLs are established in such a way that the maximum theoretical daily intake, the TMDI, as calculated from MRLs and the food basket, does not exceed the ADI.

Once the process of safety evaluation is completed and MRLs have been derived, consideration is given to the availability of analytical detection methods suitable for use in routine monitoring. The data required for the establishment of MRLs includes the provision of a chemical analytical method for monitoring of residues. The adequacy of the method for control purposes is assessed by the CVMP and verified by the European Union Reference Laboratories. Where no appropriate method is available, no recommendation can be given for establishing MRLs.

Further to the adoption of the MRLs by the European Commission and publication in the *Official Journal of the European Union*, the Agency makes the analytical method provided by the applicant available to the Union and National Reference Laboratories for monitoring purposes.

After the publication of the MRL classification in the *Official Journal*, the European Medicines Agency publishes the European Public MRL Assessment Report (EPMAR), formerly called "Summary Report," on its website (http://www.ema.europa.eu/ema/index.jsp?curl=pages/medicines/landing/vet_mrl_search.jsp&mid=WC0b01ac058008d7ad), providing a summary of the assessment carried out of the data available and the rationale for the conclusions.

Once MRLs have been established, it is then necessary in the context of granting a marketing authorization for veterinary medicinal products to determine withdrawal periods—that is, the period following the last administration of the veterinary medicinal product, during which the animal must not be slaughtered or during which milk or eggs must not be taken for human consumption—that ensure that residues from the product concerned will not exceed the MRL. Since the duration of the withdrawal period will depend on the individual pharmaceutical formulation concerned, specific withdrawal periods will be determined as part of the process of evaluation of the application for marketing authorization.

4.4 EXTRAPOLATION OF MRLs

Since the entry into force of Regulation (EC) No 470/2009 in July 2009, the possibility to extrapolate the conclusions of the assessment of the safety of residues in one species to another species or food commodity needs to be considered during the evaluation of an application for the establishment of MRLs. Although the requirement for consideration of extrapolation is recent in the legislation, the principle has been followed by the CVMP for more than 10 years. The criteria for the extrapolation of MRLs are defined in the CVMP note for guidance on the risk analysis approach for residues of veterinary medicinal products in food of animal origin (European Medicines Agency, 2001).

On the basis of knowledge on the variation of residue depletion within classes of animals, MRLs established in certain animal species may be extrapolated to other animal species within a class of animals when certain conditions are fulfilled (European Medicines Agency, 2001). In general, MRLs established in a major species can be extrapolated to a relevant minor species provided that it has been shown that the marker residue established in the major species can be measured in the minor species and that the analytical method validated for the major species is applicable in the minor species.

The approach for extrapolation is as follows:

MRLs should be allowed to be extrapolated within classes of animals as indicated in Table 4.1.

If identical MRLs were derived in cattle (or sheep), pigs, and chicken (or poultry), which represent major species with different metabolic capacities and tissue composition, the same MRLs can also be set for ovine, equidae, and rabbits, which means an extrapolation is considered possible to all food-producing animals except fish. In addition, the extrapolation from

TABLE 4.1 Potential MRL extrapolations within classes of animals

Species for which MRLs have been set	Extrapolations to
Major ruminant (meat)	All ruminants (meat)
Major ruminant (milk)	All ruminant milk
Major monogastric mammal	All monogastric mammals
Chicken and eggs	Poultry and poultry eggs
Salmonidae	All fin fish
Either a major ruminant or a major monogastric mammal	Horses

MRLs in muscle of a major species to *Salmonidae* and other finfish is possible provided that the parent substance is acceptable as marker residue for the MRL in muscle and skin; therefore, MRLs can in that case be extrapolated to all food-producing animals. In cases where MRLs were established in cattle (or sheep), pigs, and chickens (or poultry), which were slightly different, extrapolation to further species as outlined earlier is also possible. The most relevant set of MRLs for the extrapolation should be chosen on the basis of the amount of residues likely to be ingested or the most conservative MRL.

4.5 PROHIBITED DRUGS

In the context of the evaluation for the purpose of establishing MRLs, it was concluded for 10 substances that MRLs could not be recommended as residues of the concerned substances at any level whatsoever could constitute a hazard to the health of the consumers. These substances are *Aristolochia* spp. and preparations thereof, chloramphenicol, chloroform, chlorpromazine, colchicine, dapsone, dimetridazole, metronidazole, nitrofuranes (including furazolidone), and ronidazole. As a consequence, they are forbidden for use in veterinary medicinal products intended for food-producing species and biocidal products used in animal husbandry.

The use of substances having hormonal action and beta-agonists as growth promoters is forbidden in the EU. Certain hormones and beta-agonists are authorized for therapeutic and zootechnical purposes but their use is limited to specifically defined cases and under strictly controlled conditions to prevent misuse. The use of thyrostatic substances, stilbenes, and derivatives as well as estradiol is totally banned for use in food-producing species (European Council, 1996).

Since the establishment of the European Medicines Agency in 1995, any application for the authorization of products intended for use as growth enhancers has to be submitted to the Agency for evaluation by the CVMP (centralized procedure). However, no request for authorization of such a product has been submitted up to now.

4.6 EU POLICY ON MINOR USE AND MINOR SPECIES

The policy and approach on how to deal with veterinary medicinal products for minor uses and minor species (MUMS) is described in a series of guidance documents. The European Medicines Agency has developed a MUMS policy which defines the criteria for the classification of products as MUMS and describes the incentives for its authorization with the aim to stimulate development of new veterinary medicinal products for minor species and rare diseases in major species (European Medicines Agency, 2013a).

Minor species are not formally defined, but the major food-producing species based on data on food consumption and animal numbers in the EU are considered to be bovine, sheep (excluding sheep milk), porcine, chicken, and salmon. Cats and dogs are considered major species of companion animals. All other animal species are considered by default minor species. Minor uses in major species are also not specifically defined but are generally considered to be medicines intended for the treatment of diseases that occur infrequently or occur in limited geographical areas. The decision on the classification of a product as MUMS is taken on a case by case basis taking into account the justification put forward for consideration.

Specific guidelines on data requirements for products intended for minor uses and minor species are in force defining the cases where a reduced data package can be accepted in order to assess the quality, efficacy, and safety of a medicinal product intended for minor uses or minor species and ensuring consumer safety in case of food-producing species (European Medicines Agency, 2006).

4.7 EU POLICY AND LEGISLATION ON FEED ADDITIVES

The use of feed additives in the EU is regulated by Regulation (EC) No 1831/2003 (European Parliament and Council, 2003). These substances are intended to improve the quality of the feed, or the quality of the food of animal origin or to improve the performance and health of the animals. The use of a substance as a feed additive requires an authorization by the European Commission following a scientific evaluation by the European Food Safety Authority (EFSA) concluding that the use of the substance will not have harmful effects on human and animal health and on the environment. The authorization procedure of feed additives also requires the assessment for the purpose of establishing MRLs. The approach followed for the establishment of MRLs for feed additives is similar to the one followed for the establishment of MRLs with regard to veterinary medicinal products.

Feed additives are classified according to the following categories:

1. technological additives (such as preservatives, antioxidants, emulsifiers, stabilizers, thickeners, acidity regulators, denaturants);
2. sensory additives (such as colorants, flavoring compounds);
3. nutritional additives (such as vitamins, pro-vitamins, trace elements, amino acids, their salts and analogues, urea and its derivatives);
4. zootechnical additives (such as digestibility enhancers, gut flora stabilizers, substances that favorably affect the environment); and
5. coccidiostats and histomonostats

With the exception of coccidiostats and histomonostats that are intended for the prevention of coccidiosis and histomoniasis, feed additives do not have medical claims. The classification of coccidiostats as feed additives when used for prevention purposes is justified by the fact that in modern poultry production coccidiosis is present in all commercial herds even in farms with high sanitary standards and good management, and therefore they generally need to be administered throughout the lifetime of the animals. With regard to histomonostats, no products belonging to this category are currently authorized as feed additives.

The use of additives as growth promoters is no longer authorized in the EU. Up to January 2006 some antibiotics were authorized as feed additives and used with the aim to improve the performance of the animals; however, taking into account concerns related to the risks for selecting bacterial strains resistant to human or veterinary drugs their use as feed additives has been forbidden. Two other feed additives—carbadox and olaquindox—were previously authorized as growth promoters in pigs. Their use was forbidden in 1998 due to safety concerns (European Commission, 1998).

4.8 OFF-LABEL USE

If no authorized veterinary medicinal product for the species and condition concerned is available, the off-label use of authorized medicines is allowed under specified conditions, which are described in Directive 2001/82/EC (European Parliament and Council, 2004a). These conditions for the off-label use are often referred to as the "cascade" due to their stepwise approach in identifying the medicine for the off-label use.

Under these "cascade" provisions, EU member states are obliged to take certain measures. For example, if there is no authorized veterinary medicinal product in a member state for a specific condition, the responsible veterinarian may, under his/her direct personal responsibility, and in particular to avoid causing unacceptable suffering, treat the animal concerned with a veterinary medicinal product authorized in the member state concerned for another animal species or for another condition in the same species. If there is no such product authorized, a medicinal product authorized for human use in the member state concerned, or a veterinary medicinal product authorized in another member state for use in the same species, in another species for the condition in question, or for another condition, may be used. If, however, there is no such product, a veterinary medicinal product prepared extemporaneously by a person legally authorized to do so following a veterinary prescription may be used.

For food-producing animals, these provisions apply to animals on a particular holding only, the pharmacologically active substances in the medicinal product used must be classified as (a), (b), or (c) according to the MRL Regulation (Regulation (EC) No 470/2009 (European Parliament and Council, 2009)) and listed as allowed substance in Table 1 of the Annex to Regulation (EU) No 37/2010 (European Commission, 2009). Also, the veterinarian must specify an appropriate withdrawal period, which shall be at least 7 days for eggs, 7 days for milk, 28 days for meat from poultry and mammals, including fat and offal, and 500 degree-days for fish meat.

ACKNOWLEDGMENTS

The authors thank Dr Stefan Scheid for allowing the use of the graphical presentation on how the ADI is divided between the target species presented as Figure 4.1.

REFERENCES

European Commission (1998). Commission Regulation (EC) No 2788/98 of 22 December 1998 amending Council Directive 70/524/EEC concerning additives in feeding stuffs as regards the withdrawal of authorisation for certain growth promoters (EC OJ L347, pp. 31–32). Available at http://eur-lex.europa.eu/LexUriServ/LexUriServ.do?uri=OJ:L:1998:347:0031:0032:EN:PDF. Accessed May 15, 2014.

European Commission (2005). Volume 8 of the Rules Governing Medicinal Products in the European Union, Notice to Applicants and Note for Guidance, Establishment of maximum residue limits (MRLs) for residues of veterinary medicinal products in foodstuffs of animal origin. Available at http://ec.europa.eu/health/documents/eudralex/vol-8/index_en.htm. Accessed May 15, 2014.

European Commission (2009). Commission Regulation (EU) No 37/2010 of 22 December 2009 on pharmacologically active substances and their classification regarding maximum residue limits in foodstuffs of animal origin (EC OJ L15/1, 20.01.2010, pp. 1–72). Available at http://eur-lex.europa.eu/LexUriServ/LexUriServ.do?uri=OJ:L:2010:015:0001:0072:EN:PDF. Accessed May 15, 2014.

European Council (1996). Council Directive No 96/22/EC of 29 April 1996 concerning the prohibition on the use in stock farming of certain substances having hormonal or thyrostatic action and of beta-agonists and repealing Directives 81/602/EEC, 88/146/EEC and 88/299/EEC (EU OJ L 125, 23.05.1996, pp. 3–9), as amended by Directive 2003/74/EC of the European Parliament and of the Council of 22 September 2003 amending Council Directive 96/22/EC (EU OJ L 262, 14.10.2003, pp. 17–21), and Directive 2008/97/EC of the Parliament and of the Council of 19 November 2008 amending Council Directive 96/22/EC (EU OJ L 31, 28.11.2008, pp. 9–11). Available at http://eur-lex.europa.eu/LexUriServ/LexUriServ.do?uri=CONSLEG:1996L0022:20081218:EN:PDF. Accessed May 15, 2014.

European Medicines Agency (2001). CVMP Note for Guidance on the Risk Analysis Approach for Residues of Veterinary Medicinal Products in Food of Animal Origin (EMEA/CVMP/187/00-FINAL). Available at http://www.ema.europa.eu/docs/en_GB/document_library/Scientific_guideline/2009/10/WC500004534.pdf Accessed May 15, 2014.

European Medicines Agency (2004). Guideline on injection site residues (EMEA/CVMP/542/03-FINAL). Available at http://www.ema.europa.eu/docs/en_GB/document_library/Scientific_guideline/2009/10/WC500004429.pdf. Accessed May 15, 2014.

European Medicines Agency (2006). CVMP Guideline on safety and residues data requirements for veterinary medicinal products intended for minor uses or minor species (EMEA/CVMP/SWP/66781/2005). Available at http://www.ema.europa.eu/docs/en_GB/document_library/Scientific_guideline/2009/10/WC500004581.pdf. Accessed May 15, 2014.

European Medicines Agency (2013a). Revised policy for classification and incentives for veterinary medicinal products indicated for minor use, minor species (MUMS)/ limited market (EMA/429080/2009-Rev.1). Available at http://www.ema.europa.eu/docs/en_GB/document_library/Regulatory_and_procedural_guideline/2009/10/WC500005157.pdf. Accessed May 15, 2014.

European Medicines Agency (2013b). Revised reflection paper on injection site residues: considerations for risk assessment and residue surveillance (EMA/CVMP/520190/2007-Rev.1). Available at http://www.ema.europa.eu/docs/en_GB/document_library/Scientific_guideline/2013/10/WC500152656.pdf. Accessed May 15, 2014.

European Parliament and Council (2003). Regulation (EC) No 1831/2003 of the European Parliament and of the Council of 22 September 2003 on additives for use in animal nutrition (EU OJ L 268, 18.10.2003, pp. 29–43), as amended. Available at http://eur-lex.europa.eu/LexUriServ/LexUriServ.do?uri=CONSLEG:2003R1831:20100901:EN:PDF. Accessed May 15, 2014.

European Parliament and Council (2004a). Directive 2001/82/EC of the European Parliament and the Council of 6 November 2001 (EU OJ L 311, 28.11.2001, pp. 1–66), as amended by Directive 2004/28/EC of the European Parliament and the Council of 31 March 2004, amending Directive 2001/82/EC on the Community code relating to veterinary medical products (EC OJ L 136, 30.04.2004, pp. 58–84). Available at http://eur-lex.europa.eu/LexUriServ/LexUriServ.do?uri=CONSLEG:2001L0082:20090807:EN:PDF. Accessed April 6, 2014.

European Parliament and Council (2004b). Regulation (EC) No 726/2004 of the European Parliament and of the Council of 31 March 2004 laying down Community procedures for the authorisation and supervision of medicinal products for human and veterinary use and establishing a European Medicines Agency (EU OJ L 136/1, 30.4.2004, pp. 1–33). Available at http://eur-lex.europa.eu/LexUriServ/LexUriServ.do?uri=OJ:L:2004:136:0001:0033:EN:PDF. Accessed April 6, 2014.

European Parliament and Council (2009). Regulation (EC) No 470/2009 of the European Parliament and of the Council of 6 May 2009 laying down Community procedures for the establishment of residue limits of pharmacologically active substances in foodstuffs of animal origin, repealing Council Regulation (EEC) No 2377/90 and amending Directive 2001/82/EC of the European Parliament and Council, 2001; and Regulation (EC) No 726/2004 of the European Parliament and of the Council (EU OJ L 152/11, 16.06.2009, pp. 11–22). Available at http://eur-lex.europa.eu/LexUriServ/LexUriServ.do?uri=OJ:L:2009:152:0011:0022:en:PDF. Accessed April 6, 2014.

VICH (n.d.). International Cooperation on Harmonization of Technical Requirements for Registration of Veterinary Medicinal Products (VICH): VICH Guidelines. Available at http://www.vichsec.org/en/guidelines.htm. Accessed April 6, 2014.

VICH (2012). VICH Topic GL 36: Studies to evaluate the safety of residues of veterinary drugs in human food: General approach to establish a microbiological ADI. Available at http://www.vichsec.org/pdf/05_2004/Gl36_st7_F_rev.pdf. Accessed April 6, 2014.

5

METHODS TO DERIVE WITHDRAWAL PERIODS IN THE EUROPEAN UNION

G. Johan Schefferlie[1] and Stefan Scheid[2]

[1] *Veterinary Medicines Unit, Medicines Evaluation Board, Utrecht, the Netherlands*

[2] *Federal Office of Consumer Protection and Food Safety (BVL), Berlin, Germany*

5.1 INTRODUCTION

The legislation in place in the European Union (EU) requires that foodstuffs obtained from animals treated with a veterinary medicinal product do not contain residues that might represent a health risk to the consumer. Before a veterinary medicinal product intended for food-producing animals is authorized, the safety of the pharmacologically active substance and its potential residues in foods of animal origin must be evaluated and maximum residue limits (MRLs) established in accordance with Regulation (EC) No 470/2009 (European Parliament and Council, 2009).

During the marketing authorization process, the quality, safety, and efficacy of the veterinary medicinal products is evaluated. For veterinary medicines intended for administration to food-producing animals, withdrawal periods need to be set to ensure that residues do not exceed their MRLs when the label instructions for a veterinary medicinal product are followed (European Parliament and Council, 2004). Withdrawal periods are determined for all edible tissues for which MRLs have been set (usually muscle, fat, liver, and kidney, and if applicable, for milk, eggs, or honey).

Strategies for Reducing Drug and Chemical Residues in Food Animals: International Approaches to Residue Avoidance, Management, and Testing, First Edition.
Edited by Ronald E. Baynes and Jim E. Riviere.

The animal or its product is considered fit for human consumption only after the withdrawal period has elapsed. The MRLs also serve as a reference for residue control, and compliance with the legal MRLs is routinely checked in residue control programs.

The withdrawal period is usually determined on the basis of the depletion kinetics of a so-called marker residue (a single or combined residue component) in the animal tissues or animal products. Although the marker residue is only a part of the total drug-derived residue, the MRLs set for the marker residue are still protective for consumer health, because the ratios between the concentrations of marker residue and total residues have been taken into account in the establishment of MRLs.

5.2 WITHDRAWAL PERIODS FOR MEAT

Withdrawal periods for meat are established for veterinary medicines intended for use in food-producing species. A withdrawal period is calculated separately for each of the tissues for which an MRL exists. The withdrawal periods for the various tissue types are compared with each other and the tissue with the longest depletion time relative to the MRL finally determines the regulatory withdrawal period for the veterinary product. The implication of this regulatory withdrawal period is that treated animals must not be sent for slaughter until the withdrawal period has elapsed.

5.2.1 Residue Studies

Withdrawal periods in meat are based on the results of residue studies after administration of the intended veterinary medicinal product according to label instructions to food-producing animals. In these studies, the marker residue in tissues is monitored from the time of administration of a product to at least until the concentrations have fallen below the acceptable concentration, that is, the MRL.

A guideline on the conduct of residue depletion studies to establish withdrawal periods in meat has been developed under the VICH[1] process (VICH, 2011). This guideline provides recommendations on critical study design elements such as the number of animals to be used for residue depletion studies, the minimum number of time points, and sampling of tissues.

[1] International Cooperation on Harmonization of Technical Requirements for Registration of Veterinary Medicinal Products.

5.2.2 Data Evaluation

According to current EU guidelines (European Medicines Agency, 1996), there are basically two methods for the determination of a withdrawal period in animal meat:

The recommended and preferred method is a statistical method based on a linear regression model in which the upper 95% tolerance limit of the 95% percentile of the residue depletion curve is used to determine the withdrawal period. The withdrawal period is derived at the point of intersection of the tolerance limit and the MRL line (see Fig. 5.1); that is, at the end of the withdrawal period, one can be 95% sure that in 95% of tissue samples the residue is below the MRL.

If the statistical assumptions underlying the regression model are not met, and with practice this roughly happens in about 30% of data sets, then there is an alternative approach based on a decision criterion (safety factor). In this case, the time in which the residues in all tissues of all observed animals are below the respective MRLs, plus a safety span, is considered the withdrawal period.

Withdrawal times for meat from mammals and poultry are expressed as days. Where calculations result in fractions of days, these are rounded up. In fish, residue depletion is a function of time and water temperature and the withdrawal period is expressed in degree-days, derived by multiplying the withdrawal period in days by the mean daily water temperature used.

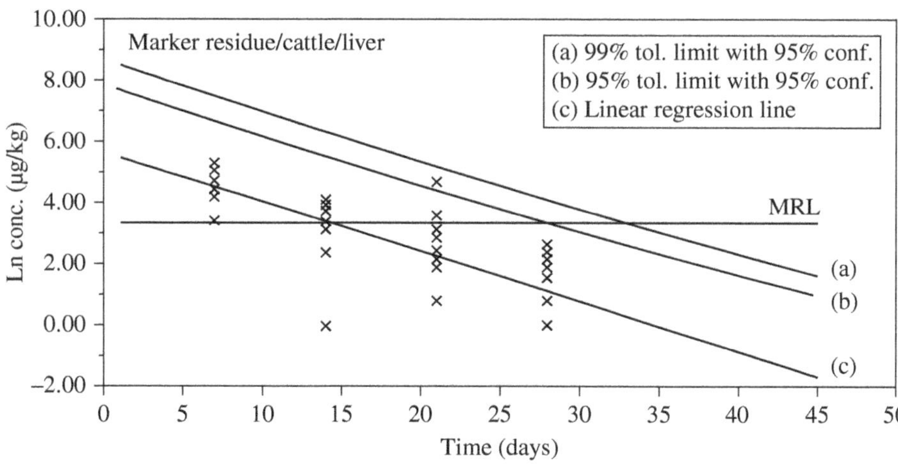

FIGURE 5.1 Example plot of withdrawal period calculation.

5.3 STATISTICAL METHOD

According to well-known pharmacokinetic models, the time course of concentration changes of residues in tissues following administration of veterinary drugs to animals can be described by a sum of n (n equal to the number of phases) exponential terms typically comprising absorption, distribution, and elimination phases.

$$C_t = \sum_{i}^{n} C_{o,i} e^{-k_i t}$$

The linear regression model for the withdrawal period calculation involves a reasonable simplification of this general equation by assuming that the withdrawal time is falling in the terminal phase of elimination, which can adequately be described by a single exponential term. The equation can then be reduced to:

$$C_t = C'_o e^{-kt}$$

Based on this equation, the depletion curve is modeled as a straight line following a logarithmic transformation of the data. Tolerance limits for the given 95% (or a higher) percentile are then calculated using the results of a regression analysis:

The basic assumptions underlying this model are:

1. The assumption of linearity of log (concentration) versus time. Visual inspection of a plot of the data is often used as an initial tool to ensure that there is a useful linear relationship. Statistical assurance of linearity of the regression line can be obtained from a lack of fit test.
2. The assumption of independence of the observations. Residue data for animal tissues meet this assumption because of
3. their originating from individual animals,
4. the normality of errors on a log scale, and
5. the homogeneity of variances of the data at all relevant times.

It may be assumed that the underlying assumptions are approximately valid in the simple pharmacokinetic scenario under investigation.

The EU guideline recommends testing the validity of each of the assumptions using appropriate standard statistical tests prior to the data analysis. Violations of the assumptions are generally not acceptable. However, given that the typically small number of data points (animals) in a withdrawal time experiment may be insufficient to fully verify the statistical assumptions in each case, not too severe deviations (e.g., for the normality assumption) may be dealt with on a case-by-case basis, depending on expert judgment.

Ideally, depletion of residues is monitored until residues are well below the concentration of the MRL so that extrapolation of the depletion curve beyond the range of measured values is not required to derive the withdrawal period. Slight extrapolations have occasionally been used based on the assumption of a monotonic linear depletion over time and the fact that the hyperbolic shape of the tolerance limits most likely results in an over- rather than an underestimation of the extrapolated withdrawal period.

A convenient calculation equation for tolerance limits has been published by Stange (1971) and modified by Graf et al. (1987):

$$\hat{y} = a + bx + ks_{y.x}$$

$\hat{y}$ in the equation represents the upper limit of the $(1-\alpha)$ tolerance interval for the proportion of $(1-\gamma)$ of residue concentrations predicted for the population of animals treated in the same way. Its significance is that at least the fraction $(1-\gamma)$ of the residue concentrations found in the animals of this population at the given time points can be expected to be below this numerical value with a confidence of $(1-\alpha)$. In the EU it is suggested that 0.95 be used for both $(1-\gamma)$ and $(1-\alpha)$ (i.e., 95% of the population and 95% statistical confidence). The respective percentage points of the standard normal distribution $(u_{1-\gamma})$ and $(u_{1-\alpha})$, which are used in the Stange equation, are 1.6449 if 0.95 is chosen for $(1-\alpha)$ (or $1-\gamma$) and 2.3263 if 0.99 is chosen for $(1-\alpha)$ (or $1-\gamma$). Experience has shown that the use of 99% percentiles often requires extreme extrapolation and, thus, limits the use of otherwise sound and acceptable data sets.

Withdrawal periods are set at the time when the upper one-sided tolerance limit with a given confidence is below the MRL. For the statistical method to be performed, theoretically, a minimum number of three animals at each of a minimum of three slaughter times in the log-linear phase of the terminal elimination of residues are required to allow a meaningful regression analysis.

MRLs are typically set close to the LOQ of the analytical residue control method (the minimum value for the MRL is required to be ≥2× LOQ). Consequently, a typical observation with residue data is that an increasing number of values at or smaller than the LOQ may occur as depletion of residues proceeds toward the withdrawal period. These data are usually named "left-censored data" and in a pragmatic approach they are replaced by half the

LOQ in the calculations. When all or most of the data at a time point are "less than" values, however, it is recommended that the entire time point be excluded to avoid a too strong bias in the regression parameters. This, in turn, can lead to data sets that are too small for statistical analysis.

The European Medicines Agency offers a computer program that analyzes meat residue data according to the statistical method (available at http://www.ema.europa.eu).

Where departures from basic model assumptions are evident, the EU guideline leaves the option to consider other statistical models and data transformations, provided they ensure equivalent statistical confidence.

5.4 ALTERNATIVE APPROACH: DECISION RULE

With the alternative approach, the point of departure for the withdrawal period is the time point at which all residues in tissues of observed animals have fallen below the respective MRLs. A safety span is then added to the depletion time to compensate for the uncertainties of biological variability. The dimension of the safety span can depend on various factors, which are decided by the study design, the quality of the data, and finally by the pharmacokinetic properties of the active substance and its formulation. As a result, an overall recommendation for a fixed rule is not possible and the relative weight given to each factor depends on expert judgment. As a rough point of reference for the safety span, a value of 10–30% of the time period is usually used. Alternatively, the safety span might be calculated from pharmacokinetic parameters such as the tissue depletion half-life, for example, a value of 1–3 times the elimination half-life, $t_{1/2}$. The recommended minimal safety span is 1–2 days.

The advantage of this method is that it is simple to perform and does not require stringent model assumptions. The disadvantage is, however, that it is not possible to provide a quantitative estimate of the uncertainty associated with the results. Long-term experience and comparison with the statistical method seems to suggest, however, that the alternative approach yields roughly comparable results but may under- or overestimate the "true" withdrawal period in some cases (Schefferlie and Hekman, 2009).

5.4.1 Injection Site Residues

Injectable formulations may exhibit a depletion profile at the site of injection that is significantly different from the kinetic profile observed in "normal" muscle or other edible tissues. Injection site residues generally tend to persist longer and at higher levels, and the time course of their release from the dose site can depend on various not entirely predictable factors: the physico-chemical properties of the substance and its formulation (slow release or depot product), the route of injection, whether subcutaneous or into the muscle, tissue reactions such as fibrosis, encapsulation, or necrosis, and others. Thus, residues in injection site muscle can be highly variable from animal to animal and may still be higher than MRLs by order(s) of magnitude at the end of the withdrawal period for noninjection site tissues. In addition, as a result of potentially nonhomogeneous dispersion of a substance at the dose site, the size and exact location of samples taken for analysis can have a considerable impact on the residues found.

There is specific guidance in the EU on assessment of withdrawal periods for injection site residues, which also includes recommendations for the conduct of residue trials and sampling (European Medicines Agency, 2004). Guidance on injection site tissue sampling is also included in the VICH GL 48 (VICH, 2011).

The approach to calculate the withdrawal period applies principally the same method as for other edible tissues. The references points for the withdrawal periods are the MRL in muscle, or, in the absence of a muscle MRL, the acceptable daily intake (ADI). Under current EU guidelines, injection site consumption is conservatively considered a "chronic" event and compared to the MRL or ADI, assuming that injection sites may be part of the daily meat consumption. As a worst-case approach, it is considered that the daily portion of muscle may originate from an injection site. Scenarios relevant to a (predominantly) short-term or acute exposure risk associated with injection site consumption, as used by some regulatory authorities outside the EU, are not applied due to the considerations as discussed in a reflection paper by the Committee for Medicinal Product for Veterinary Use (CVMP) of the European Medicines Agency (European Medicines Agency, 2013a).

As mentioned earlier, there are two assessment scenarios, depending on whether there is an MRL in muscle or not: For substances with an MRL, the injection site is treated as muscle tissue and the assessment is made relative to the MRL and the marker residue for muscle. In this case, the withdrawal period will ensure that the marker concentration has depleted below the muscle MRL at the injection site. For substances where there is no MRL for muscle, the reference value for the assessment of injection site residues is the ADI. Depending on the type of ADI, residues of concern may be either the total drug related residues or a specific toxicologically, pharmacologically, and/or microbiologically active residue fraction. Estimates of dietary exposure to residues are based on the standard food basket, which in this case includes a 300 g portion of injection muscle. For certain compounds where a spare portion of the ADI is available, a so-called injection site residue reference value (ISRRV) has been proposed. This value, which is set in addition to the MRL in muscle, is not intended for use in routine residue surveillance but provides a reference point to be used by competent authorities when setting withdrawal periods specifically at the injection site. The ISRRV is derived in a manner that allows for residues in 300 g of muscle to correspond to the unused portion of the ADI while the intake from a food basket containing the injection site should not exceed the ADI.

As the method of tissue sampling at the injection site can affect the apparent concentration, it is recommended to use a standardized sampling approach: The sample (core sample) of the injection site should be an approximately 500-g piece centered on the point of injection and taking the form of a cylinder the approximate dimensions of which should be as follows:

10-cm diameter and 6-cm depth for intramuscular injections.
15-cm diameter and 2.5-cm depth for subcutaneous injections.

It is recommended to take a second sample including tissue from the region immediately surrounding the excised core sample (surrounding sample). Experience has shown that analysis of the surrounding sample can provide valuable information on the quality of sampling and thus give confidence that the core actually contains the spot with the highest residue. In case of multiple injection sites, it is recommended that the sampling includes the site of the last injection and, at least, an injection site from the site where most of the injections were given. Where the withdrawal period can be clearly determined by residue depletion at the site of injection, there is an option of collecting data from two (or more) injection sites per animal for the determination of the withdrawal time.

5.5 WITHDRAWAL PERIODS FOR MILK

Withdrawal periods for milk need to be established for all veterinary medicinal products authorized in the EU and intended for animal species that produce milk for human consumption, irrespective of their production phase. The established withdrawal periods ensure that the marker residues of pharmacologically active substances in milk from individual animals are below the MRL.

The implication of the milk withdrawal period is that milk from all milkings before the specified withdrawal period be discarded. The milk from all milkings taken at or after the specified withdrawal period is considered fit for human consumption. In the EU, milk withdrawal periods are established for individual animals and not for tank milk because milk from individual or few animals is used for consumption and for small-scale production of dairy products on the farm level.

Milk withdrawal periods are based on residue studies for milk from animals treated with the candidate veterinary medicinal product at the maximum dose and maximum duration of treatment. Guidance on the conduct of such studies is given in the VICH GL 48 in a CVMP guideline on the calculation of withdrawal periods (European Medicines Agency, 2000; VICH, 2011).

5.5.1 Residue Studies for Milk

Residue studies for determining the withdrawal period for milk are performed with the candidate veterinary medicinal product in all target animals that produce milk for human consumption. In these studies, the concentrations of the marker residue(s) of the pharmacologically active substance(s) are monitored in all subsequent milkings starting from the first milking after treatment until the marker residue(s) is below the MRL(s). A twice-a-day milking scheme should be followed, preferably using 12-h intervals. It is recommended that the last administration of the veterinary medicinal product be given just after a milking and 12 h before the next milking.

The minimum number of animals in a milk residue study is 20, based on the statistical requirements for the calculation of the withdrawal period. This number is the minimum sample size required to allow empirical estimates of a 95th percentile. A sample size higher than 20 is advisable to give the desired confidence level of 95% at least some empirical basis. A particular stratification of the sample is not required, but the sample should be representative of the target population; therefore selection of animals should take into account factors known to be important, such as milk yield.

5.5.2 Data Evaluation

In order to estimate the time point at which the 95th percentile of the population will have milk concentrations below the MRL, several statistical methods have been developed: the time to safe concentration (TTSC), the safe concentration per milking (SCPM), and the safe concentration from linear regression (SCLR). In the EU, the TTSC method is the harmonized method, since comparisons showed that this method was applicable in the largest number of realistic cases and resulted in withdrawal periods comparable to those from the other methods. The European Medicines Agency offers a computer program that analyzes milk residue data according to the TTSC (available at http://www.ema.europa.eu/htms/vet/vetguidelines/safety.htm). The program also includes the SCPM and SCLR methods.

The TTSC denotes the first milking at which the milk contains residues below the MRL. The TTSC is therefore expressed as the number of milkings. In the TTSC method, first individual TTSCs are derived for each animal in the study. Monotonic regression is used to correct data in those cases where an increase in the concentration is observed over time. At the next step, a tolerance limit is calculated from the number of milkings based on all individual TTSCs. This tolerance limit is the number of milkings necessary for the residue concentration in the milk of most animals to reach the MRL. This method assumes a log-normal distribution of the TTSCs. The withdrawal period is calculated as the 95/95 tolerance limit, that is, the upper 95% confidence interval of the 95th percentile of the distribution of TTSCs. The calculated withdrawal period is then rounded up to the first higher full number of milkings. Considering a twice-a-day milking scheme, this number is then converted to multiples of 12 h or whole days.

An advantage of the TTSC method is that one does not need to assume a particular model to describe the depletion pattern of the residues in milk over time. Furthermore, the TTSC method does not need a strategy to deal with samples having a concentration below the LOQ of the analytical method. A disadvantage is that one needs to assume a particular distribution of the number of milkings (TTSCs). The choice for a log-normal distribution is made empirically.

In the following example (European Medicines Agency, 2000), the individual TTSC values of 25 lactating cows in a residue experiment ranged between three and eight milkings. The distribution of TTSC values is as given in Table 5.1.

Using these TTSC values, a 95/95 tolerance limit of (rounded) nine milkings can be calculated. The resulting withdrawal period for milk is $9 \times 12 = 108$ h. Milk from milkings at or after 108 h is considered safe. This implies that with a regular 12-h milking scheme the first safe milk is from the 9th milking if the

TABLE 5.1 Distribution of TTSC values per animal

TTSC	Ln (TTSC)	Frequency
3	1.099	3
4	1.386	9
5	1.609	5
6	1.792	4
7	1.946	3
8	2.079	1

treatment is given 12 h before the first milking, and from the 10th milking if the treatment is given less that 12 h before the first milking.

5.6 WITHDRAWAL PERIODS FOR EGGS

Withdrawal periods for eggs need to be established for all veterinary medicinal products authorized in the EU and intended for animal species that produce eggs for human consumption. The established withdrawal periods ensure that the marker residues of pharmacologically active substances in eggs from individual animals are below the MRL.

The implication of the withdrawal period is that all eggs laid before the specified withdrawal period should be discarded.

5.6.1 Residue Studies for Eggs

The withdrawal periods for eggs are based on residue studies for eggs from birds treated with the candidate veterinary medicinal product at the maximum dose and maximum duration. There is no specific EU guidance on the conduct of such studies, but the VICH GL 48 (VICH, 2011) specifies that the number of birds should be sufficient to collect 10 eggs per time point, over a sufficient period of time. Although the period has not been specified, it appears logical to maintain a minimum of up to 12 days, in view of the 10-day rapid growth period of yolk development.

5.6.2 Data Evaluation

In practice, whenever possible, the MRL for eggs is established in a way that allows a zero-day withdrawal period for eggs. This is considered important for economical reasons and, related to that, for reasons of compliance. Nevertheless, withdrawal periods may be necessary in a number of cases.

No formal guidance on the calculation of withdrawal periods for eggs is available in the EU at present. The type of data from residue studies for eggs resembles that from residue studies for milk. Therefore, the TTSC method described for milk can also be applied for eggs. However, for this one would need at least 20 birds, and eggs would need to be collected separately for each bird in order to establish the TTSC per animal. It is notable that the TTSC method, there being no need to assume a particular residue depletion model, is particularly useful for eggs, in view of the nonlinear depletion in (whole) eggs due to the different concentration–time profiles in yolk and albumen.

5.7 WITHDRAWAL PERIODS FOR HONEY

Establishment of withdrawal periods for residues in honey does not follow the same rules as for the withdrawal periods in mammalian or avian tissues. In the honey matrix, there is no time-dependent elimination of the residues as a result of pharmacokinetics: residues, once present in honey, tend to remain there.

Apart from possible chemical (or thermal) degradation of a substance in the honey matrix over time, the main variable responsible for the level of residues at harvest is the honey yield (dilution effect), which in large part depends on the production site (geographical area) and weather conditions at flowering time, or on crops on which bees forage, and so on. These variables are largely unpredictable and, therefore, for most products the only feasible withdrawal period, in most cases, is a "zero" withdrawal period. This would be based on results from residue studies covering a reasonable range of field conditions and provided there is acceptable statistical confidence that there are no nonconforming residues in honey (i.e., above the MRL) under conditions of good bee-keeping practice.

5.8 EXTRAPOLATION OF WITHDRAWAL PERIODS

Under certain conditions, withdrawal periods can be extrapolated between species, in particular from a so-called major species to a minor species (European Medicines Agency, 2006). The information relating to use of the veterinary medicinal product in the major species may then be used in support of the withdrawal period in the minor species, thus obviating the need for residue studies. According to animal population data and total consumption figures, based on available data from the EU, cattle (sheep), pigs, chicken, and salmon are currently considered as major food-producing species (European Medicines Agency, 2001, 2013b). All other animal species are, therefore, by default, classified as "minor." Extrapolations are performed on a case-by-case basis, with regard to the active ingredient, the formulation of the veterinary medicinal product, the dose/dosing regimen, and the route of administration and other possibly relevant factors. Each product is assessed on its own respective merits, but there are a few general principles to be observed.

5.8.1 Identical Products

In case of the same veterinary medicinal product with the same MRL in the major/minor species, it is possible to directly extrapolate the withdrawal period from the major to the minor species. Exemptions are products having a potential to leave local residues (in particular intramuscular and subcutaneous injectables and products for dermal application). In this case, information on the behavior of residues at the site of administration needs to be assessed before the withdrawal period is extrapolated. Limited residue depletion studies may be helpful in this case (e.g., at two time points, one just before the reference withdrawal period and one after it). Alternatively, an extra uncertainty (safety) factor to compensate for uncertainties in the extrapolation could be considered (multiplication of the withdrawal period in the major species by an uncertainty factor of 1.5).

5.8.2 Products with Identical Active Ingredient but with Different Formulation/Different Dosing Regimen/Routes of Administration

Extrapolation between different pharmaceutical formulations, routes of administration, or dosing regimens can be more problematic, and a more cautious approach is recommended. Limited pharmacokinetic or residue data in the minor species may be required to support the extrapolation between the major and minor species (e.g., comparison of elimination half-lives), and additional uncertainty factors could be taken into consideration. When the product for the minor species is to be used at a significantly higher dose level/dosing regimen, conventional residue studies will normally be requested to confirm the withdrawal period. Also, for injectable products (intramuscular or subcutaneous) or products for local applications, residue data at the dose site are needed to support the withdrawal period.

5.8.3 Products Not Authorized Previously for Major Species

Where no similar product was authorized for a major species, standard residue studies are required in the minor species.

REFERENCES

European Medicines Agency (1996). Note for Guidance: Approach towards harmonisation of withdrawal periods (EMEA/CVMP/036/95). Available at http://www.ema.europa.eu/docs/en_GB/document_library/Scientific_guideline/2009/10/WC500004428.pdf. Accessed April 6, 2014.

European Medicines Agency (2000). Note for Guidance for the determination of withdrawal periods for milk (EMEA/CVMP/473/98-FINAL). Available at http://www.ema.europa.eu/docs/en_GB/document_library/Scientific_guideline/2009/10/WC500004496.pdf. Accessed April 6, 2014.

European Medicines Agency (2001). CVMP Note for Guidance on the Risk Analysis Approach for Residues of Veterinary Medicinal Products in Food of Animal Origin (EMEA/CVMP/187/00-FINAL). Available at http://www.ema.europa.eu/docs/en_GB/document_library/Scientific_guideline/2009/10/WC500004534.pdf. Accessed April 6, 2014.

European Medicines Agency (2004). Guideline on injection site residues (EMEA/CVMP/542/03-FINAL. Available at http://www.ema.europa.eu/docs/en_GB/document_library/Scientific_guideline/2009/10/WC500004429.pdf. Accessed April 6, 2014.

European Medicines Agency (2006). CVMP guideline on safety and residue data requirements for veterinary medicinal products intended for minor uses or minor species (EMEA/CVMP/SWP/66781/2005). Available at http://www.ema.europa.eu/docs/en_GB/document_library/Scientific_guideline/2009/10/WC500004581.pdf. Accessed April 6, 2014.

European Medicines Agency (2013a). Revised reflection paper on injection site residues: considerations for risk assessment and residue surveillance (EMEA/CVMP/520190/2007-Rev.1). Available at http://www.ema.europa.eu/docs/en_GB/document_library/Scientific_guideline/2013/10/WC500152656.pdf. Accessed April 6, 2014.

European Medicines Agency (2013b). Revised policy for classification and incentives for veterinary medicinal products indicated for minor use, minor species (MUMS)/limited market (EMA/429080/2009-Rev.1). Available at http://www.ema.europa.eu/docs/en_GB/document_library/Regulatory_and_procedural_guideline/2009/10/WC500005157.pdf. Accessed April 6, 2014.

European Parliament and Council (2004). Directive 2001/82/EC of the European Parliament and the Council of 6 November 2001 (EU OJ L 311, 28.11.2001, pp. 1–66), as amended by Directive 2004/28/EC of the European Parliament and the Council of the 31 March 2004 amending Directive 2001/82/EC on the Community code relating to veterinary medical products (EU OJ L 136, 30.04.2004, pp 58–84). Available at http://eur-lex.europa.eu/LexUriServ/LexUriServ.do?uri=CONSLEG:2001L0082:20090807:EN:PDF. Accessed April 6, 2014.

European Parliament and Council (2009). Regulation (EC) No 470/2009 of the European Parliament and of the Council of 6 May 2009 laying down Community procedures for the establishment of residue limits of pharmacologically active substances in foodstuffs of animal origin, repealing Council Regulation (EEC) No 2377/90 and amending Directive 2001/82/EC of the European Parliament and Council, 2001; and Regulation (EC) No 726/2004 of the European Parliament and of the Council (EU OJ L 152/11, 16.06.2009, pp. 11–22). Available at http://eur-lex.europa.eu/LexUriServ/LexUriServ.do?uri=OJ:L:2009:152:0011:0022:en:PDF; http://eur-lex.europa.eu/LexUriServ/LexUriServ.do?uri=OJ:L:2004:136:0001:0033:en:PDF. Accessed April 6, 2014.

Graf U, Henning HJ, Stange K, Wilrich PT (1987). Formeln und Tabellen der angewandten mathematischen Statistik, 3rd ed. Berlin, Heidelberg, New York, London, Paris, Tokyo: Springer Verlag.

Schefferlie GJ, Hekman P (2009). The size of the safety span for pre-slaughter withdrawal periods. J Vet Pharm Ther 23(Suppl 1):249.

Stange K (1971). Angewandte Statistik, vol. II. Berlin, Heidelberg, New York: Springer Verlag.

VICH (2011). VICH Topic GL 48: Studies to evaluate the metabolism and residue kinetics of veterinary drugs in food-producing animals: Marker residue depletion studies to establish product withdrawal periods. Available at http://www.vichsec.org/en/guidelines2.htm. Accessed April 6, 2014.

6

POPULATION PHARMACOKINETIC MODELING TO PREDICT WITHDRAWAL TIMES

Sharon E. Mason

Department of Biological Sciences, Campbell University, Buies Creek, NC, USA

6.1 INTRODUCTION

Population pharmacokinetic (PK) modeling is a statistically based PK approach to modeling typically sparse amounts of data from a large number of subjects (Sheiner et al., 1977). This technique uses the variability inherent in the population to model subgroups that may not be accurately characterized by traditional PK models (del Castillo et al., 2006). This modeling technique is ideal for use in herd settings as it emphasizes the processes in the population rather than in one specific individual. However, when determining safe withholding times, we are most concerned about those animals that are sick, very old, or very young, because they tend to have different physiology. Unfortunately, most PK studies are performed in healthy, uniform groups of animals, which likely do not represent the above "at-risk" animals. Therefore, new methods to better model sources of variability in a population will prevent loss of marketable animals and minimize the risk of drug residue exposure to consumers. Population PK modeling is one such technique that may be applied to "at-risk" populations.

Strategies for Reducing Drug and Chemical Residues in Food Animals: International Approaches to Residue Avoidance, Management, and Testing, First Edition.
Edited by Ronald E. Baynes and Jim E. Riviere.

A population-based PK model is composed of a structural PK model and a covariate model. The covariate structures, or subject-specific groupings, are a system of equations overlaid on a traditional compartmental PK model. These covariate equations help to define the PK parameters within the PK structural model. The statistical components of the model allow, typically, for a normal distribution of the error terms that account for intrasubject variability and variability in PK parameter distributions (Davidian and Gallant, 1992; Park et al., 1997). These parameter distributions can be parametric, semiparametric, or nonparametric, but most modeling algorithms employ parametric distributions (Vonesh et al., 1996). These distributions are applied to the PK parameters population mean by error terms or equations based on multiplicative, additive, or exponential relationships. The determination of the error modeling scheme is best defined by the nature of the data and the type of relationships the covariates have to the PK structural model parameters (Davidian and Gallant, 1992; Davidian and Giltinan, 1993; Park et al., 1997; Parke et al., 1999). The benefit of this type of modeling is evidenced when data is scarce for individuals and there are differences in PK parameters or measured outcomes across treatments or groups of individuals. If there are no differences in the parameters or groups, then a population model may not be as useful as other PK methods.

6.2 APPLICATIONS OF POPULATION MODELING TO PRESLAUGHTER WITHDRAWAL TIMES

Population modeling techniques have been applied to veterinary medicine effectively for clinical problems (del Castillo et al., 2006; Kawai et al., 1994; Martín-Jiménez and Riviere, 2001, 2002; Mason et al., 2012; Peyou et al., 2004). However, the potential can also be applied to preslaughter withdrawal times. In calculating withdrawal times, a shift in the focus must occur from the therapeutics to the elimination of the drug and its kinetics. This change requires that the model fits be more accurate on later time points than the initial time points. Therefore, models that do not predict the absorption phase as accurately, but can correctly determine factors important to the drug's elimination, are preferred (Martín-Jiménez and Riviere, 1998). This requires a paradigm shift for most modelers as the focus is often on peak and therapeutic concentrations (Fig. 6.1) rather than for the elimination information (Fig. 6.2).

Although the statistical portion of the model is defined by parameter distributions, a traditional PK structural (i.e., compartmental) model is still the basis of the design. *A priori* information for PK parameters is helpful but not necessarily required for a population model (Wade et al., 1994). In determining which model type to use (compartmental type (micro vs. macro) and

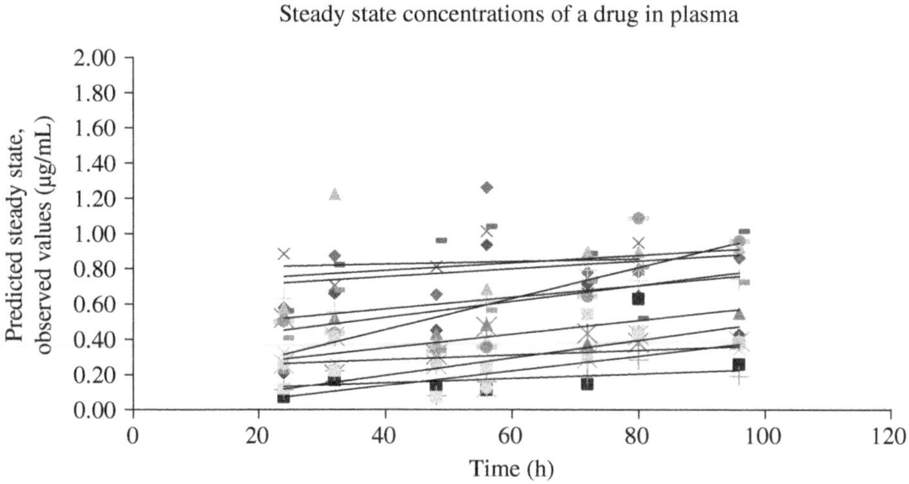

FIGURE 6.1 Model of a population approach (Phoenix WinNonMix 6.0) to estimate the steady state plasma concentrations of a medication given to a population of swine medication via at will dosing.

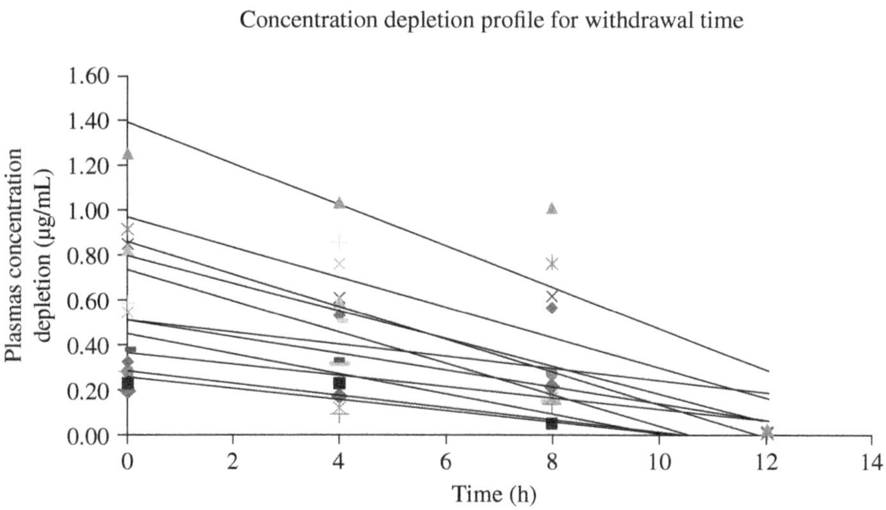

FIGURE 6.2 Simulation of a population of animals when the pharmacokinetic elimination profile is emphasized over the therapeutic profile in Figure 6.1 (Phoenix WinNonMix 6.0).

number, sampling schemes, etc.), there are many approaches in the literature, and these often involve different algorithms for crafting studies including D-optimal design (Green and Duffull, 2003; Gueorguieva et al., 2006; Tod et al., 1998); however there are two basic designs, a two-stage model approach (data intense sampling) (Tam et al., 2003; Yeap and Davidian, 2001) and a non-linear mixed effects approach with sparse data (Davidian and

Giltinan, 1993; FDA, 1999; Gobburu and Lawrence, 2002). There are also pharmacoepidemiological (Clewell and Andersen, 1996; Dorr et al., 2009) and pharmacogenetic approaches (Borges et al., 2006) that may be able to provide information on specific selection of modeling types.

In population PK models, it is useful to understand the physiology of the drug, as with other PK modeling types (Karlsson and Sheiner, 1994; Mager and Goller, 1995; Martín-Jiménez and Riviere, 2001, 2002; Vinks, 2002; Wahlby et al., 2004). However, population PK modeling is not based on a physiologic model but rather statistical theory of population distributions. To define the individual variability seen in PK models, random error, individual error (interoccasional error), and finally interindividual (or between-subject) variability are addressed typically with covariates (del Castillo et al., 2006; FDA, 1999; Merle et al., 2004). This can be explained visually by looking at the differences in Figure 6.3, where the first graph represents the predicted concentrations versus time for the PK model without covariates compared to the full covariate model comparison.

Before covariates can be used to determine interindividual differences, other statistical analysis must be performed. Statistical programs can be used to look at the limits of quantification and detection (Bressolle et al., 1996) and determine the amount of variability based on the intraindividual error, which is found by repeated measures from several individuals. However, this error term will include both individual and random error. This error is often denoted by many texts in statistical format for the concentration of the ith individual at the jth time point. This is used to look for the random (ε) and individual (η), often interoccasional, errors (Lindstrom and Bates, 1990). Once the intraindividual variability is defined or better characterized, the interindividual error can be addressed. The intersubject error is the method used to assess the covariate distributions and statistical significance in the model comparisons (Mager, 2004; Wahlby et al., 2004). Therefore, we can say that the identifiable sources of variability (parameter variability, covariate effects, and measurement error) for the model are what allow this method to work (Godfrey et al., 1980, 1994). These are notated as follows:

η for intersubject error (i.e., individual differences)
ω^2 for intersubject variability
ε for random error for each observation for the ith individual

Each of these types of errors is set for the parameters involved in the model in order to simulate the concentration–time profiles and other directly measured data. With population PK modeling, it is essential that differences among at least the intraindividual error term be specified, as the random error can be lumped in with it. Without defining the error types, model covariates or bias

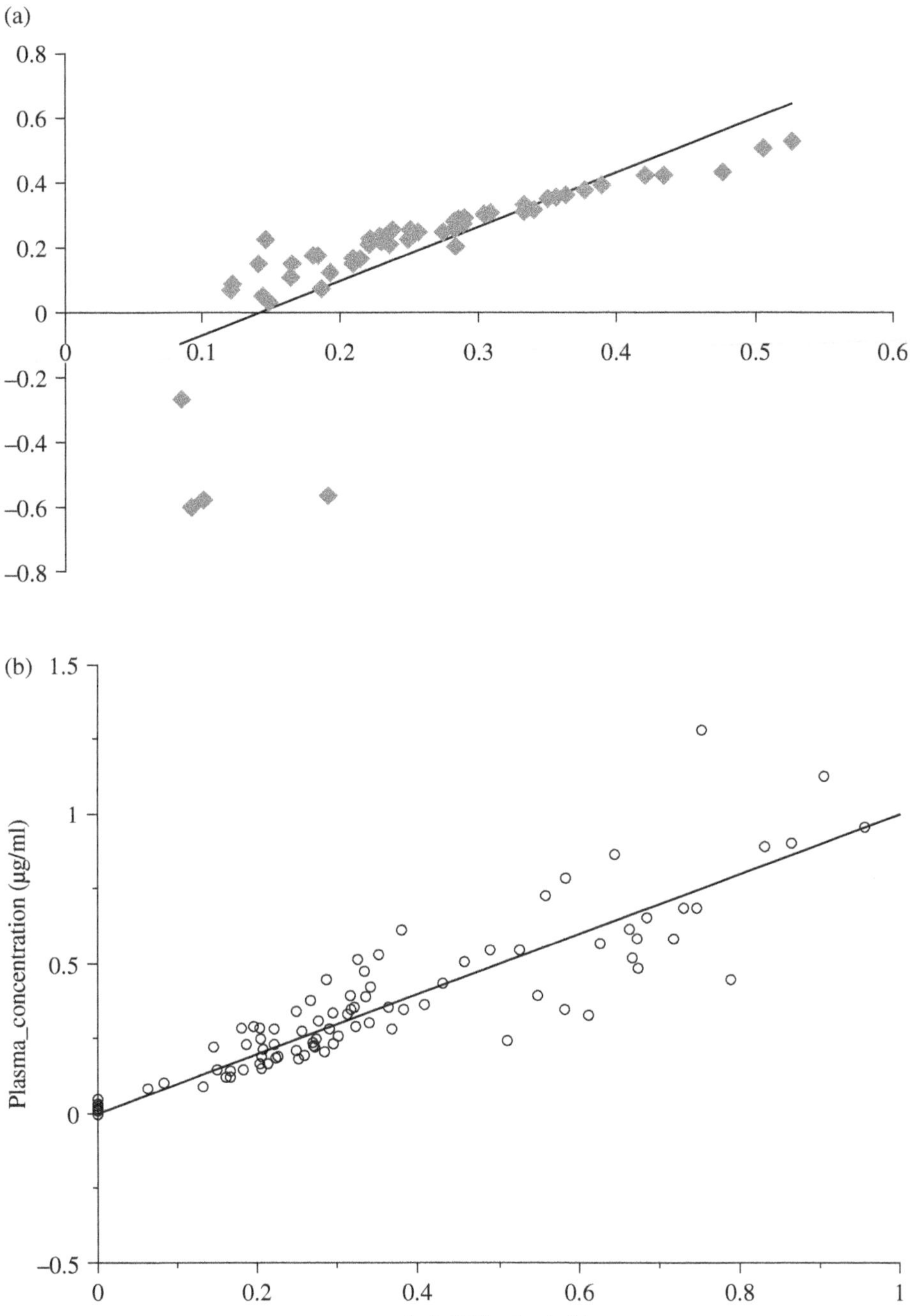

FIGURE 6.3 Graphical results of a PK modeling including covariates (a) compared to model without covariates (b).

may not be fully elucidated (Wade et al., 1994; Wahlby et al., 2004; Williams, 1990). It is possible to develop a model without defining each error type; however, the coefficient of variation across the PK parameters in the model becomes very high and is often incalculable. Therefore, the accuracy and precision of PK parameters are defined by the error specification within a model.

6.3 COVARIATE ANALYSIS

As stated earlier, covariate analysis allows definition of the sources of interindividual errors. Thus, emphasis is often placed on covariates to provide information on differences in PK parameters among subgroups of a population. These differences typically relate to or determine physiologic parameters of metabolism or clearance of a drug (such as enzyme polymorphisms or disease states) or physiologic changes due to a disease process (Martín-Jiménez and Riviere, 2001; Wahlby et al., 2004), which often affect volume of distribution and protein binding. Covariates may include continuous variables or values, such as serum creatinine, liver-specific enzyme levels, age, blood pressure, or body weight; or they may be categorical variables, such as gender or health status (Mager, 2004; Martín-Jiménez and Riviere, 2001; Wahlby et al., 2004). These covariates may provide insight into the PK related to a drug by providing a physiologic basis for variability present in the population.

Potential covariates should be investigated from the available collected data (Mager, 2004). In general, there are covariates that often have relationships to the PK parameters such as serum creatinine as a marker for clearance and body weight as a surrogate for volume of distribution. The final covariates chosen for the model should account for interindividual variability in the model and decrease the overall intrasubject and random error significantly.

However, not all covariates that are commonly important will be useful. Therefore, a selection process is necessary. There are multiple methods of performing statistical analysis to select for covariates. However, graphical comparisons of the data to potential covariates are often the easiest method for nonstatisticians. By graphically plotting residuals of a model as a function of the covariate values for each subject, a wealth of information on that covariate and its relationship to the data may be gained. The graphical method allows one to determine the effects of each variable to the results of either a linear or compartmental model (Davidian and Gallant, 1992). The graphical technique, if used properly, can also reveal nonlinear relationships of covariates that ANOVA or other linear methods cannot detect. The residuals' plot

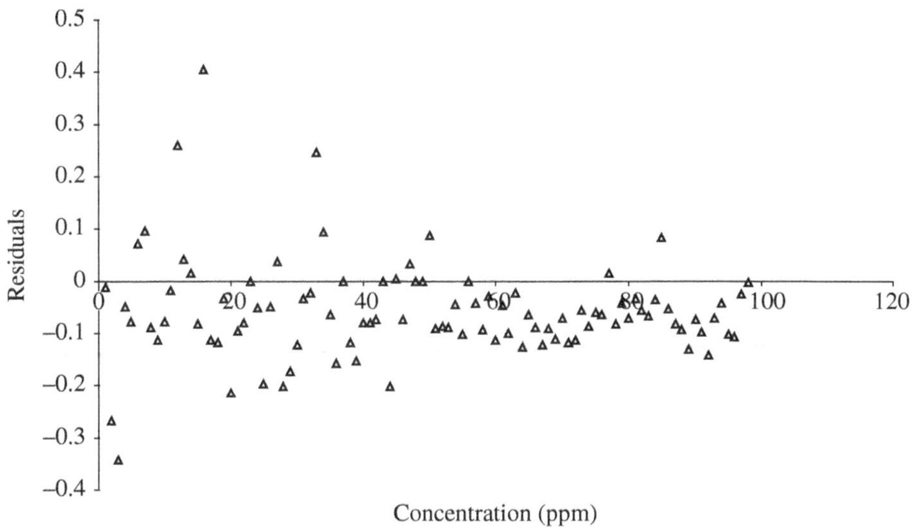

FIGURE 6.4 Plot of residuals for a model versus the concentration for a covariate. The residuals from this model are systematically too low for the data, suggesting that the covariate should be investigated further.

also helps explain the relationship of these covariates. When no relationship or bias to the model is present from a covariate, the residuals are evenly spaced above and below the estimated value for the model. However, if covariates skew the residuals in nonrandom patterns, then this covariate should be considered in the model design as it has a disproportionate effect on the data. This technique is often used with linear regression model evaluations. This highly useful tool provides some insight into variables of importance and can also suggest the type of relationship (linear, exponential, polynomial, etc.) that the covariate may have to the model. For instance, Figure 6.4 shows that there is a mild (linear) skewing of the residuals, and therefore, this covariate may be important for the model.

Another method of selecting covariates is to use stepwise additions of suspected covariates to a population compartmental analysis (Gobburu and Lawrence, 2002). Most modeling programs will plot graphical relationships between PK parameters and the covariate being tested across the populations, so differences can be detected. This is another method for selecting covariates without performing statistical assessments prior to model inclusion (Phoenix WinNonMix 6.0, Pharsight Co., Mountain View, CA). However, many models include several covariates to make the model fit more complete. With multiple covariates defined, the final model variability is decreased, as can be seen in Figure 6.3, where the model better predicts the data than prior to the covariate additions.

6.4 BENEFITS TO POPULATION-BASED MODELING TECHNIQUES

In using this type of modeling technique, there are benefits and drawbacks. One nice benefit is the ability to set the model structure based on previous data or studies. This allows the PK model structure to be defined and any physiologic changes to be seen, instead of having to develop separate models for different populations and then compare them. This modeling technique also allows populations to be compared directly, instead of, with traditional modeling, having multiple models developed, sometimes with different structures. However, the biggest benefit is the ability to define and account for various error types. This statistical definition of population variability has allowed population PK modeling to be used in safety studies on humans for drug approval and labeling by the FDA, unlike some other modeling techniques where the risk cannot be fully defined for some "diseased" populations (FDA, 1999). Another major benefit of population modeling is the ability to employ bootstrapping, resampling, and data splitting to the original data sets, if necessary, to increase sample population sizes. The aforementioned techniques along with varying the type of parameter distributions can help modelers avoid misspecification of parameter variability and improve model fit (Bustad et al., 2006; Davidian and Gallant, 1992; Oberg and Davidian, 2000; Park et al., 1997; Parke et al., 1999; Sprandel et al., 2006). Various studies have been performed in the literature in both human and veterinary medicine applying various PK modeling programs, including NONMEM, WinBUGS, and WinNonMix to population PK modeling (Anderson et al., 2006; del Castillo et al., 2006; Chervoneva et al., 2007; Concordet and Toutain, 1997; Davidian and Gallant, 1992; Gupta et al., 2001; Karlsson and Sheiner, 1994; Martín-Jiménez and Riviere, 1998, 2001, 2002; Peyrou et al., 2004; Vinks et al., 1996; Yuh et al., 1994).

6.5 LIMITATIONS OF POPULATION-BASED MODELING TECHNIQUES

Unfortunately, as with any modeling, there are assumptions and limits to the usefulness of a technique. With population-based modeling, the modeling structures tend to be more simplistic than those of physiologically based PK modeling or hybrid computational models that use combinations of compartmental and mechanistic modeling (Riviere, 1997). This modeling technique tends to use basic compartmental modeling techniques for the structural models without the complexity of intercompartmental exchange rates unless they can be defined by the available data. Due to the limitations of modeling some (often sparse) data, the use of a more simplistic model structure (two-compartmental over a three-compartmental model) may be necessary. Finally, this modeling technique is greatly affected by the population size and repeated sampling. Some of the statistical algorithms ignore repeated samples from one individual for a simpler approach and can therefore skew model parameters if appropriate algorithm changes are not made. Differences in algorithms can only be appreciated by running a model with known parameters and comparing algorithm outputs directly. In general, for normal or fairly normal populations, first order (FO) sum of squares (Bustad et al., 2006; Gobburu and Lawrence, 2002; Jonsson et al., 2000; Tanigawara et al., 1994) or first order conditional estimate (FOCE–LB) with Lindstrom Bates (Lindstrom and Bates, 1990; Yuh et al., 1994) work well. However, if there is a bimodal or non-normal population parameter, algorithms that do not assume normality and nonparametric techniques are better at fitting the data (Bustad et al., 2006). An algorithm may be chosen based on expected outcome or on model simulations. Therefore, the algorithm chosen should be appropriate for the type of data collected to have the correct statistical significance and be valid. This includes choosing parametric or nonparametric modeling schemes and appropriate algorithms for the data analysis and implemented study design (Bustad et al., 2006; Davidian and Gallant, 1992; Oberg and Davidian, 2000; Park et al., 1997; Parke et al., 1999; Sprandel et al., 2006).

6.6 FUTURE APPLICATIONS

Unfortunately, the majority of PK studies and therefore PK information is available only in healthy animals. As mentioned earlier, the use of a known or defined population model (for one species for a specific drug) can be determined from available data, and then information from ill or sick animals, often with elevated tissue or plasma concentrations, can be compared to this model. In this fashion, "sick animal" PK parameters can be defined from available sparse data, which could be obtained from different sources and exposures. As mentioned earlier, population PK modeling can be combined with Bayesian forecasting, bootstrapping, and potentially meta-analysis (for various sources of data from ill animals) to gather information on the PK changes that are seen in animals with illnesses. With these changes, either the data can be simulated for known disease processes or expected physiologic differences can be defined in the models, such as kidney failure with drastically decreased clearance for affected medications. Unfortunately, one change in the body will often impact multiple systems and PK parameters: such as volume of distribution, tissue permeabilities, and protein binding.

The benefit of just modeling these processes with data obtained from real-life sources is that the actual data can be used to develop model differences rather than having to specify large numbers of parameter changes in mechanistic based models. The ability to obtain and then model data from ill food animals will hopefully become a reality in the not so distant future.

6.7 CONCLUSIONS

The application of population-based PK models is important for explanations and predictions of therapeutic drug concentrations in veterinary patients. Previous studies have applied population PK modeling to veterinary medicine in general with few studies on food animal medicine (del Castillo et al., 2006; Martín-Jiménez and Riviere, 2001, 2002; Martín-Jiménez et al., 2002; Mason et al., 2012). The use of this modeling technique has merit for use in predicting preslaughter withdrawal times due to its flexibility (FDA, 1999; Martín-Jiménez and Riviere, 1998). Its application to veterinary medicine in general and to herd medicine specifically is warranted, as this is a technique based in statistics, which can use less intensive data collection than other modeling techniques and still maintain the ability to predict population parameters across subgroups (FDA, 1999; Wahlby et al., 2004). Fortunately, for determining preslaughter withhold information, the population is the unit of measure, not the individual, which is the same standard set by the FDA for determining safe withholding times. FDA has advocated the use of population-based PK for human and some animal studies when those studies follow appropriate sampling design and extrapolation guidelines (FDA, 1999). The current U.S. tolerance limit detection is based on statistical interpolation from slaughter data (FDA, 2006; Martinez et al., 2000). Therefore, in the future, veterinarians may use population-based PK techniques to estimate appropriate withholding times for international markets where their minimum residue limits or tolerance limits are different from those in country.

REFERENCES

Anderson BJ, Allegaert K, Holford NH (2006). Population clinical pharmacology of children: Modelling covariate effects. Eur J Pediatr 165(12):819–829.

Borges S, Desta Z, Li L, Skaar TC, Ward BA, Nguyen A, Jin Y, Storniolo AM, Nikoloff DM, Wu L, Hillman G, Hayes DF, Stearns V, Flockhart DA (2006). Quantitative effect of CYP2D6 genotype and inhibitors on tamoxifen metabolism: Implication for optimization of breast cancer treatment. Clin Pharmacol Ther 80(1):61–74.

Bressolle F, Bromet-Petit M, Audran M (1996). Validation of liquid chromatographic and gas chromatographic methods. Applications to pharmacokinetics. J Chromatogr B Biomed Appl 686(1):3–10.

Burnham KP, Anderson DR (2004). Multimodel inference: Understanding AIC and BIC in model selection. Sociol Methods Res 33(2):261–304.

Bustad A, Terziivanov D, Leary R, Port R, Schumitzky A, Jelliffe R (2006). Parametric and nonparametric population methods: Their comparative performance in analysing a clinical dataset and two Monte Carlo simulation studies. Clin Pharmacokinet 45(4):365–383.

del Castillo JR, Laroute V, Pommier P, Zemirline C, Keita A, Concordet D, Toutain PL (2006). Interindividual variability in plasma concentrations after systemic exposure of swine to dietary doxycycline supplied with and without paracetamol: A population pharmacokinetic approach. J Anim Sci 84(11):3155–3166.

Chervoneva I, Hyslop T, Hauck WW (2007). A multivariate test for population bioequivalence. Stat Med 26(6):1208–1223.

Clewell HJ, 3rd, Andersen ME (1996). Use of physiologically based pharmacokinetic modeling to investigate individual versus population risk. Toxicology 111(1–3):315–329.

Concordet D, Toutain PL (1997). The withdrawal time estimation of veterinary drugs revisited. J Vet Pharmacol Ther 20(5):380–386.

Davidian M, Gallant AR (1992). Smooth nonparametric maximum likelihood estimation for population pharmacokinetics, with application to quinidine. J Pharmacokinet Biopharm 20(5):529–556.

Davidian M, Giltinan DM (1993). Some general estimation methods for nonlinear mixed-effects models. J Biopharm Stat 3(1):23–55.

Dorr PM, Nemechek MS, Scheidt AB, Baynes RE, Gebreyes WA, Almond GW (2009). Water-flow variation and pharmacoepidemiology of tetracycline hydrochloride administration via drinking water in swine finishing farms. J Am Vet Med Assoc 235(3):299–304.

FDA (1999). Guidance for industry on Population Pharmacokinetics; availability. Food and Drug Administration, HHS. Notice. Fed Regist 64(27):6663–6664.

FDA (July 25, 2006). Guidance for industry no. 3: General principles for evaluating the safety of compounds used in food-producing animals. Rockville: US FDA.

Gobburu JV, Lawrence J (2002). Application of resampling techniques to estimate exact significance levels for covariate selection during nonlinear mixed effects model building: Some inferences. Pharm Res 19(1):92–98.

Godfrey KR, Jones RP, Brown RF (1980). Identifiable pharmacokinetic models: The role of extra inputs and measurements. J Pharmacokinet Biopharm 8(6): 633–648.

Godfrey KR, Chapman MJ, Vajda S (1994). Identifiability and indistinguishability of nonlinear pharmacokinetic models. J Pharmacokinet Biopharm 22(3):229–251.

Green B, Duffull SB (2003). Prospective evaluation of a D-optimal designed population pharmacokinetic study. J Pharmacokinet Pharmacodyn 30(2):145–161.

Gueorguieva I, Aarons L, Ogungbenro K, Jorga KM, Rodgers T, Rowland M (2006). Optimal design for multivariate response pharmacokinetic models. J Pharmacokinet Pharmacodyn 33(2):97–124.

Gupta S, Banfield C, Kantesaria B, Marino M, Clement R, Affrime M, Batra V (2001). Pharmacokinetic and safety profile of desloratadine and fexofenadine when coadministered with azithromycin: A randomized, placebo-controlled, parallel-group study. Clin Ther 23(3):451–466.

Jonsson EN, Wade JR, Karlsson MO (2000). Nonlinearity detection: Advantages of nonlinear mixed-effects modeling. AAPS PharmSci 2(3):E32.

Karlsson MO, Sheiner LB (1994). Estimating bioavailability when clearance varies with time. Clin Pharmacol Ther 55(6):623–637.

Kawai R, Lemaire M, Steimer JL, Bruelisauer A, Niederberger W, Rowland M (1994). Physiologically based pharmacokinetic study on a cyclosporin derivative, SDZ IMM 125. J Pharmacokinet Biopharm 22(5):327–365.

Lindstrom ML, Bates DM (1990). Nonlinear mixed effects models for repeated measures data. Biometrics 46(3):673–687.

Mager H (2004). Covariate selection in population pharmacokinetics/pharmacodynamics. Int J Clin Pharmacol Ther 42(3):187–188.

Mager H, Goller G (1995). Analysis of pseudo-profiles in organ pharmacokinetics and toxicokinetics. Stat Med 14(9–10):1009–1024.

Martinez M, Friedlander L, Condon R, Meneses J, O'Rangers J, Weber N, Miller M (2000). Response to criticisms of the US FDA parametric approach for withdrawal time estimation: Rebuttal and comparison to the nonparametric method proposed by Concordet and Toutain. J Vet Pharmacol Ther 23(1):21–35.

Martín-Jiménez T, Riviere JE (1998). Population pharmacokinetics in veterinary medicine: Potential use for therapeutic drug monitoring and prediction of tissue residues. J Vet Pharmacol Ther 21(3):167–189.

Martín-Jiménez T, Riviere JE (2001). Mixed effects modeling of the disposition of gentamicin across domestic animal species. J Vet Pharmacol Ther 24(5):321–332.

Martín-Jiménez T, Riviere JE (2002). Mixed-effects modeling of the interspecies pharmacokinetic scaling of oxytetracycline. J Pharm Sci 91(2):331–341.

Martín-Jiménez T, Baynes RE, Craigmill A, Riviere JE (2002). Extrapolated withdrawal-interval estimator (EWE) algorithm: A quantitative approach to establishing extralabel withdrawal times. Regul Toxicol Pharmacol 36(1):131–137.

Mason SE, Almond, GW, Riviere JE, Baynes RE (October 2012). Variables important to the population pharmacokinetic modeling of water medication in swine. *AJVR* 73(10):1641–1649.

Merle Y, Aouimer A, Tod M (2004). Impact of model misspecification at design (and/or) estimation step in population pharmacokinetic studies. J Biopharm Stat 14(1):213–227.

Oberg A, Davidian M (2000). Estimating data transformations in nonlinear mixed effects models. Biometrics 56(1):65–72.

Park K, Verotta D, Blaschke TF, Sheiner LB (1997). A semiparametric method for describing noisy population pharmacokinetic data. J Pharmacokinet Biopharm 25(5):615–642.

Parke J, Holford NH, Charles BG (1999). A procedure for generating bootstrap samples for the validation of nonlinear mixed-effects population models. Comput Methods Programs Biomed 59(1):19–29.

Peyrou M, Doucet MY, Vrins A, Concordet D, Schneider M, Bousquet-Melou A (2004). Population pharmacokinetics of marbofloxacin in horses: Preliminary analysis. J Vet Pharmacol Ther 27(5):283–288.

Riviere JE (1997). Basic principles and techniques of pharmacokinetic modeling. J Zoo Wildl Med 28(1):3–19.

Sheiner LB, Rosenberg B, Marathe VV (1977). Estimation of population characteristics of population pharmacokinetic parameters from routine clinical data. J Pharmacokinet Biopharm 5: 445–479.

Sprandel KA, Drusano GL, Hecht DW, Rotschafer JC, Danziger LH, Rodvold KA (2006). Population pharmacokinetic modeling and Monte Carlo simulation of varying doses of intravenous metronidazole. Diagn Microbiol Infect Dis 55(4): 303–309.

Tam VH, Preston SL, Drusano GL (2003). Comparative pharmacokinetic analysis by standard two-stage method versus nonparametric population modeling. Pharmacotherapy 23(12):1545–1549.

Tanigawara Y, Yano I, Kawakatsu K, Nishimura K, Yasuhara M, Hori R (1994). Predictive performance of the Bayesian analysis: Effects of blood sampling time, population parameters, and pharmacostatistical model. J Pharmacokinet Biopharm 22(1):59–71.

Tod M, Mentre F, Merle Y, Mallet A (1998). Robust optimal design for the estimation of hyperparameters in population pharmacokinetics. J Pharmacokinet Biopharm 26(6):689–716.

Vinks AA (2002). The application of population pharmacokinetic modeling to individualized antibiotic therapy. Int J Antimicrob Agents 19(4):313–322.

Vinks AA, Mouton JW, Touw DJ, Heijerman HG, Danhof M, Bakker W (1996). Population pharmacokinetics of ceftazidime in cystic fibrosis patients analyzed by using a nonparametric algorithm and optimal sampling strategy. Antimicrob Agents Chemother 40(5):1091–1097.

Vonesh EF, Chinchilli VM, Pu K (1996). Goodness-of-fit in generalized nonlinear mixed-effects models. Biometrics 52(2):572–587.

Wade JR, Beal SL, Sambol NC (1994). Interaction between structural, statistical, and covariate models in population pharmacokinetic analysis. J Pharmacokinet Biopharm 22(2):165–177.

Wahlby U, Thomson AH, Milligan PA, Karlsson MO (2004). Models for time-varying covariates in population pharmacokinetic-pharmacodynamic analysis. Br J Clin Pharmacol 58(4):367–377.

Williams PL (1990). Structural identifiability of pharmacokinetic models – compartments and experimental design. J Vet Pharmacol Ther 13(2):121–131.

Yeap BY, Davidian M (2001). Robust two-stage estimation in hierarchical nonlinear models. Biometrics 57(1):266–272.

Yuh L, Beal S, Davidian M, Harrison F, Hester A, Kowalski K, Vonesh E, Wolfinger R (1994). Population pharmacokinetic/pharmacodynamic methodology and applications: A bibliography. Biometrics 50(2):566–575.

7

PHYSIOLOGICALLY BASED PHARMACOKINETIC MODELING

Jennifer Buur

College of Veterinary Medicine, Western University of Health Sciences, Pomona, CA, USA

7.1 INTRODUCTION

Physiologically based pharmacokinetic (PBPK) models predict the absorption, distribution, metabolism, and elimination of xenobiotics based on anatomic parameters and physiological and biochemical mechanisms. The mechanistic basis allows for flexibility in the application of these models as they can be scaled across doses, routes of administration, species, and molecules of interest. Additionally, this modeling tool allows for direct correlation of tissue concentrations to specific targets such as tolerance limits, therapeutic ranges, or toxic concentrations. Finally, the mechanistic nature of PBPK allows for prediction of PK for specific individuals as well as over entire populations (Krishnan and Andersen, 2001).

Because of their mechanistic nature, PBPK models provide a complement to other pharmacokinetic modeling techniques. This is readily apparent if you compare PBPK modeling to classical compartmental analysis (Table 7.1). Compartmental analysis focuses on the rates of entry and exit of xenobiotics into aggregate compartments with theoretical volumes. The mathematical constructs are exponential equations that are fit to observe data using commonplace computer programs. Compartmental parameters make no

Strategies for Reducing Drug and Chemical Residues in Food Animals: International Approaches to Residue Avoidance, Management, and Testing, First Edition.
Edited by Ronald E. Baynes and Jim E. Riviere.

TABLE 7.1 Comparison between PBPK models and classical compartmental analysis

Concept	PBPK models	Compartmental pharmacokinetic models
Parameters	Mechanistic	Theoretical
Equations	Mass balance	Exponential
Predictive ability	Highly predictive	Descriptive
Tissue kinetics	Tissue specific	Central compartment (plasma)
Data required	Large amounts	Limited amounts
Flexibility	High	Limited
Time required for creation	Large	Limited
Software	Complicated and expensive	Commonplace and well understood
General acceptance	Low	High

claim to physiological relevance, and very little data is known a priori to running the in-silica experiment. Compartmental models are rarely tested for accuracy and precision of predictive capability and are generally limited to the dose, species, and route of administration used to generate the curve-fitting data. This makes them relatively quick and easy to perform, and they generate clinically useful data that is important in designing dosing regimens, is well established in regulatory agencies, and is widely understood in the research literature (Riviere, 1999). PBPK models, on the other hand, are based on mass balance equations that require large amounts of a priori data. They provide pharmacokinetic information in multiple tissue compartments rather than being limited to plasma or other central compartment. In addition to calibration against a limited number of data points, they may be also subjected to validation against external data sets to evaluate their predictive abilities. The equations and parameters require multiple *in vivo* studies and complicated software programs to compile and simultaneously solve the multiple differential equations involved (Chiu et al., 2007). PBPK models are data and time intensive to create. However, robust PBPK models are flexible and not limited to specific doses, routes of administration, species, or molecule of interest. Combination of compartmental analysis and PBPK models using the same experimental data yield a more complete picture of the underlying physiology that determines the pharmacokinetics of xenobiotics than either of the techniques accomplish alone. Both techniques have clinical application in different aspects of veterinary medicine.

In recent years, there has been a surge in the number of PBPK models in the literature. This is due not only to the ubiquitous nature of personal computers that are capable of the calculations but also in the acceptance of this modeling technique for regulatory as well as research purposes (Charnick et al., 1995). PBPK models are most commonly used in toxicology for human health risk assessment (Chiu et al., 2007). In fact, the U.S. Environmental Protection

Agency (US-EPA) allows the use of this modeling technique in the application and renewal processes (US-EPA, 2006). The U.S. Food Animal Residue Avoidance Databank (US-FARAD) also uses multiple PBPK models for the estimation of meat withdrawal intervals after extralabel drug use in food animal species (Riviere et al., 1997). In both human and veterinary clinical medicine, PBPK models have been developed for patient subpopulations (i.e., pregnancy) and for molecules with narrow therapeutic ranges or high individual variability (Bjorkman, 2004). In basic science research, PBPK models are used to explore relationships between molecules in mixtures, enzyme kinetics, and the physiology underlying various routes of administration (van der Merwe et al., 2006; Teeguarden et al., 2005). The variety of uses for this modeling technique is testament to the flexibility of a robust modeling system.

7.2 MODEL DEVELOPMENT AND VALIDATION

Due to their inherent flexibility and the ability to add multiple layers of complexity, all PBPK models start by defining the purpose. The purpose determines which tissues and the subsequent volumes, blood flows, and tissue-to-blood partition coefficients needed to answer specific questions. A simplistic model may only contain tissue blocks for the route of administration and elimination specifically with the rest of the animal simplified into large surrogate blocks with similar physiology. These surrogate compartments often include richly or poorly perfused tissues (Fig. 7.1). In contrast to these simplified models,

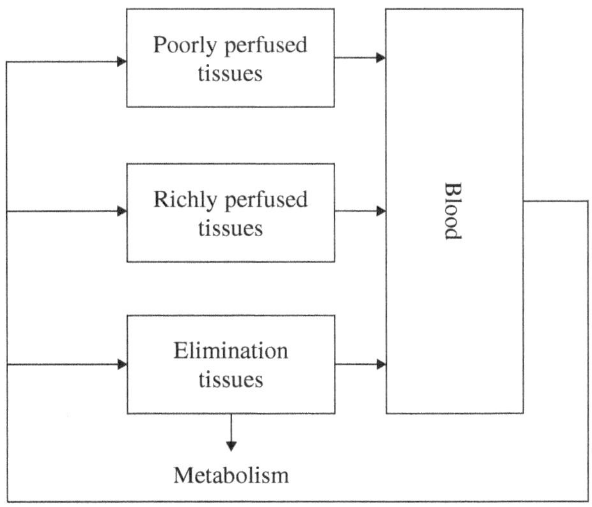

FIGURE 7.1 Schematic diagram of a simplified PBPK model. Arrows represent mass transfer via blood flow or elimination.

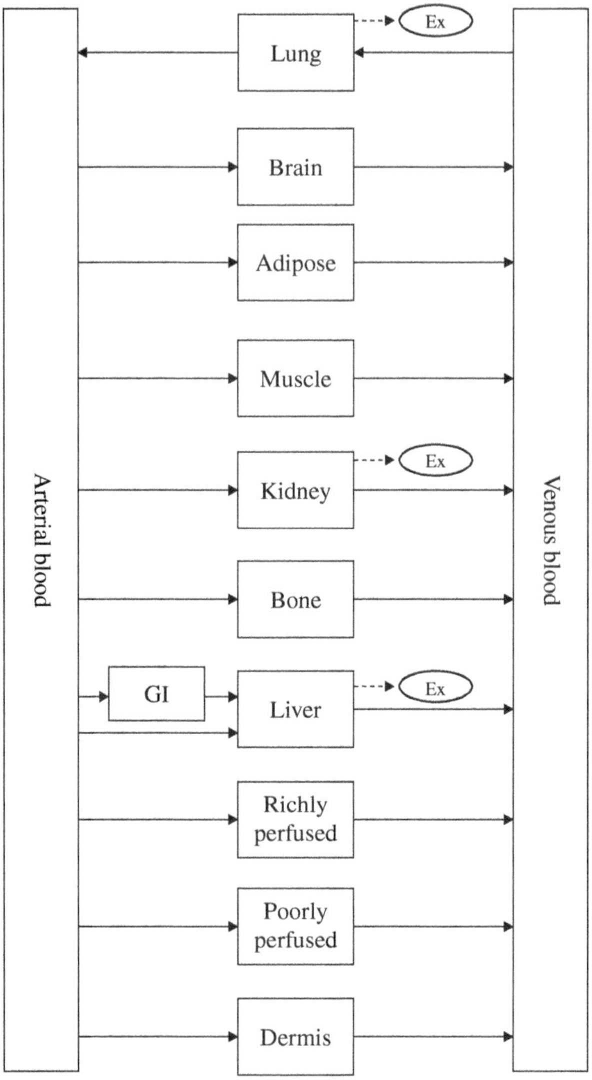

FIGURE 7.2 Schematic diagram of a more complicated PBPK model. Arrows represent mass transfer via blood flow or elimination.

PBPK has the ability to increase in complexity to include compartments for any tissue of interest. For example, models may include target organ of toxicosis or protected spaces such as the brain (Fig. 7.2). In the prevention of food residues, the tissues of interest are edible tissues, and model blocks generally consist of the kidney, liver, muscle, fat, and plasma (Fig. 7.3).

For most PBPK models used in food safety, the tissue compartments are flow limited and well mixed. The rate-limiting step for small, lipophilic drug

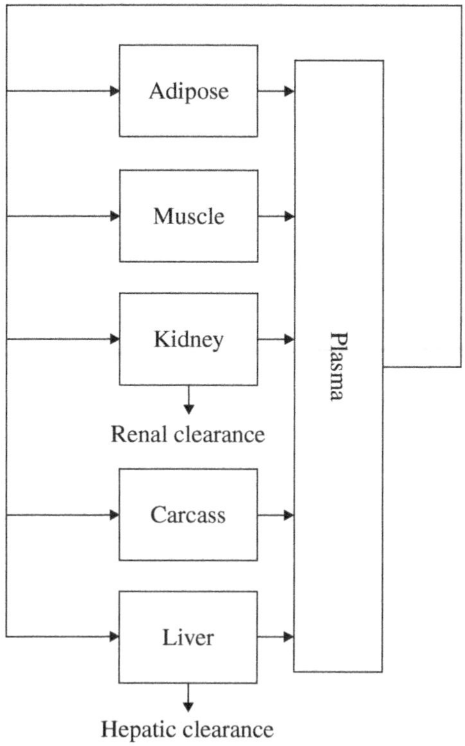

FIGURE 7.3 Schematic diagram of a representative PBPK model designed for food residue avoidance. Tissue blocks represent common edible tissues. Arrows represent mass transfer via blood flow or elimination.

distribution into the compartment is assumed to be blood flow rather than diffusion across the lipid membrane. It is also assumed that the drug is instantaneously in equilibrium and homogeneous throughout the tissue compartment. For tissue compartments where that is not the case, such as physiological protected spaces, further subdivisions can be created within each block. These most often reflect extracellular space, intracellular space, as well as tissue binding (Fig. 7.4) (Ritschel and Banerjee, 1986). Tissue binding in particular has importance in the blood compartment as multiple xenobiotics have large protein binding capacities that will limit specific distribution into other tissue compartments.

Mathematical representation of the pharmacokinetics is done through the use of differential mass balance equations. The mass balance equations detail the change in xenobiotic mass over time due to tissue distribution into and out of the compartment and clearance processes including metabolism or elimination. Clearançe mechanisms can be described simply using first-order rate constants. If needed, more complicated scenarios including Michaelis–Menten enzyme

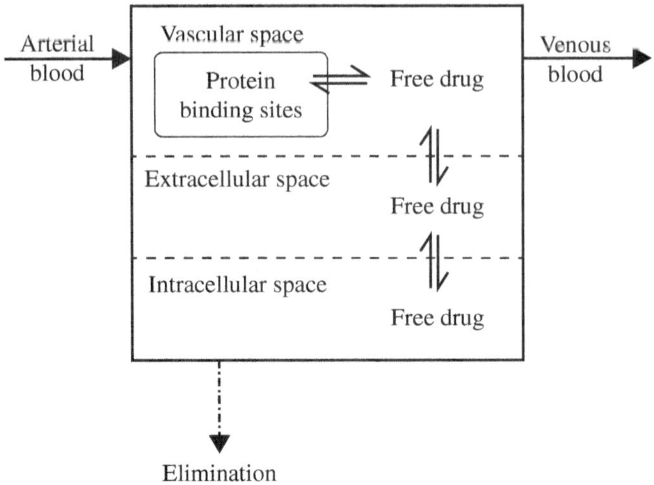

FIGURE 7.4 Schematic diagram of a tissue block containing multiple subcompartments. Arrows represent mass transfer either via blood flow, in equilibrium within the subcompartment, or through elimination.

kinetics or tissue binding kinetics can be added. For flow-limited tissue blocks, tissue-to-blood partition coefficients reflect the steady-state ratio of xenobiotic concentration in the blood and in the tissue. This is a combination parameter that incorporates the passive diffusion of the xenobiotic from the vasculature into the tissue cells along with tissue binding and active transport mechanisms. In diffusion-limited models, multiple differential equations are written to encompass the multiple subcompartments in the tissue block. Examples of common equations are found in Table 7.2. A detailed explanation of the mathematics behind these equations can be found in multiple review articles and is beyond the scope of this chapter (Colburn, 1988; Gerlowski and Jain, 1983; Krishnan and Andersen, 2001).

Simulation of PBPK models requires the simultaneous solving of multiple differential equations and thus requires a large amount of physiologic and physiochemical information a priori. Physiological parameters such as organ weight, organ density, organ volume, hematocrit, and blood flow can often be found in the literature as there are many peer-reviewed resources that contain comprehensive physiologic data for laboratory and food animal species. Other parameters including enzyme kinetics, protein binding, and absorption rates can often be directly measured using a variety of *in vivo*, *in vitro*, and *ex vivo* techniques. Physiochemical parameters such as tissue-to-blood partition coefficients can be harder to quantitate and are often estimated using quantitative structure activity relationships (QSARs), a variety of *in vitro* equilibrium models, and other in silica techniques (Beliveau et al., 2003; DeJongh and Blaauboer, 1996; US-EPA, 2006).

TABLE 7.2 Common mass balance equations found in both flow-limited and diffusion-limited PBPK models

Tissue block	Equation
Partition coefficient	$C_v = \frac{C_t}{P_t}$
Flow-limited mass balance	$V_t \cdot \frac{dC_t}{dt} = Q_t \cdot (C_a - C_v) - R_{el}$
Diffusion-limited mass balance	$V_e \cdot \frac{dC_e}{dt} = Q_t \cdot (C_a - C_e) - K_t \cdot (C_e - C_i)$
	$V_i \cdot \frac{dC_i}{dt} = K_t \cdot (C_e - C_v)$
Central tissue compartment (plasma) mass balance	$V_p \cdot \frac{dC_p}{dt} = \sum (Q_t \cdot C_t) - (Q_p \cdot C_p)$
Elimination kinetics	$R_{el} = K_{el} \cdot C_t$
Protein binding kinetics	$C_t = \frac{A}{V_p}$
	$C_f = \frac{K_d \cdot C_t}{B_{max} + K_d}$
Tissue homogenization parameter	$C_{th} = (V_{bct} \cdot C_{tb}) + [(1 - V_{bct}) \cdot C_t]$

A, amount; B_{max}, maximum binding occupancy; C, concentration; K_d, dissociation constant; K_{el}, first-order rate of elimination constant; K_t, membrane permeability coefficient for tissue t; P, blood-to-tissue partition coefficient; Q, blood flow; R_{el}, rate of elimination; V, volume of tissue compartment. Subscripts: a, arterial circulation; bct, vascular space for tissue t; e, extracellular space; f, free drug; i, intracellular space; p, plasma or central compartment; t, tissue compartment t; tb, tissue blood for tissue t; th, homogenization for tissue t; v, venous circulation.

Given the large numbers, wide variety, and often difficult to measure parameters needed for a robust PBPK model, parameters are often estimated using curve-fitting techniques to data obtained from *in vivo* experimentation. This process is collectively known as calibration. Calibration techniques can include bootstrapping, Markov chain Monte Carlo, and other "best fit" statistical calculations. Tissue-to-blood partition coefficients are best estimated using steady-state *in vivo* conditions, while absorption rates and elimination rates may be best estimated using single-dose observational data (Price et al., 2003). Care must be taken to preserve identifiability within the model and thus assure that each parameter has a single unique value. Data points from each tissue compartment are helpful in achieving identifiability while keeping the total number of points to a minimum (Williams, 1990). No matter the source of the parameter value, the ultimate applicability of PBPK models is directly dependent upon the quality of the parameters used in the equations. Thus, all parameters should be from quality resources and represent biologically plausible values or from curve-fitting procedures that utilize appropriate numbers and sources of tissue data. Additionally, uncertainty analyses can help quantitate confidence in the value of a specific parameter or parameter distribution (US-EPA, 2006).

During the calibration phase of model development, sensitivity analyses are often performed to identify parameters of importance. In the most basic situation, a sensitivity analysis relates the relative contribution of a parameter to the variable of interest. Like all aspects of PBPK modeling, the modeler must take into account the conditions of the observed calibration data points, and the conditions anticipated in the final model application as parameter contributions can change over time and concentrations. Figure 7.5 is an example sensitivity analysis looking at relative changes in plasma concentration when various parameters are altered individually. In this example, renal clearance has the greatest negative effect throughout the simulation, while hepatic clearance has an early negative effect but a late positive effect on plasma concentration. Determining what parameters are sensitive depends on the purpose of the model. If the model in Figure 7.5 was used to predict maximum concentration after an oral dose, then renal clearance may be the most sensitive parameter. However, if the model was used to determine steady-state concentrations 40 hours post dosing, then both hepatic clearance and renal clearance would be sensitive.

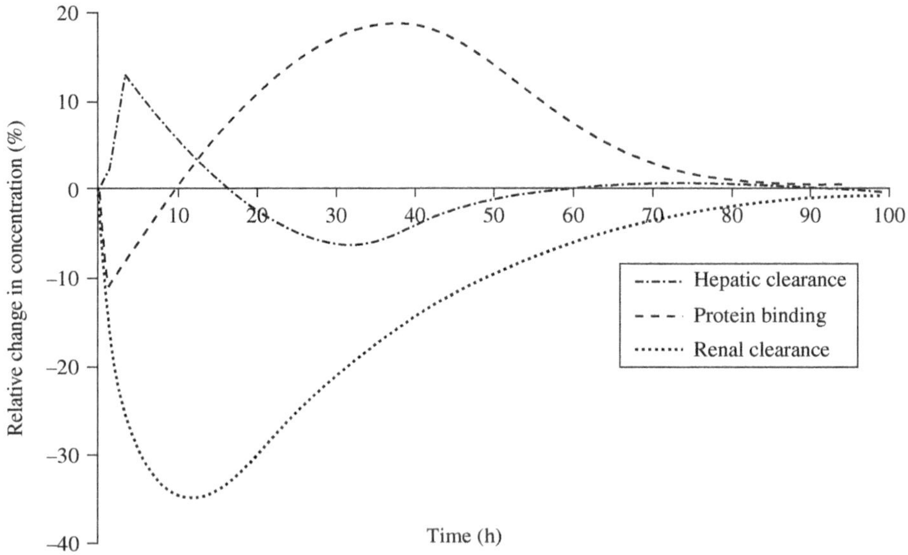

FIGURE 7.5 Example sensitivity analysis for relative change in plasma concentration over time when hepatic clearance, protein binding, and renal clearance are individually altered.

Beyond single point estimates representing the average of a population, PBPK model parameters can be represented as statistical distributions representing entire populations. Distribution methods of parameter estimation can be used to simulate individuals within a population, allowing for study of population diversity. Monte Carlo simulations are one way to implement this concept. Alternatively, Bayesian methods take into account both variability within the population and uncertainty in the distributions and can contribute to more accurate simulations (Buur et al., 2006).

After calibration, the predictive value of the PBPK model is evaluated during the validation process. Validation involves assessment of the accuracy and precision of the predictive capacity of the PBPK model. This is most often accomplished by comparing model-derived simulations with experimentally observed data sets not used in the calibration process. The validation data set should represent the correct study population. Species, breed, strain, dose, route of administration, and analytical method used for analysis should all be taken into account when developing the external data set. Ideally, the validation data set should cover the anticipated scope of model applications and thus reduce the uncertainty associated with being outside of the inference space.

There is no single method for validation of a PBPK model. In most cases, a cadre of techniques is used to provide information on both accuracy and precision of the predictive capability. Classical statistical methods are difficult to apply to these model simulations. Instead, the most common method for validation is visual inspection of the simulation compared to the validation data set. Visual inspection includes not just the magnitude of values (e.g., plasma drug concentrations) but also the general shape of the curve. To test for this, residuals are often plotted and are visually inspected for random distribution scattered equally around zero (Fig. 7.6). The closer the residual

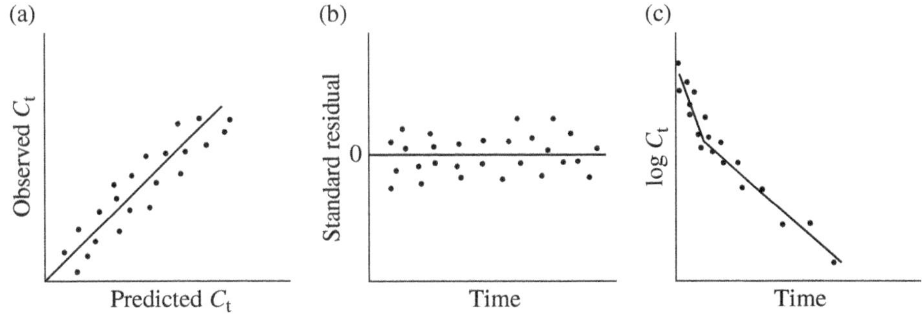

FIGURE 7.6 Examples of various different validation techniques including predicted versus observed concentration regression (a), standard residual plot (b), and direct comparison of simulated concentration–time curves to observed data (c). Dots represent observed data points. Lines represent model simulations.

values are to zero, the more accurate the model prediction. Other forms of validation include the use of Bayesian analysis and cross-validation techniques including subsampling of all data sets. No matter what method is used, the modeler must keep in mind that PBPK models have inherent uncertainty and inconsistency as not every biological process is represented in the model (US-EPA, 2006).

7.3 PBPK APPLIED TO PREDICTION OF DRUG RESIDUES

Many PBPK models have been developed and validated for use in the protection of the food supply through accurate prediction of tissue residues. These models take advantage of the flexibility and predictive nature of PBPK modeling that is not found using classical compartmental techniques. Models in the literature include sulfamethazine, sulfathiazole, and melamine in swine; oxytetracycline in sheep and Atlantic salmon; and midazolam in chickens, turkeys, quail, and pheasants (Brocklebank et al., 1997; Buur et al., 2005, 2008; Cortright et al., 2009; Craigmill, 2003; Duddy et al., 1984). Models have been applied to questions of individual animals and herd populations as well as in situations with limited available data (Buur et al., 2006).

7.3.1 Extralabel Drug Use in Individuals and Populations: Sulfamethazine in Swine

The Animal Medicinal Drug Use Clarification Act (1994) allows veterinarians in the United States to prescribe medications in an off-label manner assuming that there is a valid veterinary–client–patient relationship; there is no approved drug for use; the alternative medication is approved for either human or veterinary use and not specifically prohibited from extralabel use; use is based on medical need; and the safety of the food supply is protected by the establishment of appropriately extended withdrawal intervals such that tissue residues will be below set tolerance limits. PBPK models are uniquely suited to adapt to the wide range of doses and routes of administration that occur as a part of extralabel drug use. Additionally, PBPK models allow for the direct correlation of tissue concentrations to tolerance levels, which leads to a higher degree of confidence in the estimated withdrawal intervals.

Using PBPK models for prediction of withdrawal intervals after extralabel drug use is illustrated using a PBPK model for sulfamethazine in swine (Buur et al., 2005, 2006). The model was developed and validated with terminal tissue residues as the point of focus. Thus, it simulates the individual concentration–time curves for the edible tissues of the kidney, liver, muscle, fat, and plasma. The rest of the animal was lumped into the surrogate carcass compartment. Both intravenous and oral routes of administration were included to reflect the diversity of possible dosing regimens (Fig. 7.7). The model assumes flow-limited, well-mixed compartments and linear metabolism and elimination kinetics. The N4-acetyl metabolite is also included in the model as it undergoes deacetylation back into the parent compound and thus serves as a depot that further prolongs tissue drug concentrations. Model validation was against a data set that encompassed multiple dosing regimens. This helps provide robustness in tissue residue prediction since extralabel use is accompanied by a diversity of dosing regimens. The prime source of physiological data was the published literature. Application of this model to extralabel use was accomplished by simulating a concentration–time curve for each tissue after an IV dose of 50 mg/kg. The withdrawal interval was calculated by finding the time when all tissue concentrations were below the established tolerance limit of 0.1 ppm. In this case, it is at 120 h postinjection. The same model yielded a withdrawal interval of 160 h after an oral dose of 100 mg/kg (Fig. 7.8).

While extralabel drug use is most often on an individual level, there are times when entire populations will be treated in this manner. In this case, it is not enough to predict the average kinetics. Instead, population variability should be estimated. The goal is to predict the upper limits of tissue residues such that the probability of violative tissue residues is minimized. The same PBPK model for sulfamethazine in swine that was used for individuals was also used for populations (Buur et al., 2006). In this instance, parameters identified via sensitivity analysis as important were described by log normal distributions (Table 7.3). Simulations were implemented using Monte Carlo analysis. The predictive capability of the population model was validated by comparison of 100 Monte Carlo simulations to an external data set. Since the representative simulations encompassed the entirety of the validation set, the model was deemed satisfactory. Application of this model looked at percentiles of the population. In this case, the authors explored how long for the 99th percentile of a given population of 1000 individuals simulated by the Monte Carlo analysis for each tissue to be below tolerance. Using the label oral dose of 237.6 mg/kg/day (loading dose) followed by 118.8 mg/kg/day for 3 days, the model simulated that 99% of the herd would be below tolerance after 20 days (Fig. 7.9). This transparent methodology also allows for the creation of confidence intervals that further reflect uncertainty in the model system. The 1000-simulation Monte Carlo analysis was repeated multiple times to create a distribution of estimated withdrawal intervals. A 95% confidence interval was determined for the withdrawal interval population. Using the upper limit of the 95% confidence interval for the 99th percentile of the population resulted in the estimation of a withdrawal interval of 21 days.

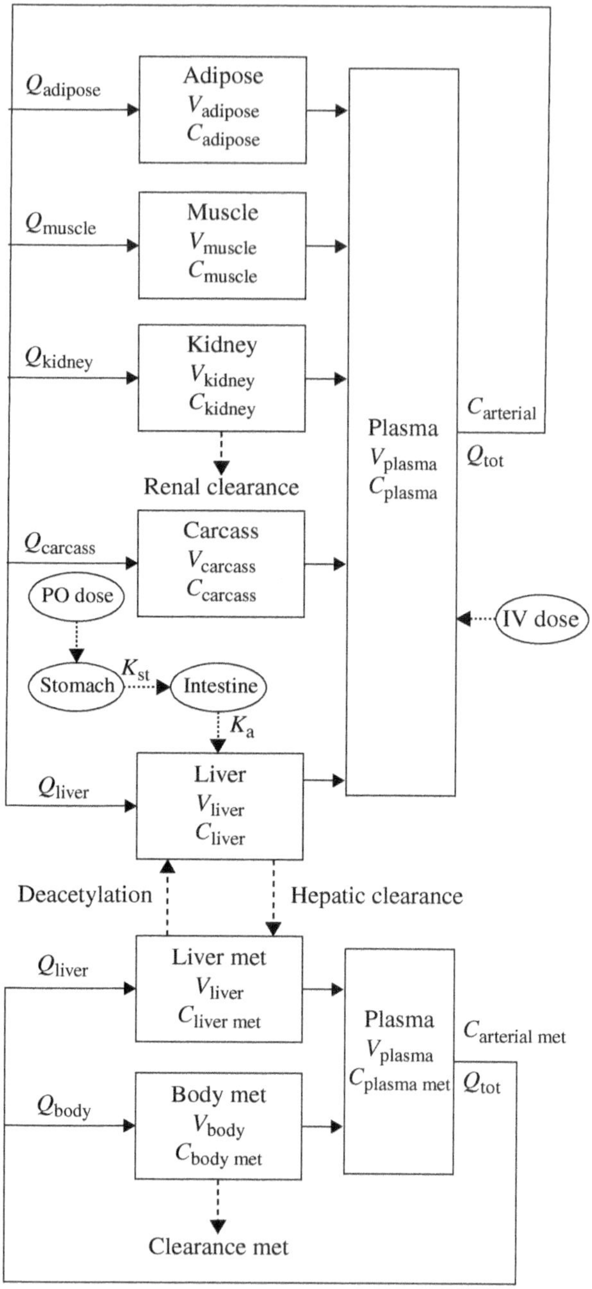

FIGURE 7.7 Schematic diagram of a PBPK model designed for the prediction of drug residues of sulfamethazine in swine. C, sulfamethazine concentration; IV, intravenous dose; K_a, rate of absorption; K_{st}, rate of gastric emptying; Met, metabolite; PO, oral dose; Q, blood flow; V, tissue volume.

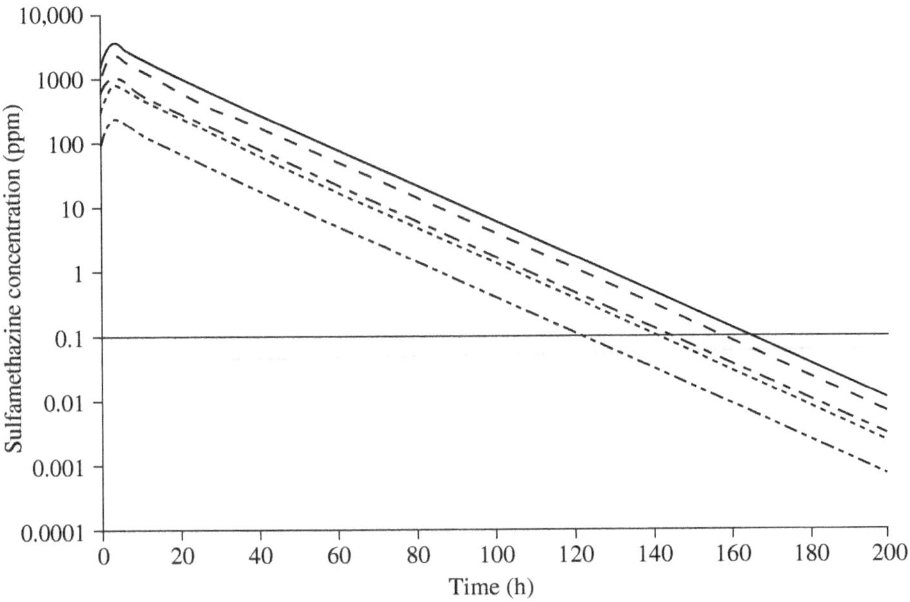

FIGURE 7.8 PBPK model simulations for sulfamethazine in edible tissues after an oral dose of 100 mg/kg given to swine. Solid line, kidney; large dash, plasma; dot–dash, liver; dot, adipose; dot–dot–dash, muscle; horizontal line, tolerance limit of 0.1 ppm.

TABLE 7.3 Comparison of parameter values in a sulfamethazine PBPK model for swine for both individual and population estimation

	Individual	Population (log normal distribution)	
Parameter	Mean	Mean	Variance
K_a (1/h)	0.1	−1	0.88
K_{st} (1/h)	0.1	−1	0.4
Hepatic clearance (ml/min/kg)	0.62	−0.4	0.32

The sulfamethazine in swine PBPK model showcases the flexibility of PBPK modeling that allows for practical application in the estimation of meat withdrawal intervals after extralabel drug use. Not only can dose and route of administration be updated as needed, but estimations can be made for target populations of either individuals or herds. Furthermore, as more information is gathered about the physiology underlying the kinetics of drugs or population variability, the physiological parameters (such as protein binding) can be updated or incorporated into the modeling system yielding a perpetual cycle of improvement.

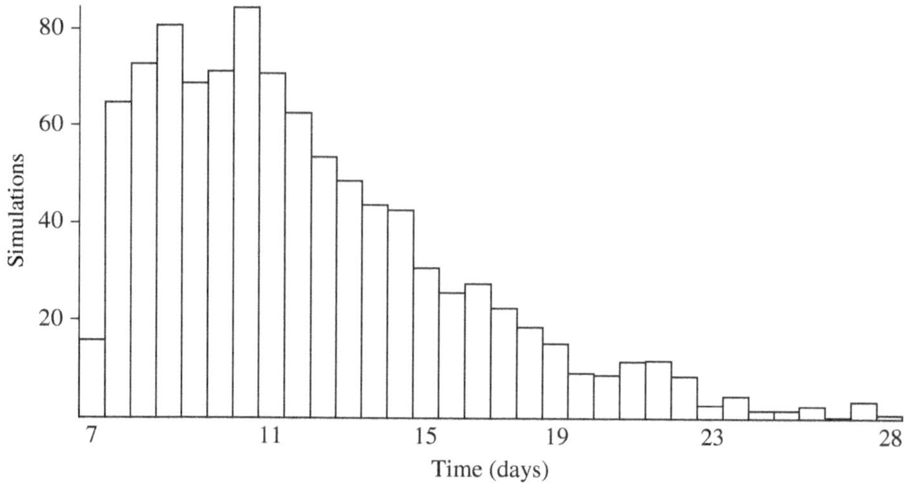

FIGURE 7.9 Population distribution of time for muscle concentrations of sulfamethazine to fall below tolerance (0.1 ppm) after the label oral dose (237.6 mg/kg on day 1 followed by 118.8 mg/kg for 3 more days) using a Monte Carlo analysis of 1000 simulations.

7.3.2 Scaling between Species: Melamine in Rats and Swine

The lack of pharmacokinetic data for all veterinary species is a major limitation in the prediction of food residues. PBPK models are uniquely qualified to help fill in gaps in the knowledge base and provide scientific evidence in the estimation of withdrawal intervals in minor species by scaling between species of interest (Cortright et al., 2009). This technique can also be applied in the estimation of withdrawal intervals in the face of little to no data in the target species (Buur et al., 2008).

Scaling between species using PBPK models is accomplished by the creation, calibration, and validation of a robust model in a major species of interest for a specific molecule. To scale between species, the modeler updates the physiological parameters to those of the species of interest. If known, the partition coefficients are also updated at this time. Absorption, elimination, and binding kinetics are generally assumed to be conserved across species unless there is evidence to the contrary. This technique is used regularly in human health risk assessment (US-EPA, 2006). However, it has been recently adapted to the prediction of tissue residues in veterinary medicine.

PBPK models were developed to explore the consequences of accidental contamination of the food supply to melamine (Buur et al., 2008). In the spring of 2007, pet food was found to be contaminated by melamine and resulted in the deaths of thousands of companion animals. Concurrently, the feed supply of pigs, sheep, and chickens were also found to be contaminated. While most of the exposed food animals did not progress into renal failure, there was justifiable concern over the levels of melamine in the edible tissues of these animals. At the time, there was no published data looking at the pharmacokinetics of melamine in species other than specific strains of laboratory rats.

A PBPK model for melamine in rats was designed, calibrated, and validated with the kidney being the tissue of focus (Fig. 7.10). Calibration of the rat melamine model was accomplished using data from multiple sources. By calibrating using urine data and plasma data from isolated gastrointestinal studies, identifiability of the multiple parameters was preserved. Due to the lack of data, validation of this model was not as robust as what was described in the sulfamethazine swine model. Correlations of 0.59–0.76 were reported for plasma and kidney tissues, respectively.

The PBPK melamine rat model was then altered to utilize porcine physiology data. The resulting simulation was compared with a single plasma pharmacokinetic study. Terminal elimination was different between simulation and observed plasma profiles most likely reflecting altered elimination physiology between rats and pigs (Fig. 7.11). However, the correlation was 0.89 for porcine plasma. When the model was applied to the issue of tissue residues, it was estimated that a 20 h withdrawal interval would be sufficient for edible tissues to be below tolerance for melamine (Fig. 7.12). A study conducted by the U.S. Food Safety Inspection Service (US-FSIS) was unable to detect melamine 24 h post exposure (US-FDA, 2007).

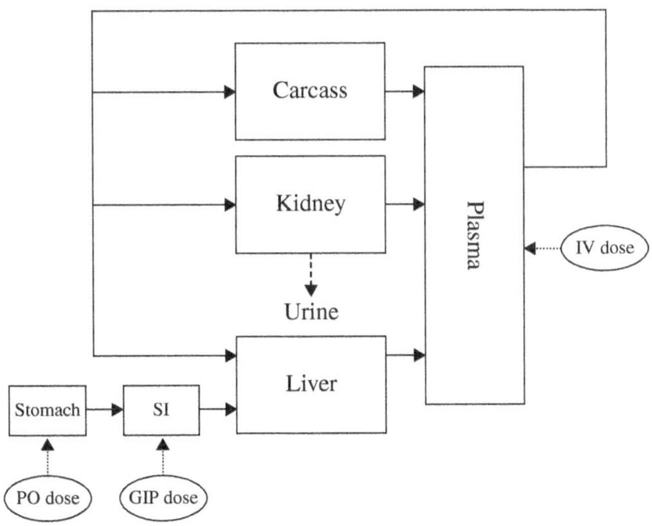

FIGURE 7.10 Schematic diagram of a PBPK model of melamine in rats and swine. Arrows represent mass transfer through blood flow or elimination into urine. GIP, isolated gastrointestinal perfusion dose; IV, intravenous dose; PO, oral dose; SI, small intestine.

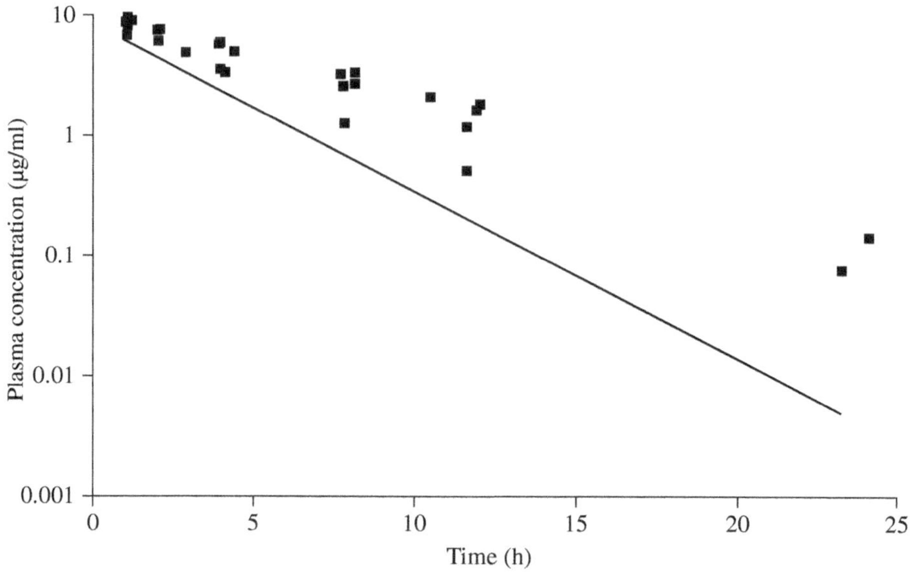

FIGURE 7.11 Plasma concentration–time simulation for melamine in swine after a single bolus IV dose. Squares represent observed data from an independent study. Reproduced with permission from Buur et al. (2008). © Elsevier.

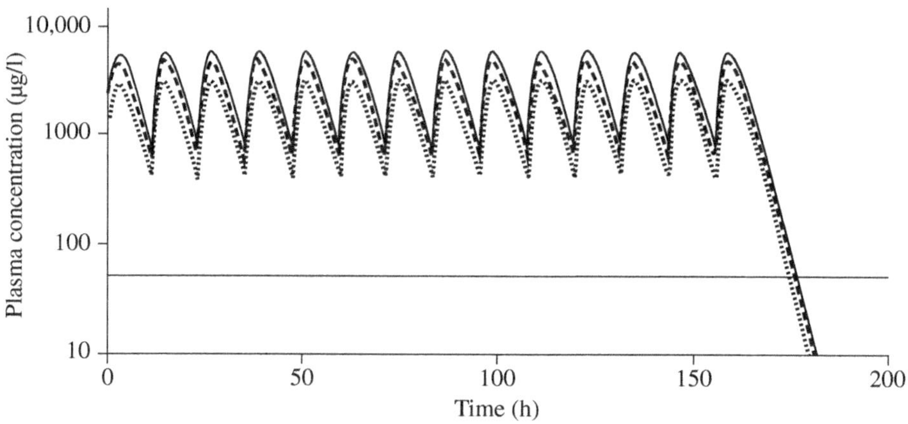

FIGURE 7.12 Concentration–time curves for edible tissues of the kidney (solid line), liver (dashed line), and plasma (dotted line) after twice daily administration of 5.12 mg/kg orally for 7 days. Horizontal line represents safe level of 50 ppb. Reproduced with permission from Buur et al. (2008). © Elsevier.

Like any other modeling tool, care must be taken in interpreting simulations in the face of limited data. Robust models that are well validated using a variety of possible scenarios provide the best predictive capability. Even then application of PBPK models to estimation of withdrawal intervals requires evaluation of model assumptions and the validity of model parameters to the practical scenario in order to have confidence in the model simulations.

7.4 CONCLUSIONS

Robust PBPK models can be used to predict tissue drug concentrations over time and thus to estimate appropriate withdrawal intervals. Due to the mechanistic nature of this modeling tool, PBPK models are flexible in terms of dose, route of administration, species, and molecule. Acceptable accuracy and predictive capability require large amounts of high-quality data. As a result, PBPK models are expensive in both time and animal studies. However, once validated, they provide scientific evidence for estimation of withdrawal intervals after extralabel drug use for both individuals and populations, and in the face of accidental exposure to contaminants. PBPK models have a place in the reducing chemical residues in our food supply.

REFERENCES

Animal Medicinal Drug Use Clarification Act (1994). Code of federal regulations – Title 21. United States Food and Drug Administration, Part 530.

Beliveau M, Tardif R, Krishnan K (2003). Quantitative structure-property relationships for physiologically based pharmacokinetic modeling of volatile organic chemicals in rats. Toxicol Appl Pharmacol 189(3):221–232.

Bjorkman S (2004). Prediction of drug disposition in infants and children by means of physiologically based pharmacokinetic (PBPK) modelling: Theophylline and midazolam as model drugs. Br J Clin Pharmacol 59(6):691–704.

Brocklebank JR, Namdari R, Law FC (1997). An oxytetracycline residue depletion study to assess the physiologically based pharmacokinetic (PBPK) model in farmed Atlantic salmon. Can Vet J 38(10):645–646.

Buur JL, Baynes RE, Craigmill AL, Riviere JE (2005). Development of a physiologic-based pharmacokinetic model for estimating sulfamethazine concentrations in swine and application to prediction of violative residues in edible tissues. Am J Vet Res 66(10):1686–1693.

Buur JL, Baynes RE, Smith G, Riviere JE (2006). Use of probabilistic modeling within a physiologically based pharmacokinetic model to predict sulfamethazine residue withdrawal times in edible tissues in swine. Antimicrob Agents Chemother 50(7):2344–2351.

Buur JL, Baynes RE, Riviere JE (2008). Estimating meat withdrawal times in pigs exposed to melamine contaminated feed using a physiologically based pharmacokinetic model. Regul Toxicol Pharmacol 51(3):324–331.

Charnick SB, Kawai R, Nedelman JR, Lemaire M, Niederberger W, Sato H (1995). Perspectives in pharmacokinetics. Physiologically based pharmacokinetic modeling as a tool for drug development. J Pharmacokinet Biopharm 23(2): 217–229.

Chiu WA, Barton HA, DeWoskin RS, Schlosser P, Thompson CM, Sonawane B, Lipscomb JC, Krishnan K (2007). Evaluation of physiologically based pharmacokinetic models for use in risk assessment. J Appl Toxicol 27(3):218–237.

Colburn WA (1988). Physiologic pharmacokinetic modeling. J Clin Pharmacol 28(8):673–677.

Cortright KA, Wetzlich SE, Craigmill AL (2009). A PBPK model for midazolam in four avian species. J Vet Pharmacol Ther 32(6):552–565.

Craigmill AL (2003). A physiologically based pharmacokinetic model for oxytetracycline residues in sheep. J Vet Pharmacol Ther 26(1):55–63.

DeJongh J, Blaauboer BJ (1996). Simulation of toluene kinetics in the rat by a physiologically based pharmacokinetic model with application of biotransformation parameters derived independently in vitro and in vivo. Fundam Appl Toxicol 32(2):260–268.

Duddy J, Hayden TL, Bourne DW, Fiske WD, Benedek IH, Stanley D, Gonzalez A, Heierman W (1984). Physiological model for distribution of sulfathiazole in swine. J Pharm Sci 73(11):1525–1528.

Gerlowski LE, Jain RK (1983). Physiologically based pharmacokinetic modeling: Principles and applications. J Pharm Sci 72(10):1103–1127.

Krishnan K, Andersen ME (2001). Physiologically based pharmacokinetic modeling in toxicology. In: Hayes AW, editor. Principles and Methods of Toxicology. Philadelphia: Taylor & Francis; p 193–241.

van der Merwe D, Brooks JD, Gehring R, Baynes RE, Monteiro-Riviere NA, Riviere JE (2006). A physiologically based pharmacokinetic model of organophosphate dermal absorption. Toxicol Sci 89(1):188–204.

Price PS, Conolly RB, Chaisson CF, Gross EA, Young JS, Mathis ET, Tedder DR (2003). Modeling interindividual variation in physiological factors used in PBPK models of humans. Crit Rev Toxicol 33(5):469–503.

Ritschel WA, Banerjee PS (1986). Physiological pharmacokinetic models: Principles, applications, limitations and outlook. Methods Find Exp Clin Pharmacol 8(10):603–614.

Riviere JE (1999). Comparative Pharmacokinetics: Principles, Techniques, and Applications. Ames: Blackwell Publishing, Inc.

Riviere JE, Martin-Jimenez T, Sundlof SF, Craigmill AL (1997). Interspecies allometric analysis of the comparative pharmacokinetics of 44 drugs across veterinary and laboratory animal species. J Vet Pharmacol Ther 20(6):453–463.

Teeguarden JG, Waechter JM, Jr., Clewell HJ, 3rd, Covington TR, Barton HA (2005). Evaluation of oral and intravenous route pharmacokinetics, plasma protein binding, and uterine tissue dose metrics of bisphenol A: A physiologically based pharmacokinetic approach. Toxicol Sci 85(2):823–838.

US-EPA (2006). Approaches for the Application of Physiologically Based Pharmacokinetic (PBPK) Models and Supporting Data in Risk Assessment (Final Report). Washington, DC: U.S. Environmental Protection Agency.

US-FDA (June 7, 2007). Interim Melamine and Analogues Safety/Risk Assessment Peer Review Report. Washington, DC: United States Food and Drug Administration.

Williams PL (1990). Structural identifiability of pharmacokinetic models—compartments and experimental design. J Vet Pharmacol Ther 13(2):121–131.

8

RESIDUE AVOIDANCE IN BEEF CATTLE PRODUCTION SYSTEMS

Virginia Fajt[1] and Dee Griffin[2]

[1] *Department of Veterinary Physiology and Pharmacology, College of Veterinary Medicine and Biomedical Sciences, Texas A&M University, College Station, TX, USA*

[2] *Great Plains Veterinary Education Center, University of Nebraska, Lincoln, NE, USA*

8.1 INTRODUCTION

Beef cattle producers in the United States have had an excellent record of keeping beef safe with very low rates of drug and chemical residues at harvest. A concerted effort since the early 1980s to reduce and prevent residues by beef producers, their trade associations such as the National Cattlemen's Beef Association, and the U.S. Department of Agriculture (USDA) has brought about this success story. While sampling strategies are not necessarily designed to estimate the total incidence of residues within the entire population of harvested animals (Cordle, 1988), an analysis of the most recent data available from the USDA-FSIS Red Books (2003–2008) showed a rate of 0.002–0.005% for beef cows and 0.0007–0.0001% for feeder cattle (D. Griffin, unpublished data), based on the number of violative residues and the estimates of total animals harvested each year. Efforts in other countries such as Australia have also resulted in the production of safe beef: From June 2008 to June 2009, 2 samples out of 5732 samples tested had greater than allowable residues (0.03% of samples)

Strategies for Reducing Drug and Chemical Residues in Food Animals: International Approaches to Residue Avoidance, Management, and Testing, First Edition.
Edited by Ronald E. Baynes and Jim E. Riviere.

(Anonymous, 2009c). The two main approaches to reducing residues are to reduce the need for application of drugs and chemicals and to insure proper management of cattle before harvest to maintain adequate withdrawal times. This chapter highlights how beef production methods keep residues low and discusses potential sources of violative residues in beef cattle, including disease incidence, typical pharmacotherapeutic programs, and risky practices. The chapter concludes with descriptions of programs that reduce those risks and makes recommendations on best practices to lower the risk of residues in edible beef.

8.2 BEEF CATTLE PRODUCTION SYSTEMS

Methods of raising cattle to produce beef differ among regions of the world and even among regions within countries. This variety is due to the resourcefulness of beef cattle producers in optimizing production through their use of available land, feedstuffs, and natural resources. The typical beef production cycle in North America is described here as a preface to understanding the areas of risk for the occurrence of residues in beef cattle. The cycle may include three major phases, with the transition between the phases being the most common time for disease challenges with resultant application of drugs and chemicals necessitating concerns about residues. The first phase is the nursing calf, the second phase is the weaned calf, and the third phase is harvest. The weaned calf phase may include a period of backgrounding (acclimating to eating out of a bunk, possible comingling of new groups of animals) and subsequent placement on pasture in a stocker operation or may only include a trip from the cow-calf production site to the feedlot. The transitions and physical movement from one phase to the next occur in the United States because the land and feed requirements for cows and nursing calves are different from the requirements for feeding cattle (Fig. 8.1 and Fig. 8.2).

The disease challenges that occur in each phase determine the need for drugs and chemicals. In cow-calf operations, calves that are born in a clean and dry environment, that receive adequate colostrum, and that are quickly moved to areas with low pathogen burden are likely to move through the first phase relatively free of disease challenges. When weather is inclement, when feedstuff availability requires close congregation of animals, or when personnel availability impacts management, any of the aforementioned may be compromised, resulting in disease in individuals or groups of animals. In the transition from nursing to weaning, weaned calves that enter a backgrounding or stocker operation may be exposed to disease challenges from transport, handling, and comingling of new animals, from lack of

FIGURE 8.1 Beef cow and calf in Midwestern United States. Photo by Dee Griffin.

FIGURE 8.2 Beef feedlot in Nebraska, United States. Photo by Dee Griffin.

immunity, from parasites and other infectious organisms, and from exposure to new surroundings, feedstuffs, and weather. If calves are taken directly to the feedlot from the cow-calf operation, similar challenges are present. Chemicals and drugs commonly used to treat these diseases are discussed in the next sections.

TABLE 8.1 Bacterial and parasitic conditions for which at least one injectable, oral, or feed additive drug is approved in the United States for beef cattle

Label claim	Organism generally associated with the condition or listed on the label (not all organisms may be included on all drug labels)	Example drug molecules included in the approved products (other drugs may also be approved)
Bacterial conditions		
Anaplasmosis	*Anaplasma marginale*	Oxytetracycline, chlortetracycline
Acute bovine interdigital necrobacillosis (foot rot, pododermatitis)	*Fusobacterium necrophorum*, *Bacteroides melaninogenicus*	Ceftiofur, tulathromycin, tylosin, florfenicol
Bovine respiratory disease (shipping fever, pneumonia)	*Mannheimia haemolytica*, *Pasteurella multocida*, *Histophilus somnus*	Ceftiofur, oxytetracycline, tilmicosin, ampicillin, enrofloxacin, tulathromycin, florfenicol
Calf diphtheria	*F. necrophorum*	Tylosin, oxytetracycline, sulfamethazine
Colibacillosis (bacterial enteritis)	*E. coli*	Sulfamethazine, oxytetracycline, neomycin, ampicillin
Infectious bovine keratoconjunctivitis (pinkeye)	*Moraxella bovis*	Oxytetracycline
Leptospirosis	*Leptospira pomona*	Oxytetracycline
Mastitis	*Streptococcus* spp.	Sulfamethazine
Metritis	*Arcanobacterium pyogenes*, *staphylococci*, *streptococci*	Tylosin, oxytetracycline, sulfamethazine
Wooden tongue	*Actinobacillus lignieresi*	Oxytetracycline
Parasitic conditions		
Gastrointestinal worms	*Bunostomum phlebotomum*, *Nematodirus helvetianus*, *Cooperia punctata*, *Cooperia oncophora*, *Trichostrongylus colubriformis*, *Oesophagostomum radiatum*	Fenbendazole, albendazole, levamisole, ivermectin, morantel, moxidectin, eprinomectin
Grubs	*Hypoderma lineatum*, *Hypoderma bovis*	Eprinomectin, fenthion, doramectin, moxidectin
Lice	*Linognathus vituli*, *Haematopinus eurysternus*, *Solenopotes capillatus*, *Bovicola (Damalinia) bovis*	Fenthion, doramectin, moxidectin
Liver abscesses	*F. necrophorum*	Tylosin, chlortetracycline
Liver flukes	*Fasciola hepatica*	Albendazole, clorsulon
Lungworm	*Dictyocaulus viviparus*	Fenbendazole, albendazole, levamisole, ivermectin, morantel, moxidectin, oxfendazole, doramectin

TABLE 8.1 (*Continued*)

Label claim	Organism generally associated with the condition or listed on the label (not all organisms may be included on all drug labels)	Example drug molecules included in the approved products (other drugs may also be approved)
Mange mites	*Chorioptes bovis*, *Sarcoptes scabiei*	Eprinomectin, doramectin, moxidectin, ivermectin
Stomach worms	*Ostertagia ostertagi*, *Haemonchus contortus*, *Haemonchus placei*, *Trichostrongylus axei*, *Teladorsagia circumcincta*	Fenbendazole, albendazole, levamisole, ivermectin, morantel, moxidectin, oxfendazole, doramectin

Source: Animal Drugs @ FDA, http://www.accessdata.fda.gov/scripts/animaldrugsatfda/, last accessed May 6, 2010.

8.2.1 Common Infectious Diseases of Beef Cattle

In beef cattle, the most common conditions requiring antimicrobial therapy include respiratory disease complex, enteritis and enterocolitis, keratoconjunctivitis, pododermatitis, and mastitis. The most common conditions requiring use of parasiticides worldwide include infections with gastrointestinal nematodes, coccidia, dermatophytes, lice, and ticks. In the United States, the diseases and conditions for which at least one antimicrobial or parasiticide is labeled are listed in Table 8.1. All other bacterial or parasitic conditions require extralabel use in the United States.

8.3 USE OF ANTI-INFECTIVE AGENTS IN BEEF CATTLE PRODUCTION

8.3.1 The Need for Anti-infectives

The most important step in reducing residues of drugs and chemicals in beef is to utilize anti-infectives only when necessary. The use of drugs to treat animals with disease is a necessary and humane practice in animal production. On the other hand, the incentives for reducing the use of pharmacotherapeutics may be economic (fewer treatments means less drug cost), market related (demand for meat not treated with drugs such as organic or "natural"), or quality related (reduce risk of residues and injection site reactions). Inappropriate or profligate use is unsound for maintaining healthy animals, a healthy food supply, and a healthy agricultural business. To better understand the need of drugs in beef production, we define "treatment" or "therapeutic" in the following text. We also discuss approaches to reduce the need for antimicrobial and parasiticide use, whether the incentive to do so is economics, market demand, or quality.

8.3.2 Definition of Therapeutic

The agency in the United States that approves drugs for use in animals, the Food and Drug Administration Center for Veterinary Medicine (FDA-CVM), considers "therapy" and "therapeutic" to include uses of drugs for treatment, control, and prevention of disease. This definition also concurs with that of the Codex Alimentarius, the international standard setting body for food and food products. Some groups, such as the Pew Commission on Industrial Food Animal Production, believe that disease prevention should not be considered a therapeutic use of drugs. However, veterinarians and other health professionals would likely agree that decision-making about therapeutics involves significant understanding of the natural history of infectious diseases, and it involves making reasonable clinical judgments about the need for prevention of disease when animals have been or are likely to be exposed to disease, even though they may appear healthy. There is a lack of understanding of beef production systems in the attempt to lump all preventive therapeutic use of drugs as "prevention in otherwise healthy animals." It may also be reasonable to treat with antimicrobial drugs prior to completing diagnostic procedures or obtaining results of patient-specific antimicrobial susceptibility tests in situations in which the progression of disease is well characterized and in which animal welfare would be compromised by a lack of timely therapy and in which the spread of disease to other animals is a concern. These empirical uses of antimicrobials do require follow-up to assess drug efficacy, and they require monitoring or surveillance of population antimicrobial susceptibility data to determine the likelihood of continuing success with empirical therapy.

In addition to approving antimicrobial drugs for treatment, prevention, and control of bacterial disease, certain antimicrobial drugs may be approved for "growth promotion and feed efficiency." Certain political and other groups have decried the approvals of antimicrobials for growth promotion and feed efficiency in cattle in the United States. In particular, they apply the term "subtherapeutic" to describe these uses. However, this term has no legal or scientific definitions, so we will not use the term in these discussions. We use the phrase "for growth promotion and feed efficiency" to describe uses that have not been conclusively demonstrated to treat, control, or prevent disease but do result in improved rate of gain or feed efficiency. The majority of antimicrobial drugs approved for growth promotion and feed efficiency also have other approvals for therapeutic purposes. In addition, the activity *in vitro* of inhibiting bacterial growth easily leads to speculation that antimicrobial drugs used for growth promotion are in fact acting at some level to inhibit pathogen growth, change the dynamics in bacterial flora in the gut, or prevent subclinical disease, but these effects have simply not yet been demonstrated. The FDA will no longer be approving antimicrobials in feed for growth promotion if they are "medically important" antimicrobials, as described in Guidance for Industry #209 (http://www.fda.gov/downloads/animalveterinary/guidancecomplianceenforcement/guidanceforindustry/ucm216936.pdf), The Judicious Use of Medically Important Antimicrobial Drugs in Food-Producing Animals.

It should be stated here that using feed medications in the United States in any manner not included on the label is illegal (Anonymous, 1996). This includes putting feed medications that are not approved to be used together in the same feed. No one, including a veterinarian, can legally prescribe the use of any feed additive other than as directed on the product label. This is not the case in some countries such as Canada, where veterinarians may legally prescribe in-feed medications in an extralabel manner, as long as the feed is labeled properly (Anonymous, 2010).

8.3.3 Clinical Scenarios in Which Extralabel Drug Use May Be Required

Extralabel drug use, as defined by U.S. federal law, is the use of a drug in any manner not specified on the approved labeling. In the United States, extralabel use of drugs may be permissible if the health of the animal is threatened and suffering or death may result if it is not treated. The importance of extralabel drug use in terms of residue avoidance and prevention is that extralabel usage may result in unexpected changes in the pharmacokinetics of drugs, resulting in the potential for residues, since approved withdrawal times for a particular drug use are based on the labeled usage. The presence of any condition not listed in Table 8.1 may require extralabel drug use. In addition, some of the conditions for which drugs are approved may require extralabel drug use, if, for example, resistant organisms are present, drug labels are inconsistent with effectiveness of the drug in the field, or concurrent disease is present compromising therapeutic effectiveness.

The specific conditions permitting extralabel use include the lack of availability of a product labeled for the condition being treated or inefficacy of a labeled product. Extralabel use can only occur by or on the order of a veterinarian, within the context of a valid veterinarian–client–patient relationship, and only for therapeutic purposes when the health of the animal is threatened or suffering or death may result from failure to treat. Extralabel use is not permitted in or on animal feed, by the order of a layperson, or if a residue above safe level or tolerance occurs. Drug cost is not considered a valid reason for extralabel drug use under the Animal Medicinal Drug Use

Clarification Act (AMDUCA) or the regulations promulgated to implement the act. Extralabel use requires the determination of extended withdrawal times, for which the veterinarian takes responsibility. Medications used extralabel must have a label that contains the contact veterinarian, instructions for use, precautions including the withdrawal time, and identification of individual animals to be treated.
Veterinarian–client–patient relationships exist when:

a. The veterinarian has assumed the responsibility for making clinical judgments regarding the health of the animal and the need for medical treatment and the client has agreed to follow the veterinarian's instructions.
b. The veterinarian has sufficient knowledge of the animal to initiate at least a general or preliminary diagnosis of the medical condition of the animal. This means the veterinarian has recently seen and is personally acquainted with the keeping and care of the animal by virtue of an examination of the animal or the medically appropriate and timely visits to the premises where the animal is kept.
c. The veterinarian is readily available for follow-up evaluation in the event of adverse reactions or failure of the treatment regimen.

8.4 APPROACHES TO MINIMIZE THE NEED FOR ANTIMICROBIAL DRUGS

The goal of minimization of antimicrobial use must be accompanied by the commitment to use antimicrobials when they are necessary. This is the basis of the National Cattlemen's Beef Association's Guide for Judicious Use of Antimicrobials in Cattle (Fig. 8.3). Regardless of the management system and management goals, cattle will contract bacterial diseases requiring antimicrobial therapy. Antimicrobial therapy should not be withheld from these animals in order to maintain a "natural" or "antibiotic-free" product; animal welfare is and should be a paramount concern of beef producers and veterinarians. The best approach is to use antimicrobials in a judicious and prudent manner and to minimize use by minimizing the need for use.

One approach to reduce the use of antimicrobial drugs is the attempt to direct public policy and legislation to ban the use of antimicrobial drugs in feed, since it is assumed that this use is unnecessary for the health of the animals and that it would result in an overall decrease in use. Examples in beef production systems to demonstrate or refute these claims do not exist. However, there is an example in pig production in Denmark, where antimicrobial

BEEF USA National Cattlemen's Beef Association

A Producers Guide for Judicious Use of Antimicrobials in Cattle

1. **Prevent Problems:** Emphasize appropriate husbandry and hygiene, routine health examinations, and vaccinations.
2. **Select and Use Antibiotics Carefully:** Consult with your veterinarian on the selection and use of antibiotics. Have a valid reason to use an antibiotic. Therapeutic alternatives should be considered prior to using antimicrobial therapy.
3. **Avoid Using Antibiotics Important In Human Medicine As First Line Therapy:** Avoid using as the first antibiotic those medications that are important to treating strategic human or animal infections.
4. **Use the Laboratory to Help You Select Antibiotics:** Cultures and susceptibility test results should be used to aid in the selection of antimicrobials, whenever possible.
5. **Combination Antibiotic Therapy Is Discouraged Unless There Is Clear Evidence The Specific Practice Is Beneficial:** Select and dose an antibiotic to affect a cure.
6. **Avoid Inappropriate Antibiotic Use:** Confine therapeutic antimicrobial use to proven clinical indications, avoiding inappropriate uses such as for viral infections without bacterial complication.
7. **Treatment Programs Should Reflect Best Use Principles:** Regimens for therapeutic antimicrobial use should be optimized using current pharmacological information and principles.
8. **Treat the Fewest Number of Animals Possible:** Limit antibiotic use to sick or at risk animals.
9. **Treat for the Recommended Time Period:** To minimize the potential for bacteria to become resistant to antimicrobials.
10. **Avoid Environmental Contamination with Antibiotics:** Steps should be taken to minimize antimicrobials reaching the environment through spillage, contaminated ground run off or aerosolization.
11. **Keep Records of Antibiotic Use:** Accurate records of treatment and outcome should be used to evaluate therapeutic regimens and always follow proper withdrawal times.
12. **Follow Label Directions:** Follow label instructions and never use antibiotics other than as labeled without a valid veterinary prescription.
13. **Extralabel Antibiotic Use Must follow FDA Regulations:** Prescriptions, including extra label use of medications must meet the Animal Medicinal Drug Use Clarification Act (AMDUCA) amendments to the Food, Drug, and Cosmetic Act and its regulations. This includes having a valid Veterinary-Client-Relationship.
14. **Subtherapeutic Antibiotic Use Is Discouraged:** Antibiotic use should be limited to prevent or control disease and should not be used if the principle intent is to improve performance.

Guidelines1-13 adapted from AVMA, AABP and AVC Appropriate Veterinary Antibiotic Use Guidelines.

FIGURE 8.3 Prudent antimicrobial drug use recommendations, from the National Cattlemen's Beef Association.

drugs in feed were completely eliminated by 1999. The use of therapeutic antimicrobial drugs increased from 1999 (Anonymous, 2009b), as did the incidence of certain diseases, such as ileitis (Vigre et al., 2008). One can only speculate as to what would happen in beef production if antimicrobial drugs in feed were no longer permitted. Unpredictable changes would be likely to occur in disease prevalence, in particular respiratory disease and enteric disease; in zoonotic organism prevalence, such as *Salmonella*; and in prevalence of resistance genes in pathogenic and nonpathogenic organisms. Intriguing results from one study (Platt et al., 2008) suggest that our predictions about the impact of removal or change in antimicrobial drug exposure will not be accurate: Cattle fed chlortetracycline in feed as compared to those fed control feed for several days had increased rates of resistant *E. coli* and *Enterococcus* spp., but there was a decrease in the proportion of *E. coli* isolates resistant to a completely different drug, ceftiofur. These results suggest that the dynamics of resistance prevalence are not as predictable as once thought.

8.4.1 Evidence for Interventions That Reduce Antimicrobial Use

The promotion of interventions that reduce antimicrobial use is common but not readily supported by evidence of actual reduction in use. The majority of prudent antimicrobial use guidelines include a recommendation to use preventive products (e.g., vaccines) and production practices that minimize disease (e.g., calving systems to reduce the incidence of diarrheal disease). However, data supporting that these practices actually result in less antimicrobial use are limited in cattle. Nonetheless, common sense and data from human hospitals suggest that any product or procedure that prevents disease will result in lower antimicrobial use. And lower antimicrobial use will result in reduced potential for drug residues.

8.4.2 Evidence for Interventions That Reduce Antimicrobial Resistance

While not strictly related to residues, this is an appropriate place to discuss the need to reduce the prevalence of antimicrobial resistance, in order to maintain efficacy of antimicrobial drugs in animals as well as in people. It is presumed that reducing antimicrobial use will reduce the selective pressure of antimicrobials on resistant bacteria, thereby reducing prevalence of resistance. While there is published evidence of this phenomenon, there are also data to suggest that targeted and specific application of antimicrobial drugs may actually reduce prevalence (Platt et al., 2008). These data must be replicated and expanded to demonstrate usefulness, but they provide an interesting potential application for targeted antimicrobial therapy.

There are a number of recommendations for interventions that may reduce antimicrobial resistance, including prudent or judicious use guidelines. While these are most often consensus documents based on input from multiple professionals or professional groups, the evidence supporting individual recommendations is relatively thin as it relates to actual impact on beef production management. The American Veterinary Medical Association has published non-species-specific prudent use guidelines for animals (http://www.avma.org/issues/policy/jtua.asp), as has the American College of Veterinary Internal Medicine (ACVIM) (Morley et al., 2005). The American Association of Bovine Practitioners has published guidelines more specifically tailored to cattle (http://www.avma.org/issues/policy/jtua_cattle.asp), but these provide no further evidence than provided in the more general documents in terms of the demonstrated impact of any of the recommendations on antimicrobial resistance in the beef production setting. In general, these guidelines speak to practices that seem to make common sense: For example, the ACVIM principles (Morley et al., 2005) include the recommendation to use culture and antimicrobial susceptibility testing to guide antimicrobial drug therapy, a practice likely to restrict antimicrobial drug use to conditions with demonstrated presence of bacteria.

8.5 APPROACHES TO MINIMIZE THE NEED FOR PARASITICIDES

With the increasing concern worldwide for resistance to all chemical groups of parasiticides, considerable scientific data have been collected on controlling parasites using production practices rather than drugs. The scope of this chapter does not permit a discussion of the details of parasite control programs, but general concepts include multiple grazing sites to reduce worm burdens, feedstuffs demonstrated to reduce parasite propagation, removing physical sites that favor parasite growth, and using biological methods of parasite control outside the animal such as the use of parasite predators. Methods of reducing the amount of parasiticide used within a herd include targeted use in animals with the highest parasite burdens rather than herd-wide use.

8.6 APPROACHES TO MINIMIZE RESIDUES (RESIDUE AVOIDANCE)

The general tenets for avoiding residues in products from beef cattle are straightforward and are outlined here.

Veterinary involvement It is important for beef producers to establish a working relationship with a licensed veterinarian. Find and use a veterinarian who is willing to be involved with your Beef Quality Assurance program. Be cautious about cattle treatment advice from anyone who is not highly acquainted with your operation and the proper use of animal health products.

Labels Use products labeled for the condition being treated, and use them as labeled. If extralabel use is necessary, as outlined previously, veterinary involvement is necessary to be legal and to provide guidance on appropriate treatment regimens and withdrawal time recommendations.

Withdrawal Times When drugs are used as labeled, withdrawal times on the label should be followed strictly. Users should observe and follow label directions and ensure that cattle are not marketed until the appropriate withdrawal time has elapsed. If an animal has remaining time on the withdrawal at the time of marketing, the complete history of product use must be transferred with the group of cattle when moved to the next production unit. A residue screening test such as the Live Animal Swab Test (LAST), Pre-Harvest Antibiotic Screening Test (PHAST), or PremiTest may offer a margin of comfort if these cattle need to be shipped soon after their prescribed withdrawal date (Fig. 8.4). Aminoglycosides present a special residue withdrawal concern in cattle. Extralabel use of these antibiotics will result in residue violation in excess of 2 years from the date of use. When drugs are used extralabel, withdrawal times should be extended, as required by AMDUCA in the United States. This will allow extended time for drugs to be eliminated from cattle, so residues are prevented.

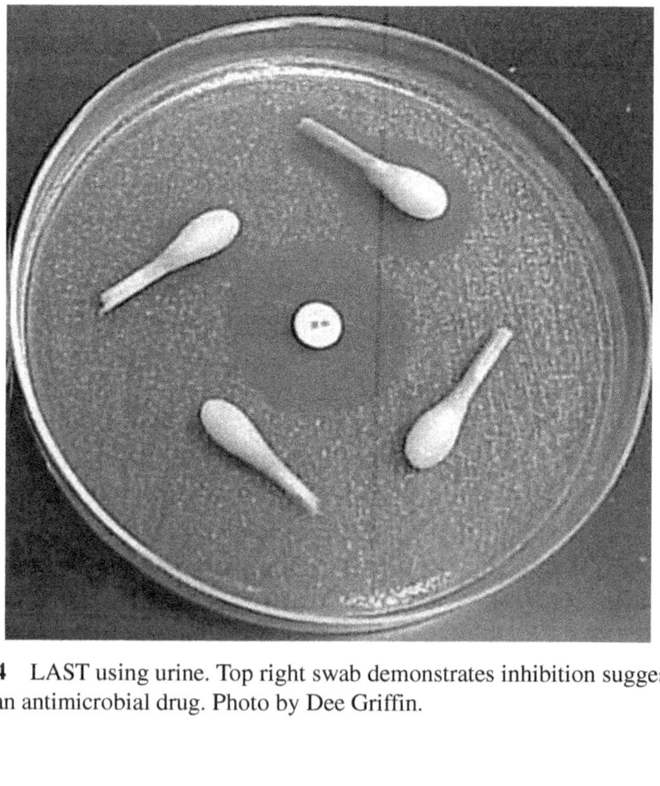

FIGURE 8.4 LAST using urine. Top right swab demonstrates inhibition suggesting the presence of an antimicrobial drug. Photo by Dee Griffin.

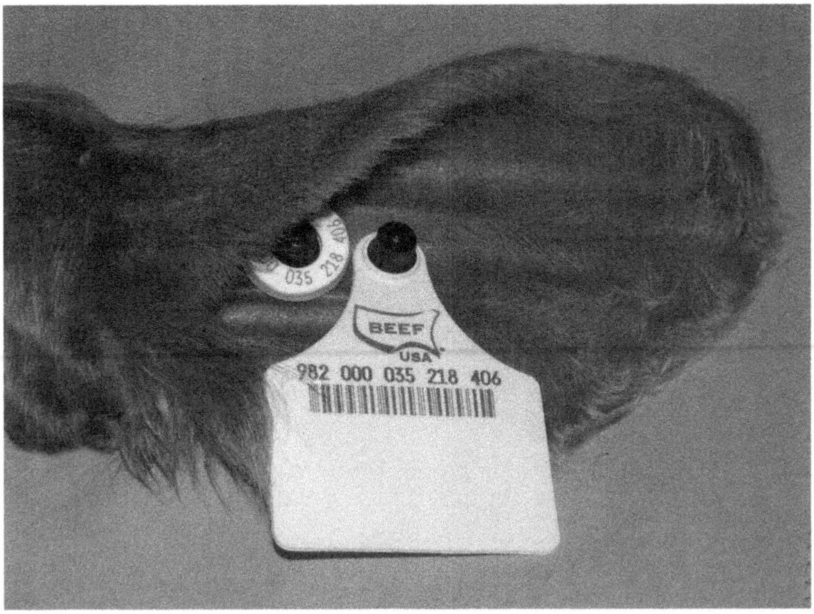

FIGURE 8.5 Combination electronic and plastic ear tag for identification of cattle. Photo by Dee Griffin.

Animal Identification All animals treated with drugs should be identified (Fig. 8.5). Identifying each animal individually is not required provided the product use information and appropriate withdrawal are applied to the entire group. As an example, several calves break with scours, and numerous calves are treated within a 10-day period. The entire group of calves would receive a withdrawal date based on the last date of administration of the product with the longest withdrawal period and the last animal treated.

Animal Records Records should be kept of all drug use, including date of treatment, amount administered, route of administration, and the identification of individual animals (Fig. 8.6).

Training All personnel who work with cattle on an operation or who can make marketing decisions about cattle should be made aware of proper use and administration of drugs and withdrawal times. They should also have the ability to check appropriate withdrawal restrictions before moving cattle to market.

Chronic Disease Animals with chronic or unresolving disease may be considered for removal from the herd by marketing for harvest. In general,

FIGURE 8.6 Computer and software for record keeping. Photo by Dee Griffin.

the best approach is to prevent unhealthy animals from entering the food supply in the first place. These nonperforming animals are at high risk for violative residues, and their presence in the food supply represents an animal welfare concern. If these animals are deemed healthy enough to enter the food supply, their records should be thoroughly reviewed by the veterinarian and manager before being released for salvaging. A residue screening program should be established for animals of this type, such as medicated cull cows and realizer feeder cattle. Records for any cattle to be marketed should be checked to ensure that treated animals have met or exceeded the label or assigned withdrawal times for all products used. Treatment records should be carefully examined for cattle with relapsed disease, in particular if the cattle are suspected to have liver or kidney disease that could result in increased clearance of drugs. If animals appear to have injection site reactions such as swelling in the neck region, treatment records should be carefully examined prior to marketing the animals.

Screening for Residues Urine can be screened for antibiotics with broad-spectrum microbial inhibition tests such as the DSN PremiTest for urine, the PHAST, or the LAST (Schneider and Lehotay, 2008). Test sensitivity relative to FDA-CVM violative residue tolerances (maximum residue limit or MRL) is listed in Table 8.2.

TABLE 8.2 A comparison of the U.S. MRL to the test detection limits of *Bacillus megaterium*, *Bacillus Stearothermophilus*, and *Bacillus subtilis* for FDA-approved cattle antibiotics

Cattle antibiotic residue tolerance and detection estimates for the STOP (*B.sub*), FAST/PHAST (*B.meg*), and Charm KIS/DSM PremiTest (*B.stearo*)					
Generic name	NADA#	MRL in cattle tissues (previously termed "tolerance")	*B.sub* detect	*B.meg* detect	*B.stearo* detect
Ampicillin	055–030	0.1 ppm edible	>0.1[b]	0.2[a]	0.005[c]
Amoxicillin	055–089	0.1 ppm edible	>0.1[b]	0.2[a]	0.005[c]
Ceftiofur sodium	140–338	0.4 ppm kidney, 1.0 ppm muscle	>0.1[b]	~0.1[b]	0.1[c]
Ceftiofur hydrochloride	140–890	0.4 ppm kidney, 1.0 ppm muscle	>0.1[b]	~0.1[b]	0.1[c]
Ceftiofur crystalline acid	141–209	0.4 ppm kidney, 1.0 ppm muscle	>0.1[b]	~0.1[b]	0.1[c]
Chlortetracycline (feed)	048–761	12.0 ppm kidney, 2.0 ppm muscle	>10[b]	~0.1[b]	0.05[c]
Danofloxacin**	141–207	0.2 ppm liver, 0.2 ppm muscle	>0.1[b]	>0.1[b]	0.6[c]
Dihydrostreptomycin	NDC	2 ppm kidney, 0.5 ppm edible	**	**	3.0
Enrofloxacin**	141–068	0.1 ppm liver, 0.1 ppm muscle	>0.1[b]	>0.1[b]	0.6[c]
Florfenicol	141–063	12.0 ppm kidney, 3.7 ppm liver	>1[b]	~5.0[b]	0.1[c]
Gentamicin**	101–862	No residue tolerance	>1[b]	0.13[b]	0.1[c]
Neomycin**	200–113	0.25 ppm edible	~10[b]	0.06[b]	0.3[c]
Oxytetracycline (LA)**	Multiple	12.0 ppm kidney, 2.0 ppm muscle	>10[b]	0.8[b]	0.05[c]
Pen G, benzathine	Multiple	0.05 ppm edible	>0.1[b]	<0.01[a]	0.005[c]
Pen G, procaine	Multiple	0.05 ppm edible	>0.1[b]	<0.01[a]	0.005[c]
Spectinomycin	141–077	4.0 ppm kidney, 0.25 ppm edible	>10[b]	6.2[a]	1.5[d]
Sulfadimethoxine (IV)	041–245	0.1 ppm edible	>100[b]	~1[b]	0.1[c]
Sulfadimethoxine (oral)	093–107	0.1 ppm edible	>100[b]	~1[b]	0.1[c]
Sulfamethazine	140–270	0.1 ppm edible	>100[b]	**	0.1[c]

(Continued)

TABLE 8.2 (*Continued*)

Cattle antibiotic residue tolerance and detection estimates for the STOP (*B.sub*), FAST/PHAST (*B.meg*), and Charm KIS/DSM PremiTest (*B.stearo*)

Generic name	NADA#	MRL in cattle tissues (previously termed "tolerance")	*B.sub* detect	*B.meg* detect	*B.stearo* detect
Tilmicosin (lung CM)	140–929	14.4 ppm kidney, 1.2 ppm liver	>10[b]	~5.0[b]	005[c]
Tulathromycin	141–244	5.5 ppm liver, 18.0 ppm kidney	>0.1[b]	>0.1[b]	18**
Tylosin	012–965	0.2 ppm kidney, 0.2 ppm liver	>1[b]	~5.0[b]	0.05[c]
Virginiamycin		0.5 ppm distiller's grains, 0.4 edible pork	**	**	2.0
			False positive	False negative	

U.S. tolerance: FDA permissible tolerance for the antibiotic in ppm (mg/kg) for target marker tissue listed.
Codex Tolerance (Max. Residue Limits (MRL)) @ http://www.codexalimentarius.net/mrls/vetdrugs/jsp/vetd_q-e.jsp.
Fast Antimicrobial Screening Test (FAST), a microbial inhibition test used by USDA-FSIS to screen for antibiotic residues.
PHAST utilizes the FAST test to screen cattle urine for antibiotic presence.
B.stearo (*Bacillus stearothermophilus* var. *calidolactis*, ATCC 10149) is the microbe used in the PremiTest and Charm KIS antibiotic screening tests.
B.meg (*Bacillus megaterium*, ATCC 9885) is the microbe used in the FAST (meat) and PHAST (urine) antibiotic screening tests.
B.sub (*Bacillus subtilis*, ATCC 6633) is the microbe used in the STOP (meat) and LAST (urine) antibiotic screening tests.
USDA-CSREES Grant: WBS # 25-6239-0098-011 (Develop Pre-Harvest Version of the USDA-FSIS Fast Antibiotic Screening Test and Education).
Note: Use the *residue detection* information only as a starting guide.
WD (withdrawal) days listed are the maximum from product labels within product class.
[a]Korsrud and MacNeil (1988).
[b]Griffin, D.D. University of Nebraska, Great Plains Veterinary Educational Center, PO Box 148, Clay Center, NE 68933.
[c]Data from FOI provided by Charm Sci (KIS) and DSM (PremiTest).
[d]Mohsenzadeh and Bahrainipour (2008).
**FARAD published estimate.

Minimize Injection Site Lesions Injection site lesions relate to food quality as well as food safety, but food quality can impact demand and therefore is an important consideration in the discussion (Fig. 8.7). Injecting drugs in an unlabeled manner or improper manner may result in a change in the pharmacokinetic behavior of the injected drug, resulting in unexpected or unpredictable residues. For the sake of quality, products that are labeled to be administered subcutaneously are preferable to intramuscular injections. For the sake of residue prevention, products should only be injected as labeled, since the labeled route of administration will result in known withdrawal times (Fig. 8.8). For the sake of quality, injectable drugs should be administered into areas with less economic value, such as the neck; if there is damage to tissues from the injection, this will minimize the economic loss. Unless labeled for large-volume injection at one site, no more than 10 ml should be administered in an intramuscular or subcutaneous injection site. Exceeding this amount will increase tissue damage and could potentially alter withdrawal time.

Nondrug Residues Careful management is necessary to prevent accidental or negligent exposure to feed, water, or soil contaminated with heavy metals, petrochemicals, polychlorinated biphenyls (PCBs), PCPs, insecticides, fungicides, herbicides, mycotoxins, or other hazardous materials. Just as with grain and forage, steps should be taken to ensure that purchased fats and oils do not contain residues. A reputable seller of fats and oils should be testing products for the following contaminants:

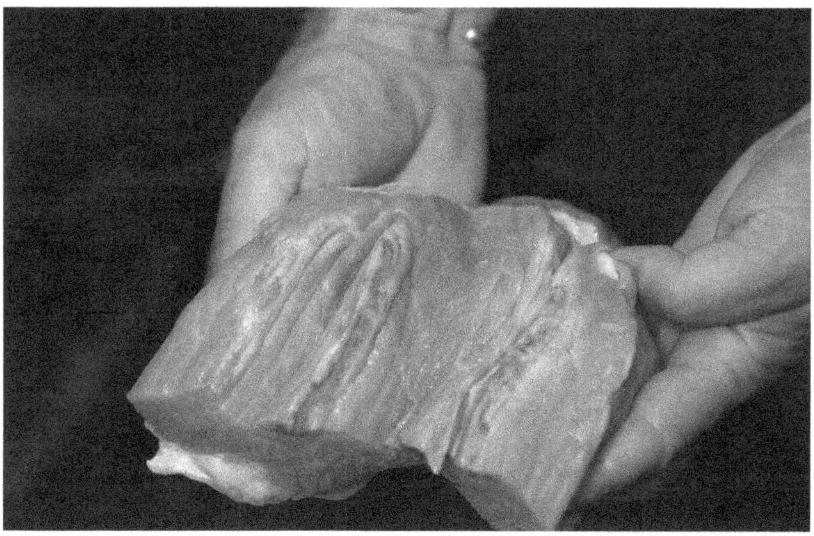

FIGURE 8.7 Lesion from intramuscular injection. Photo by Dee Griffin.

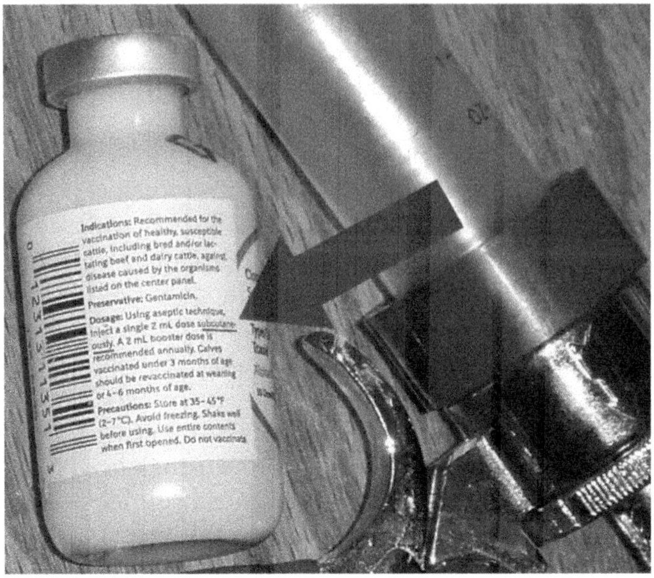

FIGURE 8.8 Read the drug label to ensure proper dose and route of administration of drugs and vaccines. Photo by Dee Griffin.

PCBs, chlorinated hydrocarbons (CHC), pesticides, heavy metals, *Salmonella*, and tall oil (hydrocarbon). Before purchasing any fat or oil, the supplier should be asked if the product is tested. The leakage of transmission and transformer fluid also poses a potential problem in residue avoidance, whether direct contamination or contamination of feed from farm equipment. Both types of fluid contain polychlorinated hydrocarbons (PCBs), resulting in a small but real risk of a violative residue in cattle. Lead and other heavy metals may be picked up through spills and leaks, batteries, paint, and other materials. To avoid accidental livestock exposure, all chemicals should be treated as potential hazards. Chemical products should be stored away from feed products and areas where cattle have access. A quality control program for incoming feed ingredients is recommended to eliminate contamination resulting from molds, mycotoxins, and chemicals. Feed should be stored in a manner that prevents development of molds and mycotoxins and prevents exposure to chemicals. Prior to usage, feed ingredients suspected of contamination should be submitted for analysis by a qualified laboratory. If a feed-related poisoning is suspected, producers and veterinarians should work with a diagnostic laboratory for assistance, and animals should be withheld from the food supply until diagnosis.

8.7 QUALITY ASSURANCE PROGRAMS

While quality assurance programs do not focus entirely on residues and residue prevention, residues are a major component of most programs, especially in the early years of adoption. Adherence to a quality assurance program may be a requirement for marketing to a particular harvest facility or it may be part of a remediation process after a violative residue. Examples of national quality assurance programs are listed in Table 8.3. Whatever the rationale for using a quality assurance program, they tend to have similar components related to residue avoidance and prevention:

- Keeping accurate records
- Keeping medications in appropriate locations under appropriate conditions (store properly)
- Following drug labels when drugs are used on label
- Tracing using identification of individual animals
- Adhering to stated withdrawal times
- Keeping unfit animals out of the food supply

TABLE 8.3 Examples of quality assurance programs websites

Title	Sponsoring organization	URL
Beef Quality Assurance	National Cattlemen's Beef Association, Denver, Colorado	http://www.bqa.org/
Dairy Beef: Maximizing Quality and Profits	Western Regional BQA Program, Washington State University Extension, Pullman, Washington	http://dairybeef.ucdavis.edu/
Livestock Production Assurance	Meat & Livestock Australia	http://www.mla.com.au/Meat-safety-and-traceability/Livestock-Production-Assurance
National Feedlot Accreditation Scheme	AUS-MEAT Limited, Murarrie, Queensland, Australia	http://www.ausmeat.com.au/audits-accreditation/nfas-feedlot-assurance.aspx
ProSafeBeef	Multiple research and industry organizations	http://www.prosafebeef.eu/
Verified Beef Production	Canadian Cattlemen's Association, Calgary, Alberta, Canada	http://www.verifiedbeef.org/

The success of these programs in the United States is evident in the rate of violative residues: Reported rates of antibiotic residues in various classes of cattle were 0.4–1.2% in 1985 (Cordle, 1988), that is, number of positive samples in tested samples. According the USDA Red Book for 2008 (Anonymous, 2009a), scheduled sampling resulted in 2/17,876 (0.01%) violative antibiotic residues, and inspector-generated sampling resulted in 885/135,552 (0.7%) violative antibiotic residues.

8.8 THE FUTURE: ANTIMICROBIAL REGULATION AND THE MARKET FOR "ANTIMICROBIAL-FREE" BEEF PRODUCTS

As mentioned previously in this chapter, there is pressure from consumer activists as well as public health groups to limit the use of so-called medically important antimicrobial drugs in food animals. In addition, the market continues to grow for "organic," "natural," and "antimicrobial-free" beef products, although legal definitions for these are limited to one for "organic." Lawmakers in the United States have introduced legislation that if passed, would severely limit the use, particularly in feed, of antimicrobial drugs with particular importance in human medicine. While the intent of this type of legislation is to reduce use by restricting it, the consequences of blanket bans on use are not well modeled or studied. It is possible that such restrictions would result in increased microbial load in the animal with a concomitant increase in prevalence in foodborne pathogens at harvest. These bans may also increase overt disease in animals, with the same result as aforementioned. We recommend further study of the potential impact of blanket restrictions before they are implemented in systems that have been shown to result in relatively low levels of foodborne illness and few unequivocal demonstrations of antimicrobial drug use in a food animal, resulting in untreatable diseases in humans.

In addition to proposed legislation, the FDA-CVM is working on the regulatory side to consider ways of restricting antimicrobial drug use in food animals through guidance documents and regulations. For example, the extralabel use of cephalosporin antimicrobial drugs has been restricted as of 2012. The FDA has published the final guidance document, Guidance 209, that outlines the future of regulation as it relates to antimicrobial drugs in food animals, including the assertion that any use as growth promotants is by definition imprudent and therefore undesirable. Beef producers and their veterinarians must continue to use antimicrobial drugs in a prudent manner, and they must support continued efforts to research the dynamics of antimicrobial drug use on foodborne pathogens and commensal organisms that can carry resistance mechanisms. This will allow the production of a safe and wholesome beef supply to feed the growing demand for beef products around the world.

REFERENCES

Anonymous (1996). Extralabel drug use in animals; Final rule, 21 CFR Part 530. Fed Reg 61:57731–57746.

Anonymous (2009a). 2008 Food safety inspection service national residue program data. Washington DC: U.S. Department of Agriculture; p 1–147.

Anonymous (2009b). DANMAP 2008 - Use of antimicrobial agents and occurrence of antimicrobial resistance in bacteria from food animals, foods and humans in Denmark. Soborg, Denmark, National Food Institute, Technical University of Denmark; p 1–132.

Anonymous (2009c). National residue survey annual report 2008–2009. Canberra, Australian Government Department of Agriculture, Fisheries and Forestry; p 1–281.

Anonymous (2010). Canada food and drug regulations C.R.C., c.870, Section C.08.012. Available at http://laws.justice.gc.ca/PDF/Regulation/C/C.R.C.,_c._870.pdf. Accessed April 6, 2014.

Cordle MK (1988). USDA regulation of residues in meat and poultry products. J Anim Sci 66:413–433.

Korsrud GO, MacNeil JD. Evaluation of the swab test on premises, the calf antibiotic and sulfa test, and a microbial inhibitor test with standard solutions of 22 antibiotics. J Food Protect 1988;51(1):43–46.

Mohsenzadeh M, Bahrainipour A. The detection limits of antimicrobial agents in cow's milk by a simple Yoghurt Culture Test. Pak J Biol Sci 2008; 11(18):2282–2285.

Morley PS, Apley MD, Besser TE, Burney DP, Fedorka-Cray PJ, Papich MG, Traub-Dargatz JL, Weese JS (2005). Antimicrobial drug use in veterinary medicine. J Vet Intern Med 19:617–629.

Platt TM, Loneragan GH, Scott HM, Norby B, Thomson DU, Brown MS, Ives SE, Brashears MM (2008). Antimicrobial susceptibility of enteric bacteria recovered from feedlot cattle administered chlortetracycline in feed. Am J Vet Res 69:988–996.

Schneider MJ, Lehotay SJ (2008). A comparison of the FAST, Premi and KIS tests for screening antibiotic residues in beef kidney juice and serum. Anal Bioanal Chem 390:1775–1779.

Vigre H, Larsen PB, Andreasen M, Christensen J, Jorsal SE (2008). The effect of discontinued use of antimicrobial growth promoters on the risk of therapeutic antibiotic treatment in Danish farrow-to-finish pig farms. Epidemiol Infect 136:92–107.

9

RESIDUE AVOIDANCE IN DAIRY CATTLE PRODUCTION SYSTEMS

Geof Smith

Department of Population Health and Pathobiology, College of Veterinary Medicine, North Carolina State University, Raleigh, NC, USA

Prevention of drug residues in the dairy industry is a major concern as both meat and milk from dairy cattle are widely consumed by humans around the world. This chapter will discuss the common prophylactic and therapeutic uses of drugs in dairy cattle. It will also address the prevalence of drug residues in both meat and milk. The majority of the discussion will focus on drug use and drug residues in the United States; however, a brief discussion on drug use and drug residues in Europe and Australia will also be included. Finally, this chapter will include a discussion on strategies to avoid drug residues and quality assurance programs available in the dairy industry.

9.1 PROPHYLACTIC USE OF DRUGS IN DAIRY CATTLE

Antibiotics are not generally fed to dairy cattle for growth promotion like is commonly done in the beef industry. Since 1997, the use of antibiotics for growth promotion has been prohibited in the European Union (EU). Therefore, drugs such as avoparcin, virginiamycin, tylosin, spiramycin, and bacitracin have been banned for several years. In addition, there are no antibiotics

Strategies for Reducing Drug and Chemical Residues in Food Animals: International Approaches to Residue Avoidance, Management, and Testing, First Edition.
Edited by Ronald E. Baynes and Jim E. Riviere.

approved in the United States that can be added to the feed of lactating dairy cattle. With the exception of ionophores, the use of antibiotics to improve growth rates or feed efficiency is not done. However, there are several other prophylactic uses of drugs in the dairy industry. One of the most significant prophylactic uses of antimicrobials in the dairy industry worldwide is the administration of intramammary antibiotics to cattle at dry-off (commonly referred to as dry-cow therapy). This is done to treat existing intramammary infections (which would be considered a therapeutic use) and to protect against intramammary infections that may develop during the dry period (prophylactic use). In the United States, over 90% of dairy farms use intramammary antibiotics at dry-off, and the vast majority of farms use dry-cow therapy on all cows (USDA, 2008). Commercially available dry-cow products in the United States include penicillin, cloxacillin, erythromycin, ceftiofur, cephapirin, and a combination of penicillin and dihydrostreptomycin. Results from the 2007 National Animal Health Monitoring Survey (NAHMS) dairy study indicated that cephapirin and penicillin/dihydrostreptomycin were by far the most commonly used antibiotics for dry-cow therapy (USDA, 2008).

The use of dry-cow therapy varies considerably in Europe. In many countries (i.e., Germany, France, Italy, Switzerland, etc.), this is still considered standard practice and is used on the majority of farms (with the exception of organic dairy farms where the use of dry-cow therapy is prohibited). For example, a study of 201 dairy herds in the Netherlands reported that 83% of herd still used intramammary antibiotics on all cows at dry-off (Barkema et al., 1998). There also seems to be some association with herd size as smaller farms are often less likely to use blanket dry-cow therapy as compared to larger farms (over 100 cows). The types of antibiotics available for intramammary administration in dry cows vary between countries; however, penicillin G, penicillin/neomycin, cloxacillin, and cefquinome appear to be the most commonly used products. There are other regions where selective dry-cow therapy is widely accepted. Selective dry-cow therapy is aimed at administering antibiotics only to cows likely to benefit from treatment. Thus, cattle at low risk of having mastitis (such as first- or second-lactation cattle with low somatic cell counts) are not treated, and cattle at low risk of successful treatment with dry-cow therapy (older cattle with high somatic cell counts or that have had multiple episodes of clinical mastitis) are recommended for culling (Østerås et al., 1999). Selective dry-cow therapy is exclusively practiced in the Scandinavian countries, and the percentage of cows that receive intramammary antibiotics at dry-off is about 17% in Finland and about 15% in Sweden (Pyörälä, personal communication). Products available in these countries include penicillin/

aminoglycoside (dihydrostreptomycin or framycetin) combinations and cloxacillin. There are actually no long-acting intramammary products available in Norway, and dry-cow therapy is not practiced in this country.

In Australia, blanket dry-cow therapy is still considered the gold standard, and the vast majority of dairy cattle receive intramammary antibiotics at the end of lactation (Brightling et al., 1998). There are multiple cloxacillin products available in Australia along with ampicillin, cefalonium, and a cloxacillin/ampicillin combination product.

Historically, another relatively common prophylactic use of antibiotics was medicated milk or milk replacers. In a 2007 survey, about 60% of dairy farms in the United States fed medicated milk replacers to preweaned heifer calves, most commonly a combination of oxytetracycline and neomycin (USDA, 2008). However, a new federal regulation that began in 2010 restricts the feeding of medicated milk replacers to a period of 7–14 days. Thus, continuous feeding of antibiotics in the milk from birth to weaning is no longer permitted. This is meant to transition the use of oral antibiotics in calves from prophylactic to therapeutic. Medicated milk replacers will now be reserved for the treatment of bacterial enteritis (diarrhea) and bacterial pneumonia in dairy calves.

Milk replacers containing antibiotics are fed as a routine practice to dairy calves in some parts of the world (such as Central and South America as well as much of Asia); however, it has become less common or even prohibited in other places. Since the late 1990s, the EU has prohibited the sale of milk replacers and other animal feeds containing antibiotics. All the feed and milk replacers for dairy cattle must be sold as nonmedicated, and then antibiotics can be added only for therapeutic use (e.g., in calves with diarrhea). Australia and New Zealand also have strict laws regarding the importation of any animal feed, and these products are generally nonmedicated as well. Overall, the conventional practice of adding antibiotics to milk or milk replacers for prophylactic use is being discouraged worldwide as research has shown that this may actually increase the severity of diarrhea and decrease growth rates in calves (Berge et al., 2009).

Anthelmintics are also frequently used in dairy cattle, both in prophylactic and therapeutic manners. There are not good data available on what percentage of dairy cattle is routinely administered anthelmintics; however, the results of two meta-analyses have demonstrated some improvement in milk yield associated with routine use of endectocides (Gross et al., 1999; Sanchez et al., 2004). Data collected as part of the national dairy survey in 2007 indicated that 70% of heifers routinely receive anthelmintics as well as 63% of lactating cows in the United States (USDA, 2007). In North America, Europe, and Australia, the most commonly used anthelmintics in dairy cattle are benzimidazoles (fenbendazole, thiabendazole) and macrocyclic lactones (ivermectin, eprinomectin, and moxidectin). Several of these products partition very poorly into milk and can be used without any milk withholding interval (i.e., eprinomectin, moxidectin). Meat or milk residues due to anthelmintic use are rare but can occur when products approved for beef cattle are accidentally used in dairy cattle. For example, ivermectin and doramectin are approved for beef cattle in the United States and are both available in either injectable or pour-on formulations. These drugs have a very long milk depletion half-life and can result in prolonged residues (45–60 days) when used accidentally in dairy cattle (Baynes et al., 2000; Chicoine et al., 2007).

Ionophores such as monensin and lasalocid are also commonly used throughout the world in dairy cattle. These are fed to preweaned heifers to prevent coccidiosis and in lactating cattle to improve feed efficiency and milk production. Although ionophores are classified as antibiotics, they are of low regulatory concern and do not normally result in milk or meat residues. Other prophylactic drugs occasionally used in the dairy industry considered to be of low importance in terms of residues include coccidiostats (decoquinate, amprolium, etc.), probiotics, and anionic salts.

9.2 THERAPEUTIC USE OF DRUGS IN DAIRY CATTLE

There are several recent surveys on therapeutic drug use in dairy cattle that have been published in the United States; however, all of these focus primarily on antimicrobials. In the dairy survey done by the 2007 NAHMS, mastitis was the most commonly treated disease with 16.5% of cows receiving intramammary antibiotics during lactation (USDA, 2008). The primary antibiotics used to treat mastitis as identified in this survey were cephalosporins (53% of cases, primarily cephapirin and ceftiofur), lincosamides (19.4%, pirlimycin), and other beta-lactams (19.1%, includes amoxicillin, cloxacillin, hetacillin, and penicillin). Other common indications for antimicrobial use are summarized in Table 9.1 but included reproductive disorders, lameness, and respiratory disease. Of the antibiotics used in cases of lameness, tetracyclines were the most common (almost half of reported cases), which represents the common use of topical oxytetracycline for digital dermatitis treatment in the United States. Cephalosporins (ceftiofur) were the next most commonly used antibiotic being used in 27% of cows treated for lameness in the dairy survey (USDA, 2008). Although treatment rates for respiratory disease in adult (lactating) cattle were fairly low, almost 12% of preweaned heifers receive antibiotics for pneumonia, and 18% of calves receive therapeutic antibiotics for the treatment of diarrhea.

TABLE 9.1 Antimicrobial use in dairy cows in the United States

	Percentage of cows treated with antibiotics for the following diseases during the previous 12 months		Percentage of farms that treated cows with any antibiotic for the following diseases during the previous 12 months	
	2002	2007	2002	2007
Mastitis	15.0	16.4	84.3	85.4
Lameness	7.0	7.1	51.6	58.6
Reproductive	4.9	7.4	42.1	52.9
Respiratory disease	2.2	2.8	49.0	55.8
Diarrhea or other digestive problems	2.0	1.9	27.9	25.0
Other	0.2	0.5	4.8	6.9

Data are summarized from the 2002 and 2007 NAHMS (USDA, 2008).

Several regional surveys have also been done during the past 10 years examining antibiotic use in dairy cows. One study looked at the use of antimicrobials on both conventional and organic dairy farms in Michigan, Minnesota, New York, and Wisconsin (Zwald et al., 2004). Use of antibiotics on conventional dairies was quite common, and the majority (85%) of producers reported treating between 1 and 10% of their cows in the previous 60-day period with about 10% reporting treatment between 11 and 25% of cattle. Ceftiofur was by far the most commonly reported antibiotic used in lactating dairy cows. Over 90% of organic dairy farms in this study reported no antibiotic treatment of dairy cows. A follow-up survey in Wisconsin showed similar results. Conventional herds commonly used antibiotics to treat diseases, most commonly mastitis, metritis, infections of the foot, and respiratory disease (Pol and Ruegg, 2007). Cephapirin was the most common antibiotic used to treat mastitis following by pirlimycin and amoxicillin. Ceftiofur was the most commonly administered antibiotic for most other diseases including foot infections, respiratory disease, and metritis. This survey attempted to estimate antimicrobial drug use by developing a daily defined dose (DDD) for antibiotic use, expressed as the number of antibiotic doses per adult cow per year. On conventional dairy farms, the estimated overall exposure was 5.43 DDD per cow per year, which was composed of 3.58 DDD of intramammary antibiotics (which included both clinical mastitis and dry-cow therapy) and 1.85 DDD of parenteral antimicrobials. Organic dairies in this study reported minimal use of antibiotics.

A survey of 381 dairy farms in Washington State showed similar results. The most common reasons for antibiotic use in this study were lameness, calf diarrhea, and mastitis; and the most commonly used drugs were cephalosporins (cephapirin and ceftiofur), penicillin, and oxytetracycline (Raymond

TABLE 9.2 Drugs currently prohibited from extralabel use in dairy cattle in the United States

Diethylstilbestrol
Chloramphenicol
Nitroimidazole (including metronidazole)
Sulfonamides (in adult dairy cattle with the exception of sulfadimethoxine, which is approved)
Nitrofurans (including topical use)
Clenbuterol
Dipyrone
Phenylbutazone
Fluoroquinolones (with the exception of enrofloxacin, which is approved for the treatment of respiratory disease in young calves)
Glycopeptides (such as vancomycin)
Cephalosporins (with the exception of cephapirin—there is a restriction on the extralabel use of cephalosporins in cattle. These drugs can be used for indications not on the label, but therapy must follow the labeled dose and duration of therapy)

et al., 2006). In this survey, 23% of producers indicated that they occasionally used drugs that were not approved for dairy cattle or that were prohibited (Table 9.2). The most commonly cited example of extralabel drug use in this study was gentamicin, which was used by 16% of farms for disease treatment. A survey of 113 dairy herds in Pennsylvania revealed that antibiotic usage was highest for calves with diarrhea (36% of animals), followed by pneumonia in calves (25%), interdigital dermatitis (foot rot) in adult cattle (16%), clinical mastitis (14%), and metritis (11%). Extralabel use of nonapproved drugs on dairy farms was common in this survey as 79% of farms reported using at least some antibiotics not approved for dairy cattle in the United States (Sawant et al., 2005).

Although no definitive data are available, flunixin is also commonly given to dairy cows in the United States. It is a nonsteroidal anti-inflammatory drug (NSAID) that is approved for the control of pyrexia associated with bovine respiratory acute bovine mastitis. It is also indicated for control of inflammation associated with endotoxemia and can only be administered intravenously. Flunixin is the only NSAID drug approved in the United States and is widely used on dairy farms, which has led to concerns about meat and milk residues (Deyrup et al., 2012; Kissell et al., 2013). Dexamethasone and isoflupredone acetate are also approved for dairy cattle (Table 9.3) but are not as widely used as flunixin.

Drug use in the EU varies considerably. There are differences in the availability and preferences of drugs between countries. There are also considerable differences in the attitude toward the use of antibiotics in food-producing animals between different countries in Europe. Some countries (such as

TABLE 9.3 Drugs approved for lactating dairy cattle in the United States along with meat and milk withholding times

Drugs	Route of administration	Meat (days)	Milk
Ampicillin	Intramuscular/SC	6	48 h
Ceftiofur sodium	Intramuscular/SC	4	0 h
Ceftiofur hydrochloride	Intramuscular/SC	3	0 h
Ceftiofur crystalline-free acid	SC (in the ear only)	13	0 h
Erythromycin	Intramuscular	14	72 h
Oxytetracycline (100 or 200 mg/ml)	IV or intramuscular	28	96 h
Penicillin (procaine)	Intramuscular	10	48 h
Sulfadimethoxine	Oral boluses	7	60 h
Sulfadimethoxine	IV	5	60 h
Amoxicillin	Intramammary—lactating cow	12	60 h
Ceftiofur	Intramammary—lactating cow	2	72 h
Cephapirin	Intramammary—lactating cow	4	96 h
Cloxacillin	Intramammary—lactating cow	10	48 h
Hetacillin	Intramammary—lactating cow	10	72 h
Penicillin	Intramammary—lactating cow	3	60 h
Pirlimycin	Intramammary—lactating cow	9	36 h
Ceftiofur	Intramammary—dry cow only	16	NA
Cephapirin	Intramammary—dry cow only	42	NA
Cloxacillin	Intramammary—dry cow only	30	NA
Erythromycin	Intramammary—dry cow only	14	NA
Penicillin–dihydrostreptomycin	Intramammary—dry cow only	60	NA
Penicillin–novobiocin	Intramammary—dry cow only	30	NA
Cloprostenol sodium	Intramuscular	0	0 days
Dexamethasone	IV or intramuscular	0	0 days
Dinoprost tromethamine	Intramuscular	0	0 days
Eprinomectin	Topical (pour-on)	0	0 days
Furosemide	IV or intramuscular	2	48 h
Fenbendazole	Oral	8	0 h (5 mg/kg dose)
Flunixin meglumine	IV	4	36 h
Gonadorelin hydrochloride	Intramuscularly	0	0 days
Gonadotropin (chorionic)	Intramuscular	0	0 days
Isoflupredone acetate	Intramuscular	0	0 days
Morantel tartrate	Oral (in feed)	14	0 h
Moxidectin	Topical (pour-on)	0	0 days
Oxytocin	IV or intramuscular	0	0 days
Sometribove zinc	SC	0	0 days
Tripelennamine hydrochloride	IV or intramuscular	4	24 h

NA, no approved milk withdrawal for dairy cows.

Denmark, Finland, and Sweden) are extremely conservative. For the most part, treatments are administered by a veterinarian, and antimicrobial use is kept to a minimum. In other countries, the use of drugs is much higher and more closely mimics what is commonly seen in the United States. The dairy industry in the EU varies considerably as well. Many of the herds are small, containing less than 50 cows. However, large, integrated dairy farms also exist, and differences in drug use between small and large operations are quite common. The diseases commonly treated in the EU are very similar to what has already been described for the United States. Calves are frequently treated for diarrhea and respiratory disease, while adult cattle are most commonly treated for mastitis, infections of the foot, and reproductive disorders (Bennedsgaard et al., 2010; Busani et al., 2004; Heuwieser et al., 2010; Menéndez González et al., 2010; Ortman and Svensson, 2004).

There are few surveys that have been done in different European countries outlining the frequency and type of drug administration to dairy cattle. One survey looked at antimicrobial use over a 1-year period on 97 dairy farms in Switzerland (Menéndez González et al., 2010). The average size of the herds included in this study was 25 cows. Penicillin and penicillin combinations (primarily with aminoglycosides) were the most commonly used intramammary drugs used to treat mastitis. The second most common therapy was cephalosporins (primarily cefquinome). Overall, the beta-lactams were by far the most common class of antibiotics used on dairy farms in this survey, followed by sulfonamides, aminoglycosides, and tetracyclines. In Italy, the use of fluoroquinolones appears to be more common. A survey of 106 Italian cattle veterinarians was done to gather information regarding their use of antibiotics (Busani et al., 2004). Most of the veterinarians included in this survey were primarily focused on dairy cattle. Fluoroquinolones and third-/fourth-generation cephalosporins were listed as "first-choice" antimicrobials for calf diarrhea and respiratory disease by a significant number of survey participants. Third- and fourth-generation cephalosporins were the most commonly reported "first-choice" treatment for mastitis treatment.

In the Netherlands, antibiotic use is monitored by a committee of scientists referred to as the Veterinary Antibiotic Usage and Resistance Surveillance Working Group (VANTURES). The committee publishes a report each year reflecting a survey of antibiotic usage in the country referred to as Monitoring of Antimicrobial Resistance and Antibiotic Usage in Animals in the Netherlands (MARAN). In 2008, the survey included 82 dairy farms, which represented over 7200 cows. Overall, dairy cattle received an average of 6.3 doses of antibiotics per year, and the most commonly used drugs were penicillins, penicillin/aminoglycoside combinations (primarily used for mastitis), and cephalosporins (VANTURES, 2010).

A survey of French cattle veterinarians indicated that a combination of penicillin and aminoglycosides was the most commonly used antibiotic, representing 25.8% of all prescriptions (Cazeau et al., 2009). Penicillin by itself was the next most commonly used (19.4% of all prescriptions), followed by cephalosporins (16.7%), fluoroquinolones (14.6%), and tetracyclines (9%). This survey included information on use of antimicrobial by type of disease treated. For digestive diseases, the penicillins and fluoroquinolones were the most commonly prescribed antibiotics representing almost 70% of the total. For locomotive disorders (including arthritis, sole abscesses, and interdigital and/or digital dermatitis), aminoglycosides were the most commonly prescribed class of antibiotics followed by penicillins and tetracycline. Fluoroquinolones were also the most commonly prescribed drug class for treatment of mastitis with the penicillins and cephalosporins following. For uterine diseases and obstetrical procedures, a combination of penicillin and aminoglycosides was by far the most frequently prescribed antibiotic. Lastly, for the treatment of respiratory disease in cattle, the fluoroquinolones were used most often (37% of cases) followed by chloramphenicol (18% of cases).

Organic dairy farms in the EU use less antibiotics as compared to conventional herds; however, their use is significantly more than organic dairies in the United States. In the United States, the organic farming regulations state that any animal receiving antibiotics be classified nonorganic for the rest of its life. Therefore, treatment with antibiotics or other drugs has a severe consequence on the fate of a dairy cow in an organic herd and is rarely used in the United States (Pol and Ruegg, 2007; Zwald et al., 2004). In contrast, under organic regulations in the EU, the consequence of using antimicrobials is only a doubling of the withdrawal time on milk and meat as compared to conventional herds (Bennedsgaard et al., 2010). So the use of most drugs is kept to a minimum; however, the majority of sick cattle still receive appropriate therapy.

There is little information on nonantibiotic drugs that are used in dairy cattle in the EU; however, NSAIDs are often used as adjunctive therapy in cows with severe endotoxemia (i.e., mastitis, metritis, respiratory disease). Carprofen is approved for dairy cattle in several European countries for control of inflammation associated with respiratory disease as well as for the control of fever associated with toxic mastitis. In a study of milk residue depletion following either IV or SC carprofen administration in dairy cows, high performance liquid chromatography revealed no drug concentrations >25 μg/kg in samples from any time point (EMEA, 2010). Therefore, this drug has been approved in the EU with no milk discard. Meloxicam is a newer NSAID that is also approved for use in dairy cattle

TABLE 9.4 Drugs currently prohibited from extralabel use in dairy cattle in the EU

Aristolochia species botanicals (common names include Aristolochia, Guan mu tong, Dutchman's pipe) have been associated with severe nephrotoxicity
Chloramphenicol
Chloroform
Chlorpromazine
Colchicine
Dapsone
Dimetridazole
Malachite green
Nitrofurans (including furazolidone and nitrofurazone)
Nitroimidazoles (including metronidazole and ronidazole)
Phenylbutazone

in the EU. It has a milk withdrawal interval of 5 days following IV or SC injection (Smith et al., 2008). Tolfenamic acid is also approved in Europe and Canada for use in cattle with acute mastitis or respiratory tract disease. The EU has set the maximum residue level (MRL) for tolfenamic acid at 50 μg/kg in muscle and milk. A dose of 4 mg/kg is approved as a single IV injection, which has a milk withdrawal time of 24 h (Smith et al., 2008).

Drugs prohibited from dairy cattle in the EU are listed in Table 9.4. In addition to those that are prohibited, there are several other drugs whose use is severely restricted. For example, β-agonists such as clenbuterol can only be used for tocolysis in both beef and dairy cattle. These drugs can only be administered by a licensed veterinarian and are not allowed to be kept on a farm. The use of clenbuterol by a producer or by a veterinarian for any reason other than tocolysis would be illegal. As previously mentioned in this chapter, the addition of antibiotics such as avoparcin, virginiamycin, tylosin, spiramycin, and bacitracin to the feed of dairy cattle is also prohibited.

A recent survey of veterinarians and producers in the state of Victoria, Australia, was done to determine the major drugs used in dairy cattle. The survey covered a wide range of topics including extralabel drug use, feed medications, compounding, and preferred treatments for specific diseases. Although the full report has not yet been released, the most commonly prescribed antibiotics were procaine penicillin, oxytetracycline, cloxacillin, ceftiofur, and cefuroxime. Survey respondents reported a fairly constant level of antibiotic use over a period of several years; however, there had been a significant increase in the frequency of NSAID administration.

9.3 PREVALENCE OF DRUG RESIDUES

Milk is one of the most heavily regulated food products in the world. Few other food products and their source of origin (the dairy farm and processing plant) are scrutinized for safety and quality quite like milk. Worldwide, the dairy industry closely monitors for residues in milk. This is done to protect consumers who may be allergic and to prevent the emergence of resistant bacteria. Antibiotic residues may also impact the manufacturing process of milk products. In the United States, the federal Food, Drug, and Cosmetic Act assigns the food industry with the primary responsibility for ensuring the safety of milk (Talley, 1999). The Food and Drug Administration's (FDA) role is to verify that the food industry is conducting its job properly and to initiate regulatory action when the food industry fails to do so. This is done primarily through cooperative agreements with state regulatory authorities. Milk quality in the United States is controlled by the Grade A Pasteurized Milk Ordinance (PMO), which assigns the responsibility for routine inspection and sampling of milk to state regulatory agencies (USHSS, 2009). Appendix N of the PMO references safe levels and tolerances for drug residues for milk in the United States.

According to the PMO, every tanker of milk in the United States must be screened for beta-lactam antibiotics prior to unloading at the milk processing plant. Screening tests used must have been evaluated and approved by the FDA and should be able to detect drug concentrations at or below the tolerance limit. All positive screening test results are reported immediately to state regulatory authorities. The farm responsible for the residue violation receives a financial penalty, and they must complete a quality assurance drug residue prevention program with a licensed veterinarian in order to have their permit to sell milk reinstated (Talley, 1999). In addition to beta-lactams, milk may be screened for other drugs by employing a random sampling program. Table 9.5 summarizes the number of milk samples tested in the United States for drug residues between 2003 and 2012 along with the number of violative residues identified. The vast majority of milk residues identified in the United States are caused by beta-lactam antibiotics (Table 9.6). However, this is by far the most common screening test performed on milk. Residues involving other classes of antibiotics (i.e., sulfonamides, aminoglycosides, macrolides, tetracyclines) are occasionally found; however, it is important to emphasize that these tests are not routinely performed. Flunixin is not routinely tested for in milk samples in the United States because a rapid screening assay has only recently been developed; however, flunixin residues in saleable milk have been identified (Kissell et al., 2013).

TABLE 9.5 Summary of drug residue data from milk samples tested in the United States between 2003 and 2009

Year	Number of milk samples tested	Number of positive tests	Percent positive
2012	3,775,440	828	0.022
2011	3,787,251	1079	0.028
2010	3,881,479	1245	0.032
2009	3,766,905	1303	0.034
2008	4,024,536	1621	0.040
2007	4,026,485	1687	0.042
2006	4,204,919	2261	0.054
2005	4,239,718	2198	0.052
2004	4,503,617	4974	0.110
2003	4,456,141	3246	0.073

Data were compiled from the National Milk Drug Residue Database Fiscal Year Annual Reports from 2003 to 2009.
Samples tested include bulk tank milk (picked up directly from farms) and pasteurized fluid milk and milk product samples (packaged products).

TABLE 9.6 Summary of drug residue data from milk samples tested in the United States between 2003 and 2012

	2007–2012		2003–2006	
	Number of milk samples tested	Total number of positive tests	Number of milk samples tested	Total number of positive tests
Beta-lactams	23,109,250	7690	17,042,811	12,473
Sulfonamides	152,481	72	246,138	157
Aminoglycosides	448	2	1,616	5
Macrolides	17,505	4	1,226	0
Tetracyclines	72,961	83	14,898	24
Enrofloxacin	18,492	5	49	0

Data were compiled from the National Milk Drug Residue Database Fiscal Year Annual Reports from 2003 to 2012.

Cull dairy cows are also frequently screened for drug residues at slaughter. The United States Department of Agriculture (USDA) and Food Safety and Inspection Service (FSIS) are responsible for collecting data on tissue residues of animal drugs. FSIS samples carcasses at slaughter on both a random basis (scheduled sampling plan) and when residues are at high risk (inspector-generated sampling plan). Under the scheduled sampling plan, inspectors collect random samples from carcasses that appear healthy and have been passed for human consumption. However, the inspector-generated sampling plan targets carcasses for sampling based on abnormal clinical signs in the live animal, lesions present in the carcass, previous known residue violations by the animals' owner, or the animal's herd history. Public health veterinarians will collect carcass samples that can be screened in the facility using the Fast Antimicrobial Screen Test (FAST) that is designed to detect a wide variety of antimicrobial classes. When animals are positive on the FAST, further testing is done to identify the exact cause of drug residue. Carcasses that test positive are also analyzed for other drugs such as flunixin meglumine and phenylbutazone. Out of 2,929,315 dairy cows slaughtered in the United States during 2011, 97,240 were tested for residues, which represent 3.3% of cattle presented for slaughter (FSIS, 2013). Major residues identified are outlined in Table 9.7 for the years 2004–2011. The data indicate that beta-lactams (especially ceftiofur and penicillin), sulfadimethoxine, and flunixin represent the primary residue concerns for cull dairy cows in the United States.

In Europe, milk testing and regulations vary between countries. MRLs are set by the European Medicines Agency (EMEA); however, the responsibility for testing and enforcing food safety regulations is left to individual countries. The number of milk and carcass samples screened for drugs and/or chemicals varies significantly across the EU. However, the European Commission states that the annual number of milk samples tested in each country within the EU should be 1 per 15,000 tons of annual milk production with a minimum of 300 samples. In the United Kingdom, residue testing in food animals is controlled by an independent group called the Veterinary Residues Committee (VRC). For a 5-year period (2005–2009), VRC tested 13,456 milk samples (an average of 2,690/year) and found five residue violations (one for each of the following: aflatoxin, amoxicillin, cefalonium, lead, and penicillin). In addition to aflatoxin and heavy metals, randomly selected milk samples are subjected to an antibiotic screen that can detect beta-lactams, chlortetracycline, doxycycline, tetracycline, dapsone, and sulfonamides. The number of milk samples tested by country in the EU and occurrence of violative residues for 2008 can be found in the European Commission report on residue monitoring data from member states (European Commission, 2008). For the entire EU, 53,333 milk samples were analyzed in 2008 and 102 residues were identified (as compared to 140 residues in 2007). The primary reason for milk residues was antibiotics; however, samples containing anthelmintics, NSAIDs, organochlorines, chloroform, and aflatoxin were also identified.

Member countries of the EU must also report carcass residue testing results to the European Commission. The minimum number of cattle to be tested each year is to be at least 0.4% of the total number of animals slaughtered in each country during the previous year. Although no distinction is made between dairy and beef cattle in meat residue reporting, data on the number of cattle tested and the number of residues identified can be found

TABLE 9.7 Summary of drug residue data from cull dairy cows tested in the United States between 2004 and 2011

Year	2004	2005	2006	2007	2008	2009	2010	2011
Ampicillin	ND	6	10	12	8	17	10	11
Aminoglycosides	106	114	129	99	86	67	82	56
Desfuroylceftiofur	ND	ND	ND	ND	69	126	69	54
Flunixin	37	107	130	259	269	194	204	91
Oxytetracycline or tetracycline	33	53	44	27	67	14	58	16
Penicillin	353	300	359	411	350	336	228	196
Phenylbutazone	ND	ND	ND	4	3	2	ND	ND
Sulfonamides	120	126	190	184	227	196	206	126
Tilmicosin	15	20	25	13	6	20	35	24
Total residues	665	727	888	1,001	1,086	1,007	896	575
Number of animals tested	84,852	94,570	57,486	93,997	80,092	81,928	96,513	97,240
Number of animals with residues	602	670	828	926	788	751	702	576
Percent of animals with residue violations (%)	0.7	0.7	1.4	1.0	1.0	0.9	0.7	0.6

Data were compiled from the Food Safety and Inspection Service National Program Residue Data ("Red Book") from 2003 to 2011 (FSIS, 2013).
The total number of residues includes compounds not specifically listed in the table including aflatoxin, heavy metals (cadmium and lead), anthelmintics, and other drugs.
ND, not done.

in the European Commission document (European Commission, 2008). The percentage of cattle tested for residues in the EU member countries averaged 0.47% and ranged from a minimum of 0.02% in Belgium to a maximum of 6.3% in Cyprus (European Commission, 2008). The primary residue identified in the EU was from antibiotics with 452 noncompliant results out of 44,069 carcasses sampled. Residues identified were from many different classes of antimicrobials. Other residues included corticosteroids (121 residues out of 31,103 samples tested), anthelmintics (9 residues out of 22,646 tests), NSAIDs (33 residues out of 4,980 tests), as well as other veterinary drugs. Additional residues identified included organochlorine compounds (PCBs), mycotoxins, and chemical elements (including cadmium, mercury, and lead). Overall in cattle from the EU in 2008, positive residue results were reported in 613 bovine carcasses. Of these 52% of the violative residues were due to antibiotics, 25% resulted from organochlorine compounds, 10% were caused by steroids, and 4% were due to NSAIDs. The remaining residues resulted from resorcylic acid lactones, prohibited substances, anthelmintics, anticoccidials, and chemicals (heavy metals).

The Australian Milk Residue Analysis (AMRA) Survey provides a national, independent residue monitoring program for the Australian dairy industry. The survey is funded by the dairy industry and coordinated by Dairy Food Victoria. In addition to providing food safety assurances to consumer, the program also facilitates the export requirements of the Australian Quarantine and Inspection Service, since about 50% of Australia's dairy products are exported to other countries. Residue results are published in the Animal Health Surveillance Quarterly reports, which are available on the Animal Health Australia website (AHA, 2010). Milk samples are routinely tested for antimicrobials, aflatoxin M1, chloramphenicol, anthelmintics, organochlorines, and organophosphates. Data from 2005 through 2012 indicate a very high degree of compliance with the Australian standards.

9.4 MINIMIZING RESIDUES IN MEAT AND MILK

The major reasons for violative drug residues in the dairy industry are as follows: (i) not following the label directions for correct treatment or dose of drug to be administered, (ii) failure to follow directions for the appropriate meat or milk withdrawal period, (iii) treatment of the animal not recorded on a written record, (iv) poor or improper animal identification, (v) long-term residues following treatment as a calf (aminoglycosides), or (vi) extralabel or illegal drug use (using a drug not approved for dairy cattle). Given the frequent use of therapeutic drug use on dairy operations and the potential involvement of farm workers in administering these drugs, veterinarians should be encouraged to set up written protocols for their herds to minimize variability in therapy and inappropriate drug selection or dosing. Unfortunately, this is not commonly done in the industry. A survey done in Washington State indicated that only about 25% of farms had written protocols in place for treating common diseases (Raymond et al., 2006). This is similar to a survey in Pennsylvania where 21% of farms had defined treatment protocols and only 32% of producers sought veterinary advice prior to treating sick cattle (Sawant et al., 2005). In addition, only about 50% of farms kept any type of written record of antimicrobial use on the farm. Another study found that the lack of adequate treatment records was the most commonly identified reason for residues in New York State (Sischo et al., 1997). Other major reasons were failure in the understanding of how to properly use drugs by farm personnel and a poor relationship between veterinarians and producers.

In addition, milk residue violations are frequently associated with the following: (i) accidentally milking a treated cow into the bulk tank, (ii) milking a cow that has received a dry-cow antibiotic formulation into the bulk tank, (iii) pipeline not diverted from bulk tank when milking cows treated with antibiotics, (iv) milk put in tank before the appropriate withdrawal period has ended, and (v) extralabel treatment (milk put into bulk tank without an appropriate withdrawal period). Farms with high somatic cell count levels have been reported to have a much higher rate of antibiotic residue violations, and larger dairy farms have also been shown to have higher rates of residues (van Schaik et al., 2002).

In the United States, there is a Milk and Dairy Beef Quality Assurance Program, which identifies 10 critical control points for residue prevention. The program is designed to be used by the dairy producer and their veterinarian as training on how to avoid drug residues. It is a voluntary program in the United States; however, once a farm has a residue violation, they may be required to complete the program in order to regain their ability to sell milk. The 10 critical control points outlined in the program are as follows:

1. Practice healthy herd management—In this part of the training, the veterinarian evaluates the housing, sanitation, nutrition and reproductive programs, biosecurity, and newborn calf care already present on the farm. Since disease prevention is often more cost effective than disease treatment, step one is designed to help the veterinarian and producer review things like milking management, hoof care, vaccination program, etc. Through the process of completing an evaluation of the current herd health management program, ways to improve herd

management and reduce the actual number of disease treatments may be identified.

2. Establishing a valid veterinarian/client/patient relationship (VCPR)—Having a valid relationship between the veterinarian and dairy producer is always helpful when drugs are being used and is mandatory in the United States if drugs are used in an extralabel manner. The American Veterinary Medical Association (AVMA) states the following requirements must be met to establish a VCPR:
   a. The veterinarian has assumed the responsibility for making clinical judgments regarding the health of the animal(s) and need for medical treatment, and the client (owner or other caretaker) has agreed to follow the instructions of the veterinarian
   b. There is sufficient knowledge of the animal(s) by the veterinarian to initiate at least a general or preliminary diagnosis of the medical condition of the animal(s). This means that the veterinarian has recently seen and is personally acquainted with the keeping and care of the animal(s) by virtue of an examination of the animal(s) and/or by medically appropriate and timely visits to the premises where the animals(s) are kept.
   c. The veterinarian is readily available or has arranged for emergency coverage or follow-up in case of adverse reactions or failure of the regimen of therapy.

   Another part of this portion of the training is to help producers understand the difference between over-the-counter drugs, approved prescription drugs, and extralabel drug use. Producers should have labels on all of their drugs stating the name of the drug, directions for use, prescribed withholding interval, and any cautionary statements.
3. Use only FDA-approved drugs with veterinarian's guidance—The veterinarian thoroughly reviews the list of prohibited drugs with the producer to ensure that these are never being used on the dairy farm (Table 9.2).
4. Maintain milk quality—This part of the training reviews the farm's milking procedures, waste management, and sanitary conditions. Since it is difficult or impossible to improve the quality of milk in the processing plant or retail locations, quality is generally determined at the dairy. The veterinarian reviews cow cleanliness, milking procedures, and milk cooling and also reviews milk quality reports (somatic cell counts, bacteria counts, etc.) with the producer.
5. Implement an effective mastitis management program—Since mastitis is frequently the most common reason cited for antibiotic use on dairy

farms, it is important for the veterinarian to review this program with the producer to help minimize the incidence of disease. The veterinarian again reviews milking sanitation (are the cows clean for milking, is the farm using an effective teat dip, etc.), how cows with clinical mastitis are identified, dry-cow program, etc. If milking management can be improved and the cases of mastitis decreased, this could potentially have a substantial impact on milk residue problems.

6. Administer all drugs properly and identify all treated animals—There are several routes of administration commonly used to administer drugs to dairy cattle including oral, topical, subcutaneous (SC), intramuscular, intravenous (IV), intramammary, and intrauterine. The veterinarian should review each of these with the producer and make sure they understand how to give drugs via each route. The veterinarian also makes sure the farm is somehow identifying animals when they are treated (using leg bands, neck bands, colored marks, etc.).
7. Maintain and use proper treatment records on all treated animals—The FDA in the United States requires that producers maintain drug treatment records for 2 years on all animals. These records should be easily accessible by anyone who works with the animals. The dairy producer should be able to show where all drug purchases were either used or disposed. The treatment record should contain the date of treatment, drug used, animal identification, dosage, route of administration, individual who administered the drug, and withdrawal period for meat and milk (Fig. 9.1).
8. Use of drug screening tests—There are various "on-farm" screening tests that are available for use by producers to screen milk for antibiotics. Examples of these rapid assays in the United States include Delvotest (DSM Food Specialties, the Netherlands), SNAP antibiotic residue test (IDEXX Labs, Inc., Westbrook, ME, USA), and various Charm II assays (Charm Sciences, Lawrence, MA, USA). Proper use of drug screening assays, particularly when a drug has been used in an extralabel manner, is strongly encouraged. In this step of the program, a veterinarian reviews how producers identify withholding intervals and assesses whether or not they are correctly using drug screening tests in certain situations. Appropriate use of milk residue test kits on farms has been associated with a significant reduction in the risk of milk residue violations (McEwen et al., 1991).
9. Implement employee/family awareness of proper drug use to avoid marketing adulterated dairy products—Many milk residues result when one person treats the animal and someone else does the milking.

Treatment Record

Herd: ______________________ Veterinarian: ______________________

COW ID	Time of Treatment			Diagnosis	Treatment Used	Dose Given	How Admin-istered	Withdrawal Time		Actual Date in Tank	Initials
	Date	AM	PM					Milk hrs.	Meat days		
				LF RF LR RR							
				LF RF LR RR							
				LF RF LR RR							
				LF RF LR RR							
				LF RF LR RR							
				LF RF LR RR							

FIGURE 9.1 A sample treatment record to be kept on dairy farms. Records should contain the date of treatment, animal identification, drug used, dosage and route of administration, individual who administered the drug, and withdrawal time for meat and milk.

In addition to maintaining accurate drug treatment records, it is important that all farm employees understand the importance and cost of drug residues and how to avoid them. The use of part-time labor to milk cows was found to be one of the most significant risk factors on dairy farms with a high risk of milk residue violations (McEwen et al., 1991). Therefore, all employees should understand how to read drug labels, how to fill in drug treatment records, and how to identify treated cattle.

10. Complete the milk and beef residue prevention protocol annually—To truly minimize residues, the training discussed earlier should be reviewed on a yearly basis to ensure everyone on the farm understands how to use drugs appropriately on the dairy farm.

Another important resource for veterinarians in the United States is the Food Animal Residue Avoidance and Depletion Program (FARAD), which is a drug and chemical database that can be accessed by veterinarians to obtain drug residue information. The ultimate goal of this program is to utilize relevant regulatory and pharmacokinetic data to provide producers and veterinarians with food safety information so that drug and chemical residues are minimal in meat and milk products. Information on extralabel drug withdrawal intervals can be obtained from FARAD via the Internet website or through telephone consultation.

Overall, the most effective way to minimize drug residues is through education between the veterinarian and farm manager and between the manager and farm employees. Establishing a valid relationship between the producer and veterinarian should be the first goal, whereby the veterinarian visits the farm regularly, has a thorough knowledge of the diseases and organisms that occur most commonly on the dairy, and establishes written treatment protocols for various diseases that may occur. The veterinarian should also educate the manager and all farm employees on proper drug storage, drug labeling, how to properly administer drugs, identification of treated animals, how to maintain and understand treatment records, and how to establish both meat and milk withdrawal times. Also the proper use of "on-farm" antibiotic screening assays can help reduce the risk of drug residues.

As we move into the future, dairy farms are becoming larger in size. This means larger numbers of cows on one facility and a greater number of employees involved in the dairy industry. We also have newer and more sensitive analytical methods that are capable of rapidly detecting even small concentrations of drugs that might be present in meat or milk samples. Globally, we are seeing a larger and larger number of milk samples tested for residues every year, which is a trend expected to continue as technology improves. As such, scrutiny of milk and milk products is at an all time high, which is expected to further increase in the future. All employees involved in the dairy industry should be reminded that drug residues are a significant public health concern and the milk and milk products get a negative image when reports of drug residue violations become public. It is in the best financial interest of dairy producers to take positive steps toward reducing and eliminating meat and milk residues.

REFERENCES

AHA (2010). Animal health surveillance quarterly report. Available at http://www.animalhealthaustralia.com.au/status/ahsq.cfm. Accessed September 30, 2010.

Barkema HW, Schukken YH, Lam TJ, Beiboer ML, Benedictus G, Brand A (1998). Management practices associated with low, medium, and high somatic cell counts in bulk milk. J Dairy Sci 81:1917–1927.

Baynes RE, Payne M, Jimenez-Martin T, Abdullah AR, Anderson KL, Webb AI, Craigmill A, Riviere JE (2000). Extralabel use of ivermectin and moxidectin in food animals. J Am Vet Med Assoc 217:668–671.

Bennedsgaard TW, Klaas IC, Vaarst M (2010). Reducing use of antimicrobials—experiences from an intervention study in organic dairy herds in Denmark. Livest Sci 131:183–192.

Berge ACB, Moore DA, Besser TE, Sischo WM (2009). Targeting therapy to minimize antimicrobial use in preweaned calves: Effects on health, growth, and treatment costs. J Dairy Sci 92:4707–4714.

Brightling P, Mein GA, Malmo J, Ryan DP (1998). Countdown Downunder: Farm Guidelines for Mastitis Control. Melbourne: Dairy Research and Development Corporation.

Busani L, Graziani C, Franco A, Di Egidio A, Binkin N, Battisti A (2004). Survey of the knowledge, attitudes and practice of Italian beef and dairy cattle veterinarians concerning the use of antibiotics. Vet Rec 155:733–738.

Cazeau PG, Botrel MA, Sala C, Chazel M, Jarrige N, Calavas D (2009). Contexte des prescriptions antibiotiques en filière bovine: résultants de l'enquête Afssa-SNGTV. Bulletin des GTV 49:55–59.

Chicoine AL, Durden DA, MacNaughton G, Dowling PM (2007). Ivermectin use and resulting milk residues on 4 Canadian dairy herds. Can Vet J 48:836–838.

Deyrup CL, Southern KJ, Cornett JA, Shultz CE, Cera DA (2012). Examining the occurrence of residues of flunixin meglumine in cull dairy cows by use of the flunixin cull cow survey. J Am Vet Med Assoc 241:249–253.

EMEA (2010) Committee for medicinal products for veterinary use: Carprofen. Available at http://www.ema.europa.eu/docs/en_GB/document_library/Maximum_Residue_Limits_-_Report/2009/11/WC500011417.pdf. Accessed September 20, 2010.

European Commission (2008). Analysis of residue monitoring data from the Member States in 2008. Available at http://ec.europa.eu/food/food/chemicalsafety/residues/workdoc_2008_en.pdf. Accessed September 28, 2010.

FSIS 2013. National residue program data. Available at http://www.fsis.usda.gov/wps/portal/fsis/topics/data-collection-and-reports/chemistry/residue-chemistry. Accessed April 18, 2014.

Gross SJ, Ryan WG, Ploeger HW (1999). Anthelmintic treatment of dairy cows and its effect on milk production. Vet Rec 144:581–587.

Heuwieser W, Iwersen M, Gossellin J, Drillich M (2010). Survey of fresh cow management practices on dairy cattle on small and large commercial farms. J Dairy Sci 93:1065–1068.

Kissell LW, Baynes RE, Riviere JE, Smith GW (2013). Occurrence of flunixin residues in bovine milk samples from the USA. Food Addit Contam A 30(9):1513–1516.

McEwen SA, Black WD, Meek AH (1991). Antibiotic residue prevention methods, farm management, and occurrence of antibiotic residues in milk. J Dairy Sci 74:2128–2137.

Menéndez González S, Steiner A, Gassner B, Regula G (2010). Antimicrobial use in Swiss dairy farms: Quantification and evaluation of data quality. Prev Vet Med 95:50–63.

Ortman K, Svensson C (2004). Use of antimicrobial drugs in Swedish dairy calves and replacement heifers. Vet Rec 154:136–140.

Østerås O, Edge VL, Martin SW (1999). Determinants of success or failure in the elimination of major mastitis pathogens in selective dry cow therapy. J Dairy Sci 82:1221–1231.

Pol M, Ruegg PL (2007). Treatment practices and quantification of antimicrobial drug usage in conventional and organic dairy farms in Wisconsin. J Dairy Sci 90:249–261.

Raymond MJ, Wohrle RD, Call DR (2006). Assessment and promotion of judicious antibiotic use on dairy farms in Washington State. J Dairy Sci 89:3228–3240.

Sanchez J, Dohoo I, Carrier J, DesCôteaux L (2004). A meta-analysis of the milk-production response after anthelmintic treatment in naturally infected adult dairy cows. Prev Vet Med 63:237–256.

Sawant AA, Sordillo LM, Jayarao BM (2005). A survey on antibiotic usage in dairy herds in Pennsylvania. J Dairy Sci 88:2991–2999.

van Schaik G, Lotem M, Schukken YH (2002). Trends in somatic cell counts, bacterial counts, and antibiotic residue violations in New York State during 1999–2000. J Dairy Sci 85:782–789.

Sischo WM, Kiernan NE, Burns CM, Byler LI (1997). Implementing a quality assurance program using a risk assessment tool on dairy operations. J Dairy Sci 80:777–787.

Smith GW, Davis JL, Tell LA, Webb AI, Riviere JE (2008). Extralabel use of nonsteroidal anti-inflammatory drugs in cattle. J Am Vet Med Assoc 232:697–701.

Talley MR (1999). The national milk safety program and drug residues in milk. Vet Clin Food Anim 15:63–73.

USDA March 2007. Dairy 2007 Part II: Changes in the US Dairy Cattle Industry, 1991–2007. Available at http://www.aphis.usda.gov/animal_health/nahms/dairy/downloads/dairy07/Dairy07_is_PartII_Highlights.pdf. Accessed April 15, 2014.

USDA October 2008. Antibiotic use on US dairy operations, 2002 and 2007. Available at http://www.aphis.usda.gov/animal_health/nahms/dairy/downloads/dairy07/Dairy07_is_AntibioticUse.pdf. Accessed April 15, 2014.

USHSS (2009). Grade "A" pasteurized milk ordinance. Available at http://www.fda.gov/downloads/Food/FoodSafety/Product-SpecificInformation/MilkSafety/NationalConferenceonInterstateMilkShipmentsNCIMSModelDocuments/UCM209789.pdf. Accessed September 9, 2010.

VANTURES (2010). Monitoring of antimicrobial resistance and antibiotic usage in animals in the Netherlands in 2008. July, 2010. Available at http://www.cvi.wur.nl/NR/rdonlyres/2D16DFE4-36C6-4140-BF4C-6F82394ED7EB/110563/MARAN__2008__definitief_corrected.pdf. Accessed September 20, 2010.

Zwald AG, Ruegg PL, Kaneene JB, Warnick LD, Wells SJ, Fossler C, Halbert LW (2004). Management practices and reported antimicrobial usage on conventional and organic dairy farms. J Dairy Sci 87:191–201.

10

RESIDUE AVOIDANCE IN AQUACULTURE PRODUCTION SYSTEMS*

Renate Reimschuessel

Veterinary Laboratory Investigation and Response Network, Center for Veterinary Medicine, FDA, Laurel, MD, USA

10.1 INTRODUCTION

During the 1900s, as chemical and drug use in crop and livestock agriculture increased, so too did concern about residues of chemicals and drugs in the foods we eat. Indeed, in the United States, government agencies such as the U.S. Department of Agriculture (USDA) and U.S. Food and Drug Administration (USFDA) were established during the twentieth century to help regulate food production and conduct inspections to assure the quality of the products (Hilts, 2004). Over the last 50 years, farming has become a more corporate industry with intensive agricultural practices. In developed countries, this meant a greater reliance on fertilizers and pesticides for crops and therapeutics for animals (Conklin, 2009; Lee et al., 2007; Mazoyer and Roudart, 2006). Third world countries have followed this lead and now may actually rely more heavily on such products than farms in developed countries (Little et al., 2008).

*The views expressed in this chapter are those of the author and may not reflect the official policy of the Department of Health and Human Services, the US Food and Drug Administration, or the US Government.

Strategies for Reducing Drug and Chemical Residues in Food Animals: International Approaches to Residue Avoidance, Management, and Testing, First Edition.
Edited by Ronald E. Baynes and Jim E. Riviere.

Aquacultural production has also increased significantly over the last 30 years (FAO, 2006, 2009). Throughout the world, there are many different forms of aquaculture, from subsistence farms in third world countries to larger production systems found in Norway, Canada, and Chile (FAO, 2009). As with land-based farms, there has been an increase in therapeutant use in fish farming and thus greater public concern about the presence of drug residues in farm-raised fish. In addition, reports that farm-raised fish may accumulate more environmental toxins than wild-caught fish (Amberg and Hall, 2008; Hites et al., 2004) brought much media attention to the potential risks of eating farm-raised fish. The risks versus benefits of consuming aquacultured fish have subsequently been examined by a number of authors, both to help understand risks and to find appropriate interventions to reduce risks (Cole et al., 2009; Mozaffarian and Rimm, 2006; Santerre, 2010; Sapkota et al., 2008; Tuomisto and Froyland, 2008).

Residues in farm-raised fish, as a result of either intentional or unintentional exposure, originate from two main sources, the aquatic environment or the feed. This chapter will first provide a broad overview of the potential residues that may be present in aquacultural products. The second section will focus on a recent international adulteration event, melamine in protein sources, as a case study for reducing residues in aquacultural species.

10.2 ENVIRONMENTAL CONTAMINANTS

In most developed countries, clean drinking water is routinely provided to terrestrial livestock. This reduces the risk of animals ingesting high concentrations of pollutants that can be in runoff water from contaminated sites or treated fields. Such is not always the case for aquaculture. Fish farms relying on water from streams can receive and retain many different chemicals, either industrial or agricultural (Jana, 1998; Kemper, 2008; Minh et al., 2006; Schuetze et al., 2008; Turner et al., 1986). In some localities, groundwater contamination with various metals or organic substances can put aquaculture facilities at risk (Adel, 2001; Barnes et al., 2004; IFEN, 2004; Scheidleder Agrath et al., 1999; Zhao et al., 2006). Aquatic species are particularly vulnerable to waterborne pollutants because they ingest water and other aquatic life forms in the water (algae, invertebrates, fish), as well as "breathe" the water via the gills. Some contaminants are absorbed through the gills directly from the water, resulting in residues in fish raised for human consumption (Li et al., 2010). More commonly, however, the route of exposure for contaminants in farm-raised fish is more indirect. Wild fish accumulate environmental pollutants via the food chain (Chu et al., 2000; Kiser et al., 2010; Streit, 1998; Varanasi and Stein, 1991). When those wild fish are used to make fish feeds for aquaculture, the accumulated environmental contaminants become an important source of residues in farmed fish (Berntssen and Lundebye, 2008; Crossland et al., 1987; Tacon and Metian, 2008a).

There are two main groups of environmental contaminants in aquaculture that pose a potential human health threat, persistent organic pollutants (POPs) and metals. Organic pollutants, such as polycyclic aromatic hydrocarbons (PAHs) and organochlorine compounds [dioxins and polychlorinated biphenyls (PCBs)], tend to accumulate in fatty tissues of fish from contaminated environments (Beyer and Biziuk, 2009; Puy-Azurmendi et al., 2010; Ruus et al., 1999; Tsapakis et al., 2010). These compounds biomagnify up the food chain and can reach rather high concentrations in wild fish, especially top predatory species, such as tuna, shark, and swordfish (Froehner et al., 2011; Serrano et al., 2008a; Shaw et al., 2006). The presence of POPs in aquaculture feeds is an unintentional consequence of using fish meal and oil from wild-caught fish (Carlson and Hites, 2005; WHO, 1999). The concentration of POPs in farmed fish has been correlated to the concentrations of POPs in feed (Easton et al., 2002; Jacobs et al., 2002) and to the type of fish grown (Berntssen et al., 2005, 2007; Karl et al., 2003). For example, feeds fed to Atlantic salmon (*Salmo salar*) and rainbow trout (*Oncorhynchus mykiss*) contain a large percent of fat in their muscles, resulting in higher POP exposure levels to these fish species. Also, because salmon and trout are oily fish, in addition to being fed higher levels of contaminated oil in their feed, they accumulate more of the lipid-soluble contaminants than less oily fish such as sea bream (*Dicentrarchus labrax*) and carp (*Cyprinus carpio*).

Residues of POPs, however, are not always higher in farm-raised fish. Johnson et al. found that PCBs and PAHs were present in higher levels in fish from the wild than in hatchery-reared fish (Johnson et al., 2007). Likewise, a study of Mediterranean gilthead sea bream found significantly higher concentrations of PCBs in tissues of the wild fish than in farmed fish from the same area (Serrano et al., 2008b). Additional concerns arise regarding wild fish PAH accumulation from environments that have had major oil spills (Cakirogullari and Secer, 2011; Froehner et al., 2011; Puy-Azurmendi et al., 2010). The use of dispersants to manage these spills (Jung et al., 2009; PWSRCAC, 2010; Ramachandran et al., 2004) may actually cause greater accumulation of PAH in the wild fish swimming near these sites. Food safety concerns about such events as the recent Deepwater Horizon oil spill in the Gulf of Mexico have stimulated new efforts on detecting PAH contaminants in fish (USFDA, 2010a). In cases of environmental spills, consumers would be sure that farm-raised fish would not have had such a direct exposure. However, instead of being the recipient of contaminants, sometimes, fish farms themselves can be point sources for contaminants in environmental marine sediments (Bustnes et al., 2010; Tsapakis et al., 2010). In these cases, however, PAHs were found primarily in the region underneath and immediately surrounding the fish cages, with minimal contamination 1000 m from the production site. Understanding the source of those PAHs, whether in the fish feed or from other farming practices, is important for any efforts to reduce both environmental contamination and potential residues in farmed fish.

Recently, brominated flame retardants, such as polybrominated diphenyl ethers (PBDEs) and hexabromocyclodecane (HBCD), have been identified as residues in a variety of foods, including wild and farmed fish (van Leeuwen et al., 2009; Schecter et al., 2010; Takahashi et al., 2010; Xian et al., 2008). Again, the highest residues in fish feed ingredients are found in the fish oil component. When fish are fed diets containing less fish oils, the residues of these compounds are significantly reduced in their edible tissues (Berntssen et al., 2010). The concentrations found in wild versus farm-raised fish vary with the species examined, and in some cases, wild-caught fish have higher PBDE residues (Hayward et al., 2007). A recent study found that outmigrant juvenile Chinook salmon from the Pacific Northwest had higher PBDE residue levels than their hatchery-raised counterparts (Sloan et al., 2010). This study indicates that these contaminants bioaccumulate in tissues of these fish during the time they are living in the open ocean and estuary rather than during the time they are fed aquaculture feeds.

Using fish oils from marine fish with naturally low levels of dioxins or PCBs has been shown to help reduce the levels of these residues in farmed fish (Isosaari et al., 2005; Lundebye et al., 2004). There is a growing movement to reduce the amount of fish oil used in aquaculture feeds, moving to plant-based oils (Berntssen et al., 2005; Friesen et al., 2008; RAFOA, 2010; Tacon and Metian, 2008b). This trend is based on the desire both to reduce contaminants in products fed to cultured fish and to develop more sustainable aquaculture practices (Li and Hu, 2009; Naylor et al., 2009). Using more plant-based ingredients in fish feeds, however, adds a greater risk for contamination with PAHs and with agricultural pesticides (Berntssen et al., 2010; Tsapakis et al., 2010).

Organochlorine pesticides, such as DDT, dieldrin, toxaphene, and endosulfan, are also environmental contaminants that persist in the aquatic environment. Pesticides are widely used to manage a variety of organisms in many different contexts. Some pesticides are primarily used to control weeds in agriculture, forests, parks, and household yards. Others are used to control invertebrate parasites on livestock and pets, and others are used for controlling larger pests such as rodents in both rural and urban settings. Thus, the introduction of pesticides to those locales is often intentional, but runoff causes contaminated waterways with unintended effects on aquatic biota and background contamination of wild fish (Barbash et al., 1999; Gilliom, 2007; IFEN, 2004; Varanasi and Stein, 1991). Pesticide contamination of groundwater has been identified in multiple aquifers worldwide (Mathys, 1994; Sampat, 2000; Scheidleder Agrath et al., 1999; Zhao et al., 2006). Absorption can be directly from waterborne exposure, but these substances also biomagnify through the food chain, ending up as contaminants in aquaculture feeds (Jacobs et al., 2002; Streit, 1998). As noted previously, more and more producers of aquaculture feeds want to use plant proteins and oils to develop more sustainable and profitable fish feeds (Hasan et al., 2007; Jackson, 2009). With this effort, however, they risk introducing higher levels of pesticides into the feed from terrestrial crop residues (Cabras et al., 1997; Holland et al., 1994; Lentza-Rizos and Avramides, 1995). Terrestrial pesticide use in agriculture has increased greatly during the last 50 years, especially in Asian countries (Little et al., 2008). Also, many of the pesticides marketed in third world countries do not meet international quality standards, which could mean that even more impurities enter the environment when those products are used (EJF, 2003). As more plant feed ingredients are sought for fish feeds, more surveillance for pesticide residues will undoubtedly be needed (FAO, 2001; Sun and Chen, 2008).

The second group of environmental contaminants in aquacultural products is metals. Naturally present in the aquatic environment due to geological activity, metals may also enter waters from man-made sources such as mining and industry. One of the most notable incidences of human toxicity due to metal residues in fish is the Minamata Bay incident in the 1950s (Eto, 1997; Harada, 1995). For over 30 years, a chemical factory released effluent containing methyl mercury into the bay, contaminating sediment and resident wildlife. Many fish died while others bioaccumulated the mercury. When consumed, the contaminated fish poisoned cats, birds, and humans, resulting in thousands of illnesses and fatalities (MOE, 2010). Globally, mercury accumulation in fish not exposed to this type of point source tends to be highest in the older, predatory fish, such as swordfish, tuna, pike, and shark (Evans et al., 2005). One strategy to reduce methyl mercury concentrations in farmed fish is to avoid using fish oils from such predatory species when making fish feeds (Bethune et al., 2006).

Other metals reported in fish filets include cadmium, lead, and arsenic (European Commission, 2008; Foran et al., 2004; Franklin et al., 2005; Hove et al., 2008). A number of recent reports of residues in wild fish indicate that the levels of these metals are generally below those considered harmful, but this can depend on the species examined and the regions they inhabit (Burger and Gochfeld, 2005; Cheung et al., 2008; IPCS-INCHEM, 2010; Storelli, 2009; Wang et al., 2010b). While most metals accumulate in the protein compartment of fish, arsenic is present in relatively high concentrations in fish oil. As a result, fish oil can be a source of arsenic in aquatic feeds (Sloth et al., 2005). The toxicity of arsenic depends greatly on its chemical form, with the inorganic species being more toxic than the organic forms (Francesconi and Edmonds, 1993; Gomez-Caminero et al., 2001; Heikens, 2006; Shiomi, 1994). In fish, the primary form is organic and thus may pose less of a food safety risk than inorganic arsenicals found in water (EFSA, 2009; Foran et al., 2004).

Other emerging environmental contaminants are pharmaceuticals and personal-care products (PPCPs) and man-made nanoparticles (Johnston et al., 2010; Kahru and Dubourguier, 2010). The occurrence of PPCPs in the environment has been recognized as a potential risk for both humans and aquatic life (Halling-Sørensen et al., 1998; Kostich and Lazorchak, 2008; Wang et al., 2010a). Although these have come primarily from human products, they also come from veterinary uses, including aquaculture (Burridge et al., 2010; Khan et al., 2008). Residues of pharmaceuticals, including fluoxetine and sertraline (antidepressants) and their metabolites, diphenhydramine (antihistamine), diltiazem (calcium channel blocker), and carbamazepine (anticonvulsants) used in human medicine, have been detected in edible fish tissues (Brooks et al., 2005; Chu and Metcalfe, 2007; Mehinto et al., 2010; Ramirez et al., 2007; Zhou et al., 2008). Likewise, residues of galaxolide and tonalide (fragrances used in soap) have also been recently reported in fish (Anonymous, 2009a; Leal et al., 2010). As methods are developed to detect these compounds in different matrices, it is likely that more reports of their presence in aquatic life forms will result.

As nanotechnology advances and more nanotech products are used by industry and consumers, it is important to assess any risks that those products pose for the environment or aquatic organisms. Studies have shown that nanoparticles can penetrate the blood–brain barrier of fish and induce toxicity in larvae (Kashiwada, 2006). Waterborne nanoparticles can also localize within the liver after penetrating the gills and inducing the gill enzyme *cyp1a2* and in some cases can have effects on growth (Gao et al., 2011; Handy et al., 2008). Along with potential ecological risks, impacts on the food supply need to be evaluated (Bouwmeester et al., 2009; Das et al., 2009; Tiede et al., 2008). Unfortunately, little is known about the risks of nanoparticles in food matrices, and precautionary principles have been recommended (Myhr and Myskja, 2011). Nanoparticle toxicity has been shown in cells and small aquatic organisms, but how relevant those toxicities are to larger organisms ingesting small quantities is unknown and an area of active research.

In summary, the main source of environmental contaminants in farm-raised fish is from the feeds they eat. Feeds using fish meal and oils from wild fish may contain more organic pollutants or metals than ingredients produced from plants. Pesticide residues from terrestrial crops may, however, be more prevalent in plant oils. Ways to decrease environmental contaminant residues in farm-raised fish include using wild-caught fish from cleaner areas, varying the fish sources depending on season, and extracting contaminants from the oils being used (Laender et al., 2010; Vives et al., 2004). Certainly, monitoring feed ingredients for contaminants is the first step toward reducing feed contamination and residues in the final product (Berntssen and Lundebye, 2008; Loutfy et al., 2007; Usydus et al., 2009).

10.3 DRUG USE AS A SOURCE OF RESIDUES

The previous section described potential sources of residues as the result of environmental contamination, presumably from unintentional exposures. In this section, we will consider intentional exposures, those due to treating fish with drugs. Unlike environmental exposures, in the case of veterinary drugs, the exposures are planned by the attending veterinarian or fish health professional, and efforts to reduce residues rely more on regulation than trying to limit exposure. Veterinary drugs are used to treat infections in farmed fish caused by organisms such as fungi, bacteria, and parasites. Drugs are also used to sedate or anesthetize fish for transport or harvesting. Sometimes, drugs are used as spawning aids and marking devices and to promote growth. Any of these therapeutic compounds could leave residues in a fish intended for human consumption.

Throughout the world, there are different regulatory agencies that oversee the use of veterinary compounds (EMEA, 2010; Fingleton, 2004; Schnick et al., 1999; Subasinghe and Alderman, 2002; USFDA, 2010b). These agencies often interact with global organizations such as the World Health Organization (WHO) and the Food and Agriculture Organization (FAO), which coordinate internationally recognized standards and codes of practice. A Web search of the FAO Fisheries and Aquaculture Department shows 77 fact sheets of regulations regarding aquaculture drug use by countries throughout the world. In general, a common theme in regulations worldwide is their goal to reduce the risk of harmful residues entering the food supply after veterinary use. Although there are still a number of countries that do not control drug use in aquaculture, many developing countries have recently begun to promote regulations for aquaculture along with the infrastructure to enforce those rules (FAO/IAEA, 2010; Lovatelli et al., 2006; Murshed-E-Jahan, et al., 2008; Yamprayoon and Sukhumparnich, 2010).

The classes of drugs used in aquatic veterinary medicine throughout the world are the same as those used in terrestrial, avian, or even human medicine (Alderman and Nichel, 1992). Most of the compounds have a rather extensive history of use in more mainstream veterinary medicine; thus, the patterns of residue accumulation and depletion are most likely known for a number of conventional species. Nevertheless, most regulatory agencies require depletion studies in the aquatic target species prior to approving any use of the specific drug. The information obtained from these studies is analyzed by regulators who set residue limits and withdrawal times to ensure that harmful residues will not be present in edible fish products. That process usually involves some type of risk assessment, exposure assessment, and risk characterization (FAO/WHO, 2006a; Sumner et al., 2004).

Worldwide standardization efforts to reduce veterinary drug residues in foods are being advanced by the Codex Alimentarius Committee on Residues of Veterinary Drugs in Foods (CCRVDF). This WHO/FAO committee seeks to establish internationally acceptable guidelines defining what residue levels are considered acceptable or legally permitted in edible tissue (CODEX, 2008; FAO/WHO, 2006a, b). These levels are called "maximum limit of residues of veterinary drugs" (MRLVD) by Codex, "maximum residue level" (MRL) by the European Medicines Agency (EMEA) (Chapter 4) and Canadian Food Inspection Agency (CFIA), or "tolerances" by the USFDA. In addition to establishing a limit on the amount of residue that may be present in a food, an "acceptable daily intake" (ADI) value may be established. ADIs are an estimate of the amount of residue considered to be without any toxicological hazard or appreciable health risk if ingested daily over a lifetime. Although ADIs and MRLs can vary between countries, there is an international effort to harmonize these values (FAO/WHO, 2004, 2010).

During the drug approval process, withdrawal periods are designated to ensure that the edible tissues will not contain residues above the MRLs or tolerances established by the regulatory agency. MRLs and tolerances are developed by considering the general risk to the consumer and balancing the potential toxicity of the compound with the potential exposure levels. The basic principles used to develop withdrawal periods for terrestrial food animals are those used for fish, but temperature is added to the equation (Chapters 2 and 5). This may be in the form of "degree days" (the °C multiplied by the number of days, e.g., 50 days at 10°C = 500 degree days, as does 25 days at 20°C) (Alderman, 2000; EMEA, 2010; USFDA, 2010b, c). The European Union (EU) regulations also have provisions for a generic withdrawal period of 500 degree days for compounds that are used when no withdrawal period has been set (EMEA, 2009).

Since there are myriad fish species, regulatory authorities have acknowledged the need to deal with uncertainties when it comes to therapeutants and fish. Aquatic animals require veterinary care, and since there are few pharmaceutical companies willing to spend the money to develop the data to support drug approvals, there are few approved drugs for fish. Some countries have exemptions that allow veterinarians to use a drug in a minor species if that drug is approved in a major species. In the EU and Australia, this is referred to as "off-label" drug use, while in the United States and Canada, it is called "extralabel drug use" (EMEA, 2009; Grignon-Boutet et al., 2008; USFDA, 2004, 2010d). In the United States, the Minor Use and Minor Species Animal Health Act of 2004 legislation provides more flexibility for veterinarians prescribing medicines to aquatic animals. The EMEA has adopted similar policies in the EU (EMEA, 2009).

Many countries are also developing provisions for using therapeutic agents in minor species even though those compounds are not approved in a major species. For example, in the United States, there are substances listed by the USFDA as "low regulatory priority." Such substances include sodium chloride, sodium bicarbonate, and urea. These chemicals might be used in conjunction with other drug treatments and could affect the depletion of those other drugs. However, in general, it is believed that use of these compounds poses no concern for public or animal health.

Another category of drug regulation by the USFDA is "indexing." In this case, the agency can add certain drugs to an index of legally marketed but unapproved new animal drugs for use in minor species. Some have voiced concern that such alternative drug use paradigms, if not carefully administered, could increase the risk of residues exceeding MRLs or tolerances in the food products (Lupin, 2009). Thus, developing generic withdrawal periods can be a valuable exercise for regulatory authorities. Knowledge of the pharmacokinetics and depuration patterns of different drugs in different fish species is essential for those establishing such withdrawal periods. A good source of such information in fish is PhishPharm (Reimschuessel et al., 2005; USFDA, 2011), a free online database that contains data from over 450 publications.

Which veterinary drugs then, are most frequently found as violative residues in edible fish products? A survey of the recent literature reveals a short list of compounds including antibiotics (furans, enrofloxacin, oxolinic acid, tetracyclines, sulfadiazine, chloramphenicol), antiparasitics (emamectin, oxfendazole), and dyes (malachite green/leucomalachite green, crystal violet) (CFIA, 2006; EFSA, 2008; Hove et al., 2008; Love et al., 2011; Tittlemier et al., 2007; Turnipseed and Andersen, 2008; VRC, 2005, 2009; Yess, 1991). Of course, only compounds for which one establishes a testing program will show up on a survey. It is thus desirable to develop multiresidue testing programs so that screens can be done, which "cast a wide net" and are more cost-effective.

Antibiotics are one of the most commonly used therapeutants in fish medicine. Of primary concern for potential adverse effects are antimicrobials such as chloramphenicol, which can cause aplastic anemia, and nitrofurans, which are carcinogenic. Many countries have prohibited the use of these antibiotics in aquacultured species (Reimschuessel and Miller, 2006). Another group of antibiotics of concern is the β-lactams (penicillin, amoxicillin, etc.), whose residues could pose problems for people with allergies to those antimicrobials (Reimschuessel, 2008; Sapkota et al., 2008). Finally, there is also increasing concern that antimicrobial residues in fish flesh could cause human gut bacteria or pathogens to develop into "superbugs" resistant to antimicrobials used in human medicine (Miller and Reimschuessel, 2006; Miranda and Rojas, 2007; Serrano, 2005).

To reduce residues of any type of antimicrobial in edible fish tissues, the mode of treatment and the appropriate withdrawal time must be adhered to. With appropriate precautions, residues will be below any MRLs or tolerances established for antimicrobials administered for therapeutic reasons. One additional way to help reduce residues is to regulate drug use by requiring veterinary prescription and oversight. Registering drugs and monitoring sales can also help regulatory agencies track trends in use and make recommendations if violative residues are identified. In addition to prescription regulations, many countries are developing guidelines for judicious use of antimicrobials in order to prevent antimicrobial resistance developing in pathogens and environmental bacteria. In the United States, such guidelines have been proposed by the American Veterinary Medical Association, FAO/WHO, and USFDA (AVMA, 2002, 2010; FAO/WHO, 2006a, b, c; Serrano, 2005; USFDA, 2009a, 2010e). The recommendations in these documents are, in general, similar to guidelines proposed for antimicrobial use in terrestrial animals (CODEX, 2005).

It is possible, however, that antibiotics can be present as contaminants either in the environment or in the feed. Any drug therapy in aquaculture must take into account potential contamination of the environment and downstream effects to reduce any chance of indirect reentry of the drugs into the food chain. This is of concern in terrestrial medicine as well. Kumar et al. (2005) showed that corn, green onion, and cabbage, when grown in manure containing chlortetracycline from treated cattle, accumulate that antibiotic. Chloramphenicol has also recently been identified as naturally occurring in herbs and grass in various parts of the world (Berendsen et al., 2010). Even trucks that haul medicated feed, when inadequately cleaned, can be a source of antimicrobial contamination for other animal feeds (Van Donkersgoed et al., 2010). One other potential source of inadvertent antibiotic administration to fish is the use of plant-based feed ingredients, which are by-products of alternative fuel production. Distillers' grains, a by-product of biofuel manufacture, may contain antimicrobials that are used to reduce bacterial growth while making ethanol. The USFDA has recently undertaken a limited survey to determine the extent and level of antibiotic residues in a limited number of domestic and import samples of distillers' grains (USFDA, 2006, 2009b). Results of this study are not yet published. In general, however, the most likely causes of violative antibiotic residues, especially higher levels, in a fish ready for market, are antimicrobial treatment followed by inadequate withdrawal time.

Similar principles apply to residues of antiparasitics and antifungals, including the prohibited triphenylmethane dyes. Unfortunately, illegal drug use, especially of malachite green and crystal violet, which continues in many parts of the world (Anonymous, 2010b; Tittlemier et al., 2007; UK Food

Standards Agency, 1999), has resulted in import alerts or bans (USFDA, 2009c). Other antiparasitics, such as emamectin and pyrethroids, have been demonstrated in sediments or surrounding crustaceans during/after treatment (Burridge et al., 2010). In the Norwegian 2008 monitoring system, emamectin was identified in 9 of 56 samples, in the 1–9 μg/kg range. The residues detected were well below the EU MRL for emamectin, 100 μg/kg (Hove et al., 2008). No residues for other veterinary compounds, such as antibacterials, anthelmintics, other sea lice agents, and dyes, were detected in that monitoring program. Even fewer drugs were detected in 2009 (Lunestad, 2010). Continued monitoring programs such as this one will greatly help assessing potential exposures and reducing risks to human health.

10.4 MELAMINE ADULTERATION OF AQUACULTURE FEEDS: A CASE STUDY

During 2007, the United States had the largest recall of pet foods due to melamine adulteration of feed ingredients, including those for fish feeds (USFDA, 2007a, 2010f). Melamine, an s-triazine, is used in the manufacture of plastics, textiles, and glues (Anonymous, 1993) and as a flame retardant when combined with cyanuric acid (Anonymous, 2010a). Melamine, because of its ability to form complexes, may actually be useful as a carrier for much more toxic drugs such as chemotherapy agents (Lim and Simanek, 2005; Neerman et al., 2004). Because melamine contains many nitrogen molecules, it has also been used as an explosive stimulant for testing landmine detectors (Viesti et al., 1999, 2005). For the same reason (high nitrogen content), melamine and other s-triazines have been used to adulterate food ingredients to fake the actual protein content. Most tests to identify protein content in foods rely on the Kjeldahl method, which actually measures nitrogen, not protein concentrations (Lynch and Barbano, 1999; Munro and Fleck, 1969). Thus, triazines in products are falsely identified as protein when tested using the Kjeldahl assay.

Melamine and other s-triazines were investigated as potential nonprotein nitrogen (NPN) feed supplements for ruminants in the 1950s–1960s (Altona and MacKenzie, 1964; Clark, 1966; Clark et al., 1965; Hatfield et al., 1959; MacKenzie, 1966; MacKenzie and van Rensburg, 1968; Newton and Utley, 1978). Ruminants, due to their fermentative gastrointestinal systems, contain microorganisms that can degrade s-triazines and substances like urea, releasing nitrogen for absorption and subsequent protein synthesis. Some of these studies reported weight loss, urinary crystals, and mortalities related to supplementing feed with triazines, especially melamine in sheep. In general, the s-triazines have not been widely used as NPN supplements (Loosli and McDonald, 1968).

One of the earliest reports of food product adulteration by melamine was in 1970 in Italy, where 70% of imported meat meal and fish meal were found to contain melamine (Cattaneo and Ceriani, 1988). With the introduction of a surveillance program, this number was reduced to 5% in 1987. Another report demonstrated melamine in potato meal in Germany in the 1980s (Bisaz and Kummer, 1983). Unfortunately, these studies were not widely recognized in the rest of the world, and melamine or other s-triazines were not considered likely food or feed adulterants. Urea and urea-formalin were, however, detected as adulterants in wheat (Folkenberg et al., 1990) and fish meal (Guo et al., 2002). The widespread use of melamine and s-triazines as adulterants was discovered during the 2007 U.S. pet food recall (Dobson et al., 2008; USFDA, 2007a, 2010f). The contamination was not limited to pet food but included fish, poultry, and hog feeds (Barboza and Barrionuevo, 2007; Fraser, 2007a; Reimschuessel et al., 2008; USFDA, 2007b, c, 2009d). Additional contamination incidents involving melamine and triazines were reported in Europe, South Africa, and Asia (Cocchi et al., 2010; González et al., 2009; Reyers, 2007; Yhee et al., 2009). In 2008, the contamination of milk and infant formula in China also caused worldwide recalls of milk products (Barboza, 2008; Bradsher, 2008; Parry, 2008; Qing, 2008; Xin and Stone, 2008). At that time, additional contamination of animal feeds came to light as melamine residues were found in eggs (Anonymous, 2008) and chickens fed soy mislabeled as "organic" (Adams, 2008).

The main feed commodities adulterated in the 2007 pet food incident, wheat gluten and rice protein concentrate, were both actually wheat flour mixed with melamine. The melamine used was a poor-quality scrap form that contained multiple triazines, including cyanuric acid and sometimes ammeline and ammelide (Dobson et al., 2008; USFDA, 2009c). The nephrotoxicity seen in the dogs and cats that consumed these feeds was due to the formation of melamine cyanurate crystals in the kidney, with resulting intratubular obstruction.

Because of the media attention, a variety of animal feeds were screened and, if tested positive for melamine, were also recalled. Many of those products contained only melamine, which, by itself, is not particularly toxic. Large quantities of melamine, however, consumed for long periods of time can induce urinary tract stone formation (Heck and Tyl, 1985; Ogasawara et al., 1995) but not acute renal failure. Thus, it is important to differentiate the contamination as either melamine alone or melamine in combination with cyanuric acid.

Melamine adulteration not associated with clinical disease was also found in fish feeds (Fraser, 2007a; Gutierrez, 2009; USFDA, 2007c, d). During the summer of 2007, in response to concern for public health raised by the contamination of livestock and fish feeds, the Center for Veterinary

Medicine (CVM) began to develop methods for detecting melamine in tissues of animals intended for human consumption. The first studies conducted used four different fish species: trout, salmon, tilapia, and catfish (Andersen et al., 2008). Fish were fed melamine (either alone or in combination with cyanuric acid), and filets were tested to demonstrate the performance of the newly developed methods. Unexpectedly, the control fish filets also contained melamine (0.04–0.12 ppm). When the feeds that the fish had been consuming prior to the study were tested, melamine was identified with trout feed containing 0.5 ppm and salmon feed containing 6.7 ppm (Reimschuessel et al., 2008). Due to those findings, CVM conducted a retrospective analysis of feeds still archived in the laboratory. Melamine was found in four different feeds received in 2006–2007 with concentrations of 1.1, 2.8, 75, and 102 ppm. In addition, archived fish filets from trout obtained in 2006 from a producer who "always used high protein fish feeds" contained 0.32 ppm melamine (Reimschuessel et al., 2010).

Very high melamine concentrations (170 ppm) were also found in shrimp feeds used during a study evaluating triazine residue accumulation in shrimp (Andersen et al., 2008; Karbiwnyk et al., 2009, 2010). This was not an isolated event. A report of several cases of melamine cyanurate crystals in shrimp antennal glands found multiple shrimp feeds containing melamine levels ranging from 112 to 183 ppm (Consumer Affairs, 2007; Lightner et al., 2009).

Considering these findings, the retrospective studies in Iberian pigs and Asian pets (González et al., 2009; Yhee et al., 2009) and the Italian reports of adulterated fish meal (Cattaneo and Ceriani, 1988), melamine was very likely present in fish feeds during the last 20 years, possibly as early as the 1970s. Without the rapid dissemination of such information via the Internet, however, reports of commodity adulteration by compounds high in nitrogen (triazines, urea, biuret) were unfortunately buried in the literature (Bisaz and Kummer, 1983; Cattaneo and Ceriani, 1988; Folkenberg et al., 1990). The common thread in all these cases was the same, economic fraud. The best deterrents to such behavior are adequate surveillance systems that cooperate on a global basis. Thus, there have been calls to test foods for potential adulterants that mimic protein and to develop tests that assess protein directly, rather than relying on the Kjeldahl reaction to test for nitrogen (AOAC, 2009; Fraser, 2007b).

The insidious effect of this long-term adulteration of fish feeds on nutrition research has not been adequately publicized. As mentioned previously, there are numerous studies exploring the use of plant proteins as an alternative for fish meal. The routine test done by researchers to evaluate protein in the test diets or ingredients has been the Kjeldahl reaction. Both the reference feeds and the test feeds could have had false protein data due to melamine adulteration. This may explain many of the conflicting reports about feed utilization after using plant protein-based diets (Kaushik and Seiliez, 2010). Unfortunately, the fact that melamine adulteration was common practice in China likely invalidates most of the fish feed utilization studies conducted over the last 20 years. Particularly disturbing is the fact that, despite the widespread press reports about melamine adulteration of animal feeds and infant formula, most fish nutrition investigators still report only testing their feeds using the Kjeldahl reaction (Bell et al., 2010; El-Husseiny et al., 2008; Emre et al., 2008; Zhou and Yue, 2010). Aquaculture feed prices have increased greatly in the past 5 years, imposing a great financial burden on fish farmers (Anonymous, 2009b; Gaung, 2010; Rana et al., 2009). Some of this increase could be due to the substitution of triazines that were used in the past with real protein. Increased vigilance and surveillance for fraudulent ingredients are essential for the health of the animals, appropriate nutrition, and the safety of the world's food supply.

Increasingly, our food supply is coming from a global market. It is imperative that we monitor the quality of the ingredients used in human food and animal feeds. This requires surveillance systems that detect contaminants that can arise from industrial chemicals, pesticides, metals from both natural and pollution sources, PCPs, pharmaceuticals (human and veterinary), and new products such as nanoparticles. In addition, developing systems that can rapidly trace food/feeds to their source can make it more difficult to successfully adulterate or substitute ingredients. Greater communication cooperation and harmonization of standards between nations via organizations such as the FAO/WHO, and increased quality assurance programs for fish farmers, especially in developing countries, will help this endeavor.

ACKNOWLEDGMENTS

I am grateful to V.M.R. Myers and J. Myers for their help in editing this chapter.

REFERENCES

Adams M (2008). Melamine found contaminating soy meal fed to organic chickens. Available at http://www.naturalnews.com/News_000571_melamine_organic_chickens_China.html. Accessed August 12, 2010.

Adel MM (2001). Effect on water resources from upstream water diversion in the Ganges basin. J Environ Qual 30(2):356–368.

Alderman DJ (2000). Antimicrobial drug use in aquaculture. In: Prescott JF, Baggot JD, Walker RD, editors. Antimicrobial Therapy in Veterinary Medicine. 3rd ed. Ames: Iowa State University Press; p 692–711.

Alderman DJ, Nichel C (1992). Chemotherapy in aquaculture today. In: Michel C, Alderman DJ, editors. Chemotherapy in Aquaculture: From Theory to Reality. Paris: Office International des Epizooties; p 3–24.

Altona RE, Mackenzie HI (1964). Observations on cyanuric acid as a source of non-protein nitrogen for sheep. J S Afr Vet Med Assoc 35:203–205.

Amberg S, Hall TE (2008). Communicating risks and benefits of aquaculture: A content analysis of U.S. newsprint representations of farmed salmon. J World Aquac Soc 39:143–157.

Andersen WC, Turnipseed SB, Karbiwnyk CM, Lee RH, Clark SB, Rowe WD, Madson MR, Miller KE (2007). Quantitative and confirmatory analyses of crystal violet (gentian violet) and brilliant green in fish, Laboratory Information Bulletin #4395. Silver Spring, MD: U.S. Food and Drug Administration.

Andersen WC, Turnipseed SB, Karbiwnyk CM, Clark SB, Madson MR, Gieseker CM, Miller RA, Rummel NG, Reimschuessel R (June 25, 2008). Determination and confirmation of melamine residues in catfish, trout, tilapia, salmon, and shrimp by liquid chromatography with tandem mass spectrometry. J Agric Food Chem 56(12):4340–4347.

Anonymous (1993). Evaluation and control of the risks of existing substances: Melamine. Official Journal of the European Communities, L84 1. Available at http://www.inchem.org/documents/sids/sids/108781.pdf. Accessed August 6, 2010.

Anonymous (2008). Melamine found in Hong Kong eggs. Available at http://www.upi.com/Top_News/2008/10/26/Melamine-found-in-Hong-Kong-eggs/UPI-45391225035386/. Accessed August 12, 2010.

Anonymous (2009a). First National Pilot Study Finds Human Medications and Personal Care Products in Fish Caught in Several U.S. Waterways. Baylor University Press. Available at http://www.baylor.edu/pr/news.php?action=story&story=57434. Accessed August 6, 2010.

Anonymous (2009b). Feed prices put Vietnamese farmers out of business. Available at http://www.allaboutfeed.net/news/feed-prices-put-vietnamese-farmers-out-of-business-3322.html. Accessed August 12, 2010.

Anonymous (2010a). Melamine derivatives. Available at http://www.specialchem4p-olymers.com/tc/Melamine-Flame-Retardants/index.aspx?id=4004. Accessed August 6, 2010.

Anonymous (2010b). Veterinary Residues Committee announcement on the results of the brand naming survey of dyes, tetracyclines, nitrofurans, chloramphenicol, florfenicol, antimicrobials and quinolones in imported farmed fish-282696. http://www.vmd.gov.uk/vrc/Reports/Brand_naming_report_2009.pdf. Accessed September 14, 2010.

AOAC (2009). Fraud in food and drug safety. Available at http://www.aoac.org/meetings1/123rd_annual_mtg/science_sessions.htm#FRAUD. Accessed August 9, 2010.

AVMA (2002). Judicious and prudent antimicrobial drug use principles for food fish veterinarians (revised 2008). Am Vet Med Assoc. Available at http://www.avma.org/issues/policy/jtua_fish.asp. Accessed August 14, 2010.

AVMA (2010). AVMA testifies in support of judicious use of antimicrobials in food animals. Available at http://www.avma.org/press/releases/100714_antimicrobials_food_animals.asp. Accessed August 14, 2010.

Barbash JE, Thelin GP, Kolpin DW, Gilliom RJ 1999. Distribution of major herbicides in ground water of the United States. Available at http://water.usgs.gov/nawqa/pnsp/pubs/wrir984245/text.html. Accessed April 25, 2014.

Barboza D (2008). Hong Kong finds tainted Chinese fish feed. http://www.nytimes.com/2008/11/13/world/asia/13china.html?_r=1. Accessed April 6, 2014.

Barboza D, Barrionuevo A (2007). Filler in animal feed is open secret in China. Available at http://www.nytimes.com/2007/04/30/business/worldbusiness/30food.html?pagewanted=1&_r=1. Accessed August 6, 2010.

Barnes KK, Christenson SC, Kolpin DW, Focazio MJ, Furlong ET, Zaugg SD, Meyer MT, Barber LB (2004). Pharmaceuticals and other organic wastewater contaminants within a leachate plume downgradient of a municipal landfill. Ground Water Monit Remediat 24:119–126.

Bell JG, Pratoomyot J, Strachan F, Henderson RJ, Fontanillas R, Hebard A, Hunter D, Tocher T, Guy DR (2010). Growth, flesh adiposity and fatty acid composition of Atlantic salmon (Salmo salar) families with contrasting flesh adiposity: Effects of replacement of dietary fish oil with vegetable oils. Aquaculture 306: 225–232.

Berendsen B, Stolker L, de Jong J, Nielen M, Tserendorj E, Sodnomdarjaa R, Cannavan A, Elliott C (2010). Evidence of natural occurrence of the banned antibiotic chloramphenicol in herbs and grass. Anal Bioanal Chem 397(5):1955–1963.

Berntssen MH, Lundebye AK (2008). Environmental contaminants in farmed fish and potential consequences for seafood safety. In: Lie Ø, editor. Improving Farmed Fish for the Consumer. Cambridge, England: Woodhead Publishing; p 39–64.

Berntssen MH, Lundebye AK, Torstensen BE (2005). Reducing the levels of dioxins and dioxin-like PCB's in farmed Atlantic salmon by dietary substitution of fish oils with vegetable oils; a life cycle study. Aqua Nutr 11:219–232.

Berntssen MH, Giskegjerde TA, Rosenlund G, Torstensen BE, Lundebye AK (2007). Predicting World Health Organization toxic equivalency factor dioxin and dioxin-like polychlorinated biphenyl levels in farmed Atlantic salmon (Salmo salar) based on known levels in feed. Environ Toxicol Chem 26(1):13–23.

Berntssen MH, Julshamn K, Lundebye AK (2010). Chemical contaminants in aquafeeds and Atlantic salmon (Salmo salar) following the use of traditional—Versus alternative feed ingredients. Chemosphere 78(6):637–646.

Bethune C, Seierstad SL, Seljeflot I, Johansen O, Arnesen H, Meltzer HM, Rosenlund G, Frøyland L, Lundebye AK (2006). Dietary intake of differently fed salmon: A preliminary study on contaminants. Eur J Clin Invest 36(3):193–201.

Beyer A, Biziuk M (2009). Environmental fate and global distribution of polychlorinated biphenyls. Rev Environ Contam Toxicol 201:137–158.

Bisaz R, Kummer A (1983). Determination of 2,4,6-triamino-1,3,5-triazine (melamine) in potato proteins. Mitt Lebensmittelunters Hyg 74(1):74–79.

Bouwmeester H, Dekkers S, Noordam MY, Hagens WI, Bulder AS, de Heer C, ten Voorde SE, Wijnhoven SW, Marvin HJ, Sips AJ (2009). Review of health safety aspects of nanotechnologies in food production. Regul Toxicol Pharmacol 53(1):52–62.

Bradsher K (2008). China begins inquiry into tainted baby formula. Available at http://www.nytimes.com/2008/09/13/world/asia/13milk.html?hp. Accessed August 6, 2010.

Brooks BW, Chambliss CK, Stanley JK, Ramirez A, Banks KE, Johnson RD, Lewis RJ (2005). Determination of select antidepressants in fish from an effluent-dominated stream. Environ Toxicol Chem 24(2):464–469.

Burger J, Gochfeld M (2005). Heavy metals in commercial fish in New Jersey. Environ Res 99(3):403–412.

Burridge L, Weis JS, Cabello F, Pizarro J, Bostick K (2010). Chemical use in salmon aquaculture: A review of current practices and possible environmental effects. Aquaculture 306:7–23.

Bustnes JO, Lie E, Herzke D, Dempster T, Bjørn PA, Nygård T, Uglem I (2010) Salmon farms as a source of organohalogenated contaminants in wild fish. Environ Sci Technol 44(22):8736–8743.

Cabras P, Angioni A, Garau VL, Melis M, Pirisi, FM, Karim, M, Minelli EV (1997). Persistence of insecticide residues in olives and olive oil. Agric Food Chem 45(6):2244–2247.

Cakirogullari GC, Secer S (2011). Seasonal variation of organochlorine contaminants in bonito (*Sarda sarda* L. 1758) and anchovy (*Engraulis encrasicolus* L. 1758) in Black Sea region, Turkey. Chemosphere 85(11):1713–1718.

Carlson DL, Hites RA (2005). Polychlorinated biphenyls in salmon and salmon feed: Global differences and bioaccumulation. Environ Sci Technol 39(19):7389–7395.

Cattaneo P, Ceriani L (1988). Melamine in animal meals. Tec Molitoria 39:28–32.

CFIA (2006). Interim Guidelines for the Presence of Malachite Green (MG) and Leucomalachite Green (LMG) in Aquaculture Fish Products. Canadian Food Inspection Agency. Available at http://www.inspection.gc.ca/english/fssa/fispoi/commun/20060329e.shtml. Accessed August 6, 2010.

Cheung KC, Leung HM, Wong MH (2008). Metal concentrations of common freshwater and marine fish from the Pearl River Delta, south China. Arch Environ Contam Toxicol 54(4):705–715.

Chu S, Metcalfe CD (2007). Analysis of paroxetine, fluoxetine and norfluoxetine in fish tissues using pressurized liquid extraction, mixed mode solid phase extraction cleanup and liquid chromatography–tandem mass spectrometry. J Chromatogr A 1163:112–118.

Chu F-LE, Soudant P, Cruz-Rodriguez LA, Hale RC (2000). PCB uptake and accumulation by oysters (Crassostrea virginica) exposed via a contaminated algal diet. Mar Environ Res 50:217–221.

Clark R. (1966). Melamine crystalluria in sheep. J S Afr Vet Med Assoc 37:349–351.

Clark R, Barratt EL, Kellerman JH (1965). A comparison between nitrogen retention from urea, biuret, triuret and cyanuric acid by sheep on low protein roughage diet. J S Afr Vet Med Assoc 36:79–80.

Cocchi M, Vascellari M, Gallina A, Agnoletti F, Angeletti R, Mutinelli F (2010). Canine nephrotoxicosis induced by melamine-contaminated pet food in Italy. J Vet Med Sci 72(1):103–107.

CODEX (2005). Code of Practice to Minimize and Contain Antimicrobial Resistance. Codex Alimentarius Commission. Available at www.codexalimentarius.net/download/standards/10213/CXP_061e.pdf. Accessed August 12, 2010.

CODEX (2008). Maximum residue limits (MRL) for veterinary drugs in foods. Updated at the 31st Session of the Codex Alimentarius Commission; July 2008. Available at www.codexalimentarius.net/vetdrugs/data/MRL2_e_2009.pdf. Accessed August 8, 2010.

Cole DW, Cole R, Gaydos SJ, Gray J, Hyland G, Jacques ML, Powell-Dunford N, Sawhney C, Au WW (2009). Aquaculture: Environmental, toxicological, and health issues. Int J Hyg Environ Health 212(4):369–377.

Conklin PK (2009). A Revolution Down on the Farm: The Transformation of American Agriculture since 1929. Kentucky: University Press of Kentucky.

Consumer Affairs (2007). Two US companies recall melamine-tainted feed. Available at http://www.consumeraffairs.com/news04/2007/05/melamine_us.html. Accessed August 12, 2010.

Crossland NO, Bennett D, Wolff CJM (1987). Fate of 2,5,4'-trichlorobiphenyl in outdoor ponds and its uptake via the food chain compared with direct uptake via the gills in grass carp and rainbow trout, Ecotoxicol Environ Saf 13(2):225–238.

Das M, Saxena N, Dwivedi PD (2009). Emerging trends of nanoparticles application in food technology: Safety paradigms. Nanotoxicology 3:10–18.

Dobson RLM, Motlagh S, Quijano M, Cambron RT, Baker TR, Pullen AM, Regg BT, Bigalow-Kern AS, Vennard T, Fix A, Reimschuessel R, Overmann G, Shan Y, Daston GP (2008). Identification and characterization of toxicity of contaminants in pet food leading to an outbreak of renal toxicity in cats and dogs. Toxicol Sci 106:251–262.

Easton MD, Luszniak D, Von der Geest E (2002). Preliminary examination of contaminant loadings in farmed salmon, wild salmon and commercial salmon feed. Chemosphere 46(7):1053–1074.

EFSA (2008). Report for 2008 on the Results from the Monitoring of Veterinary Medicinal Product Residues and Other Substances in Food of Animal Origin in the Member States. Parma: European Food Safety Authority. Available at http://www.efsa.europa.eu/en/scdocs/scdoc/1559.htm. Accessed August 9, 2010.

EFSA (2009). Scientific Opinion on Arsenic in Food. Parma: European Food Safety Authority. Available at http://www.efsa.europa.eu/en/scdocs/scdoc/1351.htm. Accessed August 9, 2010.

EJF (2003). What's Your Poison—Health Threats Posed by Pesticides in Developing Countries. London: Environmental Justice Foundation. Available at http://www.ejfoundation.org/pdf/whats_your_poison.pdf. Accessed August 9, 2010.

El-Husseiny OM, El Din G, Abdul-Aziz M, Mabroke RS (2008). Effect of mixed protein schedules combined with choline and betaine on the growth performance of Nile tilapia (*Oreochromis niloticus*) Aquac Res 39:291–300.

EMEA (2009). Policy for Classification and Incentives for Veterinary Medicinal Products Indicated for Minor Use Minor Species (MUMS) Limited Markets. London: European Medicines Agency. Available at http://www.ema.europa.eu/htms/vet/availability/availability.htm; http://www.ema.europa.eu/docs/en_GB/document_library/Scientific_guideline/2009/10/WC500004581.pdf. Accessed August 9, 2010.

EMEA (2010). London: European Medicines Agency. Available at http://www.ema.europa.eu/ema/index.jsp?curl=pages/home/Home_Page.jsp&murl=&mid=. Accessed August 9, 2010.

Emre Y, Sevgili H, Sanl M (2008). A preliminary study on the utilization of hazelnut meal as a substitute for fish meal in diets of European sea bass (*Dicentrarchus labrax* L.). Aquac Res 39:324–328.

Eto K (1997). Pathology of Minamata disease. Toxicol Pathol 25(6):614–623. Erratum in: Toxicol Pathol (1998) 26(6):741.

European Commission (2008). Results of national residue monitoring plans: 2008. Available at http://ec.europa.eu/food/food/chemicalsafety/residues/control_en.htm. Accessed August 9, 2010.

Evans MS, Muir D, Lockhart WL, Stern G, Ryan M, Roach P (2005). Persistent organic pollutants and metals in the freshwater biota of the Canadian Subarctic and Arctic: An overview. Sci Total Environ 351–352:94–147.

FAO (2001). Technical Guidelines for Responsible Fisheries—Aquaculture Development: 1. Good Aquaculture Feed Manufacturing Practice. Rome: Food and Agriculture Organization of the United Nations. Available at http://www.fao.org/docrep/005/y1453e/y1453e00.htm. Accessed August 9, 2010.

FAO (2006). State of World Aquaculture 2006, FAO Fisheries Technical Paper No. 500. Rome: Food and Agriculture Organization of the United Nations. Available at http://www.fao.org/docrep/009/a0874e/a0874e00.htm. Accessed April 6, 2014.

FAO (2009). State of World Aquaculture 2008. Rome: Food and Agriculture Organization of the United Nations. Available at http://www.fao.org/docrep/011/i0250e/i0250e00.htm. Accessed August 9, 2010.

FAO/IAEA (2010). FAO/IAEA Technical co-operation projects joint FAO/IAEA programme of nuclear techniques in food and agriculture. Available at http://www-naweb.iaea.org/nafa/fep/field-projects-fep.html. Accessed August 12, 2010.

FAO/WHO (2004). Technical workshop on residues of veterinary drugs without ADI/MRL met in Bangkok, Thailand from 24th to 26th August 2004. Rome: Food and Agriculture Organization of the United Nations/World Health Organization. Available at http://www.fao.org/docrep/008/y5723e/y5723e00.htm. Accessed August 9, 2010.

FAO/WHO (2006a). Codex Alimentarius Commission 29. Rome: Food and Agriculture Organization of the United Nations/World Health Organization. Available at http://www.codexalimentarius.net/web/index_en.jsp. Accessed August 9, 2010.

FAO/WHO (2006b). Codex Alimentarius Manual 19 Edition-English. Rome: Food and Agriculture Organization of the United Nations/World Health Organization. Available at http://www.codexalimentarius.net/web/procedural_manual.jsp. Accessed August 9, 2010.

FAO/WHO (2006c). Antimicrobial use in aquaculture and antimicrobial resistance. Joint expert consultation on antimicrobial use in aquaculture and antimicrobial resistance. Seoul, Republic of Korea, June 13–16, 2006. Geneva: Food and Agriculture Organization of the United Nations/World Organisation for Animal Health formerly International Epizootic Office/World Health Organization. Available at ftp://ftp.fao.org/ag/agn/food/aquaculture_rep_13_16june2006.pdf. Accessed August 9, 2010.

FAO/WHO (2010). Joint FAO/WHO expert committee on food additives (JECFA). Seventy-second meeting, February 16–25, 2010. Rome: Food and Agriculture Organization of the United Nations/World Health Organization. Available at http://www.who.int/foodsafety/chem/summary72_rev.pdf. Accessed August 9, 2010.

Fingleton J (2004). Legislation for Veterinary Drugs Control, legal papers on line # 38. Rome: FAO. Available at http://www.fao.org/legal/prs-ol/lpo38.pdf. Accessed August 9, 2010.

Folkenberg J, Nelson R, Snider S (1990). Spiked wheat—Schuler Grain Co. uses additive to artificially boost the protein content of its wheat. FDA Consumer. Available at http://findarticles.com/p/articles/mi_m1370/is_n10_v24/ai_9246932. Accessed August 9, 2010.

Foran JA, Hites RA, Carpenter DO, Hamilton C, Amos M, Schwager SJ (2004). A survey of metals in tissues of farmed Atlantic and wild Pacific salmon. Environ Toxicol Chem 23(9):2108–2110.

Francesconi KA, Edmonds JS (1993). Arsenic in the sea. Oceanogr Mar Biol Annu Rev 31:111–151.

Franklin NM, Glover CN, Nicol JA, Wood CM (2005). Calcium/cadmium interactions at uptake surfaces in rainbow trout: Waterborne versus dietary routes of exposure. Environ Toxicol Chem 24(11):2954–2964.

Fraser S (2007a). Uniscope and Tembec voluntary recall feed ingredients. Available at http://www.aquafeed.com/read-article.php?id=1926. Accessed August 9, 2010.

Fraser S (2007b). FAO and WHO urge all countries to be more vigilant about food safety. Available at http://www.aquafeed.com/read-article.php?id=2072. Accessed August 9, 2010.

Friesen EN, Ikonomou MG, Higgs DA, Ang KP, Dubetz C (2008). Use of terrestrial based lipids in aquaculture feeds and the effects on flesh organohalogen and fatty acid concentrations in farmed Atlantic salmon. Environ Sci Technol 42(10):3519–3523.

Froehner S, Maceno M, Machado KS (2011). Predicting bioaccumulation of PAHs in the trophic chain in the estuary region of Paranagua, Brazil. Environ Monit Assess 174(1–4):135–145.

Gao J, Wang Y, Folta KM, Krishna V, Bai W, Indeglia P, Georgieva A, Nakamura H, Koopman B, Moudgil B (2011). Polyhydroxy fullerenes (fullerols or fullerenols): beneficial effects on growth and lifespan in diverse biological models. PLoS One 6(5):e19976.

Gaung JS (2010). Bran, bean paste costs raise fish feed prices. Myanmar Times 26(513). Available at http://www.mmtimes.com/2010/business/513/b51304.html. Accessed August 12, 2010.

Gilliom RJ (2007). Pesticides in the nation's streams and ground water. Environ Sci Technol 41(10):3408–3414.

Gomez-Caminero A, Howe P, Hughes M, Kenyon E, Lewis DR, Moore M, Ng JC, Aitio A, Becking G (2001). Environmental Health Criteria 224 Arsenic and Arsenic Compounds. 2nd ed. Geneva: World Health Organization.

González J, Puschner B, Pérez V, Ferreras MC, Delgado L, Muñoz M, Pérez C, Reyes LE, Velasco J, Fernández V, García-Marín JF (2009). Nephrotoxicosis in Iberian piglets subsequent to exposure to melamine and derivatives in Spain between 2003 and 2006. J Vet Diagn Invest 21(4):558–563.

Grignon-Boutet R, Ireland MJ, Adewoye L, Mehrotra M, Russell S, Alexander I (2008). Health Canada's policy on extra-label drug use in food-producing animals in Canada. Can Vet J 49(7):689–693.

Guo J, Yang H, Liu D, Zhu T, Qian F (2002). Distinguishing ureaformaldehyde polymers in fish meal. Cereal Feed Ind 8:49.

Gutierrez D (2009). China admits melamine widely used in animal feed. Available at http://www.naturalnews.com/025836.html. Accessed August 9, 2010.

Halling-Sørensen B, Nors Nielsen S, Lanzky PF, Ingerslev F, Holten Lützhøft HC, Jørgensen SE (1998). Occurrence, fate and effects of pharmaceutical substances in the environment—A review. Chemosphere 36(2):357–393.

Handy RD, Henry TB, Scown TM, Johnston BD, Tyler CR (2008). Manufactured nanoparticles: Their uptake and effects on fish—A mechanistic analysis Ecotoxicology 17(5):396–409.

Harada M (1995). Minamata disease: Methylmercury poisoning in Japan caused by environmental pollution. Crit Rev Toxicol 25(1):1–24.

Hasan MR, Hecht T, De Silva SS, Tacon AGJ, editors (2007). Study and Analysis of Feeds and Fertilizers for Sustainable Aquaculture Development, FAO Fisheries Technical Paper No. 497. Rome: Food and Agriculture Organization of the United Nations. Available at http://www.fao.org/docrep/011/a1444e/a1444e00.htm. Accessed August 12, 2010.

Hatfield EE, Garrigus US, Forbes RM, Neumann AL, Gaither W (1959). Biuret—A source of NPN for ruminants. J Anim Sci 18:1208–1219.

Hayward D, Wong J, Krynitsky AJ (2007). Polybrominated diphenyl ethers and polychlorinated biphenyls in commercially wild caught and farm-raised fish fillets in the United States. Environ Res 103(1):46–54.

Heck HD, Tyl RW (1985). The induction of bladder stones by terephthalic acid, dimethyl terephthalate, and melamine (2,4,6-triamino-s-triazine) and its relevance to risk assessment. Regul Toxicol Pharmacol 5(3):294–313.

Heikens A (2006). Arsenic contamination of irrigation water, soil and crops in Bangladesh: Risk implications for sustainable agriculture and food safety in Asia. Available at ftp://ftp.fao.org/docrep/fao/009/ag105e/ag105e00.pdf. Accessed August 12, 2010.

Hilts PK (2004). Protecting America's Health: The FDA, Business, and One Hundred Years of Regulation. Chapel Hill: The University of North Carolina Press.

Hites RA, Foran JA, Carpenter DO, Hamilton MC, Knuth BA, Schwager SJ (2004). Global assessment of organic contaminants in farmed salmon. Science 303:226–229.

Holland PT, Hamilton D, Ohlin B, Skidmore MW (1994). Effects of storage and processing on pesticide residues in plant products (technical report). Pure Appl Chem 66(2):335–356.

Hove HT, Lunestad BT, Hannisdal R, Julshamn K (2008). Monitoring Program for Residues of Therapeutic Agents, Illegal Substance, Pollutants and Other Undesirables in Farmed Fish. Bergen: National Institute of Nutrition and Seafood Research. Available at www.nifes.no/file.php?id=1122. Accessed August 9, 2010.

IFEN (2004). The state of groundwater resources in France quantitative and qualitative aspects. Institut Français de l'Environnement. Available at http://www.stats.environnement.developpement-durable.gouv.fr/fileadmin/publications/les_dossiers/2004/et43 anglais1.pdf. Accessed August 9, 2010.

IPCS-INCHEM (2010). Arsenic. International program on chemical safety. Available at http://www.inchem.org/documents/jecfa/jecmono/v024je08.htm. Accessed August 9, 2010.

Isosaari P, Lundebye AK, Ritchie G, Lie O, Kiviranta H, Vartiainen T (2005). Dietary accumulation efficiencies and biotransformation of polybrominated diphenyl ethers in farmed Atlantic salmon (Salmo salar). Food Addit Contam 22(9):829–837.

Jackson A. (2009). The continuing demand for sustainable fishmeal and fish oil in aquaculture diets. Int Aqua Feed Sept–Oct:23–36.

Jacobs MN, Covaci A, Schepens P (2002). Investigation of selected persistent organic pollutants in farmed Atlantic salmon (Salmo salar), salmon aquaculture feed, and fish oil components of the feed. Environ Sci Technol 36(13):2797–2805.

Jana BB (1998). Sewage-fed aquaculture: The Calcutta model. Ecol Eng 11:73–85.

Johnson LL, Ylitalo GM, Sloan CA, Anulacion BF, Kagley AN, Arkoosh MR, Lundrigan TA, Larson K, Siipola M, Collier TK (2007). Persistent organic pollutants in outmigrant juvenile chinook salmon from the Lower Columbia Estuary, USA. Sci Total Environ 374(2–3):342–366.

Johnston BD, Scown TM, Moger J, Cumberland SA, Baalousha M, Linge K, van Aerle R, Jarvis K, Lead JR, Tyler CR (2010). Bioavailability of nanoscale metal oxides TiO(2), CeO(2), and ZnO to fish. Environ Sci Technol 44(3):1144–1151.

Jung JH, Yim UH, Han GM, Shim WJ (2009). Biochemical changes in rockfish, Sebastes schlegeli, exposed to dispersed crude oil. Comp Biochem Physiol C Toxicol Pharmacol 150(2):218–223.

Karbiwnyk CM, Andersen WC, Turnipseed SB, Storey JM, Madson MR, Miller KE, Gieseker CM, Miller RA, Rummel NG, Reimschuessel R. (April 1, 2009). Determination of cyanuric acid residues in catfish, trout, tilapia, salmon and shrimp by liquid chromatography–tandem mass spectrometry. Anal Chim Acta 637(1–2):101–111.

Karbiwnyk CM, Williams RR, Andersen WC, Turnipseed SB, Madson MR, Miller KE, Reimschuessel R (December 2010). Bioaccumulation of cyanuric acid in edible tissues of shrimp following experimental feeding. *Food Addit Contam Part A: Chem Anal Control Expo Risk Assess* 27(12):1658–1664.

Kahru A, Dubourguier HC (2010). From ecotoxicology to nanoecotoxicology. Toxicology 269(2–3):105–119.

Karbiwnyk CM, Andersen W, Turnipseed SB, Storey JM, Madson MR, Gieseker CM, Miller RA, Rummel NG, Miller KE, Reimschuessel R (2008). Determination of cyanuric acid residues in catfish, trout, tilapia, salmon and shrimp by liquid chromatography-tandem mass spectrometry. Anal Chim Acta 637:101–111.

Karl H, Kuhlmann H, Ruoff U (2003). Transfer of PCDDs and PCDFs into the edible parts of farmed rainbow trout, Oncorhynchus mykiss (Walbaum), via feed. Aquac Res 34:1009–1014.

Kashiwada S (2006). Distribution of nanoparticles in the see-through medaka (Oryzias latipes). Environ Health Perspect 114(11):1697–1702.

Kaushik SJ, Seiliez I (2010). Protein and amino acid nutrition and metabolism in fish: Current knowledge and future needs. Aquac Res 41:322–332.

Kemper N (2008). Veterinary antibiotics in the aquatic and terrestrial environment. Ecol Indic 8(1);1–13.

Khan SJ, Roser DJ, Davies CM, Peters GM, Stuetz RM, Tucker R, Ashbolt NJ (2008). Chemical contaminants in feedlot wastes: Concentrations, effects and attenuation. Environ Int 34(6):839–859.

Kiser T, Hansen J, Kennedy B (2010). Impacts and pathways of mine contaminants to bull trout (Salvelinus confluentus) in an Idaho watershed. Arch Environ Contam Toxicol 59(2):301–311.

Kostich MS, Lazorchak JM (2008). Risks to aquatic organisms posed by human pharmaceutical use. Sci Total Environ 389(2–3):329–339.

Kumar K, Gupta SC, Baidoo SK, Chander Y, Rosen CJ (2005). Antibiotic uptake by plants from soil fertilized with animal manure. J Environ Qual 34(6): 2082–2085.

Laender FD, Oevelen DV, Frantzen S, Middelburg JJ, Soetaert K (2010). Seasonal PCB bioaccumulation in an arctic marine ecosystem: A model analysis incorporating lipid dynamics, food-web productivity and migration. Environ Sci Technol 44(1):356–361.

Leal LH, Vieno N, Temmink H, Zeeman G, Buisman CJN (2010). Occurrence of xenobiotics in gray water and removal in three biological treatment systems. Environ Sci Technol 44(17):6835–6842. Erratum in Environ Sci Technol 2011;45(1):351.

Lee LS, Carmosini N, Sassman SA, Dion HM, Sepulveda MS (2007). Agricultural contributions of antimicrobials and hormones on soil and water quality. In: Sparks DE, editor. Advances in Agronomy. San Diego: Academic Press; 93:p 1–68.

Lentza-Rizos C, Avramides EJ (1995). Pesticide residues in olive oil. Rev Environ Contam Toxicol 141:111–134.

Li D, Hu X (2009). Fish and its multiple human health effects in times of threat to sustainability and affordability: Are there alternatives? Asia Pac J Clin Nutr 18(4):553–563.

Li Q, Zhang X, Yan C (2010). Polycyclic aromatic hydrocarbon contamination of recent sediments and marine organisms from Xiamen Bay, China. Arch Environ Contam Toxicol 58(3):711–721.

Lightner DV, Pantoja CR, Redman RM, Hasson KW, Menon JP (2009). Case reports of melamine-induced pathology in penaeid shrimp fed adulterated feeds. Dis Aquat Org 86:107–112.

Lim J, Simanek EE (2005). Toward the next-generation drug delivery vehicle: Synthesis of a dendrimer with four orthogonally reactive groups. Mol Pharm 2(4):273–277.

Little DC, Milwain GK, Price C (2008). Pesticide contamination in farmed fish: Assessing risks and reducing contamination. In: Lie Ø, editor. Improving Farmed Fish for the Consumer. Cambridge: Woodhead Publishing Limited; p 71–91.

Loosli JK, McDonald IW (1968). Nonprotein Nitrogen in the Nutrition of Ruminants, FAO Agricultural Studies No. 73. Rome: Food and Agriculture Organization of the United Nations. Available at http://www.fao.org/DOCREP/004/AC149E/AC149E00.HTM. Accessed August 9, 2010.

Loutfy N, Fuerhacker M, Tundo P, Raccanelli S, Tawfic AM (2007). Monitoring of polychlorinated dibenzo-p-dioxins and dibenzofurans, dioxin-like PCBs and polycyclic aromatic hydrocarbons in food and feed samples from Ismailia city, Egypt. Chemosphere 66:1962–1970.

Lovatelli A, Phillips MJ, Arthur JR, Yamamoto K (2006). The future of mariculture: A regional approach for responsible development in the Asia-Pacific region, FAO/NACA Regional Workshop, March 7–11, 2006, Guangzhou, China. Rome: Food and Agriculture Organization of the United Nations/World Health Organization. Available at http://www.fao.org/docrep/011/i0202e/i0202e00.htm. Accessed August 9, 2010.

Love DC, Rodman S, Neff RA, Nachman KE (2011). Veterinary drug residues in seafood inspected by the European Union, United States, Canada, and Japan from 2000 to 2009. Environ Sci Technol;45(17):7232–7240.

Lundebye A, Berntssen MHG, Lie O, Ritchie G, Isosaari P, Kiviranta H, Vartiainen T (2004). Dietary uptake of dioxins (PCDD/PCDFs) and dioxin-like PCBs in Atlantic Salmon (Salmo salar). Aquac Nutr 10:199–207.

Lunestad BT (2010). Report 2009: Analysis of drug residues and undesirable substances in farmed fish. Available at http://www.nifes.no/index.php?page_id=&article_id=3442&lang_id=2. Accessed August 12, 2010.

Lupin HM (2009). Human health aspects of drug and chemical use in aquaculture. Options Méditerranéennes, A / no. 86, 2009 The use of veterinary drugs and vaccines in Mediterranean aquaculture. Available at ressources.ciheam.org/om/pdf/a86/00801065.pdf. Accessed August 12, 2010.

Lynch JM, Barbano DM (1999). Kjeldahl nitrogen analysis as a reference method for protein determination in dairy products. J AOAC Int 82(6):1389–1398.

MacKenzie HI (1966). Melamine for sheep. J S Afr Vet Med Assoc 37:153–157.

MacKenzie HI, van Rensburg IBJ (1968). Ammelide and Ammeline as non-protein nitrogen supplements for sheep. J S Afr Vet Med Assoc 39:41–45.

Mathys W (1994). Pesticide pollution of groundwater and drinking water by the processes of artificial groundwater enrichment or coastal filtration: Underrated sources of contamination. Zentralbl Hyg Umweltmed 196(4):338–359.

Mazoyer M, Roudart L (2006). A History of World Agriculture: From the Neolithic age to the Current Crisis. New York: Monthly Review Press.

Mehinto AC, Hill EM, Tyler CR (2010). Uptake and biological effects of environmentally relevant concentrations of the nonsteroidal anti-inflammatory pharmaceutical diclofenac in rainbow trout (Oncorhynchus mykiss). Environ Sci Technol 44(6):2176–2182.

Miller RA, Reimschuessel R (2006). Epidemiologic cutoff values for antimicrobial agents against Aeromonas salmonicida isolates determined by frequency distributions of minimal inhibitory concentration and diameter of zone of inhibition data. Am J Vet Res 67(11):1837–1843.

Minh NH, Minh TB, Kajiwara N, Kunisue T, Iwata H, Viet PH, Tu NP, Tuyen BC, Tanabe S (2006). Contamination by polybrominated diphenyl ethers and persistent organochlorines in catfish and feed from Mekong River Delta, Vietnam. Environ Toxicol Chem 25(10):2700–2708.

Miranda CD, Rojas R (2007). Occurrence of florfenicol resistance in bacteria associated with two Chilean salmon farms with different history of antibacterial usage. Aquaculture 266:39–46.

MOE (2010). Minamata Disease: The History and Measures. Ministry of the Environment Government of Japan. Available at http://www.env.go.jp/en/chemi/hs/minamata2002/ch2.html. Accessed August 12, 2010.

Mozaffarian D, Rimm EB (2006). Fish intake, contaminants, and human health: Evaluating the risks and the benefits. JAMA 296(15):1885–1899. Erratum in: JAMA 2007 297(6):590.

Munro HN, Fleck A (1969). Analysis of tissues and body fluids for nitrogenous constituents. In: Munro HN editor. Mammalian Protein Metabolism, III. New York: Academic Press; p 423–525.

Murshed-E-Jahan K, Beveridge MCM, Brooks AC (2008). Impact of long-term training and extension support on small-scale carp polyculture farms of Bangladesh. J World Aquac Soc 39(4):441–453.

Myhr AI, Myskja BK (2011). Precaution or integrated responsibility approach to nanovaccines in fish farming? A critical appraisal of the UNESCO precautionary principle. Nanoethics 5(1):73–86.

Naylor RL, Hardy RW, Bureau DP, Chiu A, Elliott M, Farrell AP, Forster I, Gatlin DM, Goldburg RJ, Hua K, Nichols PD (2009). Feeding aquaculture in an era of finite resources. Proc Natl Acad Sci USA 106(36):15103–15110. Erratum in: Proc Natl Acad Sci USA 2009 106(42):18040.

Neerman MF, Chen HT, Parrish AR, Simanek EE (2004). Reduction of drug toxicity using dendrimers based on melamine. Mol Pharm 1(5):390–393.

Newton GL, Utley PR (1978). Melamine as a dietary nitrogen source for ruminants. J Anim Sci 47:1338–1344.

Ogasawara H, Imaida K, Ishiwata H, Toyoda K, Kawanish, T, Uneyama C, Hayashi S, Takahashi M, Hayashi Y (1995). Urinary bladder carcinogenesis induced by melamine in F344 male rats: Correlation between carcinogenicity and urolith formation. Carcinogenesis 16(11):2773–2777.

Parry J (2008). China's tainted milk scandal spreads around world. Br Med J. 337:a1890. Available at http://www.bmj.com/cgi/content/full/337/oct01_1/a1890. Accessed August 11, 2010.

Puy-Azurmendi E, Navarro A, Olivares A, Fernandes D, Martínez E, López de Alda M, Porte C, Cajaraville MP, Barceló D, Piña B (2010). Origin and distribution of polycyclic aromatic hydrocarbon pollution in sediment and fish from the biosphere reserve of Urdaibai (Bay of Biscay, Basque country, Spain). Mar Environ Res 70(2):142–149.

PWSRCAC (2010). Dispersants Prince William sound regional citizens' advisory council. Available at http://pwsrcac.info/dispersants/. Accessed August 11, 2010.

Qing T (2008). FDA announces recalls of more melamine tainted Chinese products. The Epoch Times [internet]. Available at http://www.theepochtimes.com/n2/content/view/4876/. Accessed August 11, 2010.

RAFOA (2010). Researching alternatives to fish oils in aquaculture. Available at http://www.rafoa.stir.ac.uk/project_publications.html. Accessed August 11, 2010.

Ramachandran SD, Hodson PV, Khan CW, Lee K (2004). Oil dispersant increases PAH uptake by fish exposed to crude oil. Ecotoxicol Environ Saf 59(3):300–308.

Ramirez AJ, Mottaleb MA, Brooks BW, Chambliss CK (2007). Analysis of pharmaceuticals in fish using liquid chromatography-tandem mass spectrometry. Anal Chem 79(8):3155–3163.

Rana KJ, Siriwardena S, Hasan MR (2009). Impact of rising feed ingredient prices on aquafeeds and aquaculture production. FAO Fisheries and Aquaculture Technical Paper 541. Available at http://www.fao.org/docrep/012/i1143e/i1143e00.htm. Accessed August 9, 2010.

Reimschuessel R (2008). Assessing the human health implications of new drugs used in fish farming. In: Lie Ø, editor. Improving Farmed Fish for the Consumer. Cambridge: Woodhead Publishing; p 128–156.

Reimschuessel R, Miller RA (2006). Antimicrobial drug use in aquaculture. In: Giguere S, Prescott JF, Baggot JD, Walker RD, Dowling T, editors. Antimicrobial Therapy in Veterinary Medicine. 4th ed. Ames: Blackwell Publishing; p 593–606.

Reimschuessel R, Stewart L, Squibb E, Hirokawa K, Brady T, Brooks B, Shaikh B, Hodsdon C (2005). Phish-Pharm—A searchable database of pharmacokinetics data in fish. Am Assoc Pharma Sci J 7(2):E288–E327. Available at http://www.aapsj.org/view.asp?art=aapsj070230; http://www.fda.gov/AnimalVeterinary/ScienceResearch/ToolsResources/Phish-Pharm/default.htm. Accessed August 11, 2010.

Reimschuessel R, Gieseker C, Miller RA, Rummel N, Ward J, Boehmer J, Heller D, Nochetto C, De Alwis H, Bataller N, Andersen W, Turnipseed SB, Karbiwnyk CM, Satzger RD, Crowe J, Reinhard MK, Roberts JF, Witkowski M (2008). Evaluation of the renal effects of experimental feeding of melamine and cyanuric acid to fish and pigs. Am J Vet Res 69:1217–1228.

Reimschuessel R, Evans E, Andersen WC, Karbiwnyk CM, Turnipseed SB, Mayer TD, Nochetto C, Rummel NG, Gieseker CM (2010). Residue depletion of melamine and cyanuric acid in catfish and rainbow trout following oral administration. J Vet Pharmacol Ther 33:172–182.

Reyers F (2007). Melamine-contaminated pet food. The South African experience. Veterinary News, May 8–12.

Ruus A, Ugland KI, Espeland O, Skaare JU (1999). Organochlorine contaminants in a local marine food chain from Jarfjord, Northern Norway, Mar Environ Res 48(2):131–146.

Sampat P (2000). Groundwater shock: The polluting of the world's major freshwater stores. World Watch Jan–Feb:10-22.

Santerre CR (2010). The risks and benefits of farmed fish. J World Aquac Soc 41(2): 250–257.

Sapkota A, Sapkota AR, Kucharski M, Burke J, McKenzie S, Walker P, Lawrence R (2008). Aquaculture practices and potential human health risks: Current knowledge and future priorities. Environ Int 34(8):1215–1226.

Schecter A, Haffner D, Colacino J, Patel K, Päpke O, Opel M, Birnbaum L (2010). Polybrominated diphenyl ethers (PBDEs) and hexabromocyclodecane (HBCD) in composite U.S. food samples. Environ Health Perspect 118(3):357–362.

Scheidleder Agrath J, Winkler G, Stärk U, Koreimann C, Gmeiner C, Gravesen P, Leonard J, Elvira M, Nixon S, Casillas J, Lack TJ (1999). Technical report No 22 Groundwater quality and quantity in Europe—Data and basic information. Copenhagen, European Environment Agency. Available at http://www.eea.europa.eu/publications/TEC22. Accessed August 9, 2010.

Schnick RA, Alderman DJ, Armstrong R, Le Gouvello R, Ishihara S, Lacierda EC, Percival S, Roth M (1999). Worldwide aquaculture drug and vaccine registration progress. Available at http://www.aquanic.org/aquadrugs/publications/world_drug_progress_9-20-99.htm. Accessed August 11, 2010.

Schuetze A, Heberer T, Juergensen S (2008). Occurrence of residues of the veterinary drug malachite green in eels caught downstream from municipal sewage treatment plants. Chemosphere 72(11):1664–1670.

Serrano PH (2005). Responsible use of antibiotics in aquaculture, FAO Fisheries Technical Paper No. 469. Rome: Food and Agriculture Organization of the United Nations. Available at ftp://ftp.fao.org/docrep/fao/009/a0282e/a0282e00.pdf. Accessed August 11, 2010.

Serrano R, Barreda M, Blanes MA (2008a). Investigating the presence of organochlorine pesticides and polychlorinated biphenyls in wild and farmed gilthead sea bream (*Sparus aurata*) from the Western Mediterranean sea. Mar Pollut Bull 56(5):963–972.

Serrano R, Blanes MA, López FJ (2008b). Biomagnification of organochlorine pollutants in farmed and wild gilthead sea bream (Sparus aurata) and stable isotope characterization of the trophic chains. Sci Total Environ 389(2–3):340–349.

Shaw SD, Brenner D, Berger ML, Carpenter DO, Hong C-S, Kannan K (2006). PCBs, PCDD/Fs, and organochlorine pesticides in farmed Atlantic salmon from Maine, eastern Canada and Norway, and wild salmon from Alaska. Environ Sci Technol 40:5347–5354.

Shiomi K (1994). Arsenic in marine organisms: Chemical forms and toxicological aspects. In: Nriagu JO, editor. Arsenic in the Environment: Part II, Human Health and Ecosystem Effects. New York: John Wiley & Sons, Inc.; p 261–282.

Sloan CA, Anulacion BF, Bolton JL, Boyd D, Olson OP, Sol SY, Ylitalo GM, Johnson LL (2010). Polybrominated diphenyl ethers in outmigrant juvenile Chinook salmon from the lower Columbia River and Estuary and Puget Sound, Washington. Arch Environ Contam Toxicol 58(2):403–414.

Sloth JJ, Larsen EH, Julshamn K (2005). Survey of inorganic arsenic in marine animals and marine certified reference materials by anion exchange high-performance liquid chromatography-inductively coupled plasma mass spectrometry. J Agric Food Chem 53(15):6011–6018.

Storelli MM (2009). Intake of essential minerals and metals via consumption of seafood from the Mediterranean Sea. J Food Prot 72(5):1116–1120.

Streit B (1998). Bioaccumulation of contaminants in fish. EXS 86:353–387.

Subasinghe R, Alderman D (2002). Basic overview of the regulatory procedures for authorization of veterinary medicines with emphasis on residues in food animal species. Available at ftp://ftp.fao.org/docrep/fao/004/AC343E/AC343E00.pdf. Accessed August 9, 2010.

Sumner J, Ross T, Ababouch L (2004). Application of risk assessment in the fish industry, FAO Fisheries Technical Paper No. 442. Rome: Food and Agriculture Organization of the United Nations. Available at http://www.fao.org/docrep/007/y4722e/y4722e00.HTM. Accessed August 12, 2010.

Sun F, Chen HS (2008). Monitoring of pesticide chlorpyrifos residue in farmed fish: Investigation of possible sources. Chemosphere 71(10):1866–1869.

Tacon AG, Metian M (2008a). Aquaculture feed and food safety. Ann N Y Acad Sci 1140:50–59.

Tacon AG, Metian M (2008b). Global overview on the use of fish meal and fish oil in industrially compounded aquafeeds: Trends and future prospects. Aquaculture 285:146–158.

Takahashi S, Oshihoi T, Ramu K, Isobe T, Ohmori K, Kubodera T, Tanabe S (2010). Organohalogen compounds in deep-sea fishes from the western North Pacific, off-Tohoku, Japan: Contamination status and bioaccumulation profiles. Mar Pollut Bull 60(2):187–196.

Tiede K, Boxall AB, Tear SP, Lewis J, David H, Hassellov M (2008). Detection and characterization of engineered nanoparticles in food and the environment. Food Addit Contam Part A Chem Anal Control Expo Risk Assess 25(7):795–821.

Tittlemier SA, Van de Riet J, Burns G, Potter R, Murphy C, Rourke W, Pearce H, Dufresne G (2007). Analysis of veterinary drug residues in fish and shrimp composites collected during the Canadian Total Diet Study, 1993–2004. Food Addit Contam Part A 24:14–20.

Tsapakis M, Dakanali E, Stephanou EG, Karakassis I (2010). PAHs and n-alkanes in Mediterranean coastal marine sediments: Aquaculture as a significant point source. J Environ Monit 12:958–963.

Tuomisto J, Froyland L (2008). The risks and benefits of consumption of farmed fish. In: Lie Ø, editor. Improving Farmed Fish for the Consumer. Cambridge: Woodhead Publishing; p 3–27.

Turner JWD, Sibbald RR, Hemens J (1986). Chlorinated secondary domestic sewage effluent as a fertilizer for marine aquaculture: III. Assessment of bacterial and viral quality and accumulation of heavy metals and chlorinated pesticides in cultured fish and prawns. Aquaculture 53(2):157–168.

Turnipseed SA, Andersen WC (2008). Veterinary drug residues. In: Barcelo D, editor. Food Contaminants and Residue Analysis, Volume 51 (Comprehensive Analytical Chemistry). Amsterdam: Elsevier Science: p 307–338.

UK Food Standards Agency (1999). Statement on Surveillance for Malachite Green and Leucomalachite Green in Farmed Fish. London: Committee on toxicity of chemicals in food, consumer products and the environment. Available at http://archive.food.gov.uk/dept_health/pdf/mala.pdf. Accessed August 11, 2010.

USFDA (2004). Minor use/minor species act 2004. Available at http://frwebgate.access.gpo.gov/cgi-bin/getdoc.cgi?dbname=108_cong_public_laws&docid=f:publ282.108.pdf. Accessed August 12, 2010.

USFDA (2006). FY 2006 nationwide survey of distillers grains for aflatoxins. Available at http://www.fda.gov/AnimalVeterinary/Products/AnimalFoodFeeds/Contaminants/ucm050480.htm. Accessed August 12, 2010.

USFDA (2007a). Melamine pet food recall of 2007. Available at http://www.fda.gov/animalveterinary/safetyhealth/recallswithdrawals/ucm129575.htm. Accessed August 12, 2010.

USFDA (2007b). Press announcements: Joint update: FDA/USDA update on tainted animal feed. Available at http://www.fda.gov/NewsEvents/Newsroom/PressAnnouncements/2007/ucm108901.htm. Accessed August 12, 2010.

USFDA (2007c). Press announcements: Tembec and Uniscope voluntary recall feed ingredients, FDA asks feed manufacturers to avoid ingredients containing melamine. Available at http://www.fda.gov/NewsEvents/Newsroom/PressAnnouncements/2007/ucm108925.htm. Accessed August 12, 2010.

USFDA (2007d). Press announcements: FDA's safety/risk assessment on interim melamine and melamine analogues is available. Available at http://www.fda.gov/NewsEvents/Newsroom/PressAnnouncements/2007/ucm108922.htm. Accessed August 12, 2010.

USFDA (2009a). Judicious use of antimicrobials for aquatic veterinarians. Available at http://www.fda.gov/AnimalVeterinary/SafetyHealth/AntimicrobialResistance/JudiciousUseofAntimicrobials/ucm095473.htm. Accessed August 12, 2010.

USFDA (2009b). FY 2010 nationwide survey of distillers grains for antibiotic residues September 30, 2009. Available at http://www.fda.gov/AnimalVeterinary/Products/AnimalFoodFeeds/Contaminants/ucm190907.htm. Accessed August 12, 2010.

USFDA (2009c). Questions and answers on FDA's import alert on farm-raised seafood from China. Available at http://www.fda.gov/Food/FoodSafety/Product-SpecificInformation/Seafood/ucm119105.htm#q3. Accessed August 12, 2010.

USFDA (2009d). Melamine pet food recall—frequently asked questions. Available at http://www.fda.gov/animalveterinary/safetyhealth/RecallsWithdrawals/ucm129932.htm. Accessed August 12, 2010.

USFDA (2010a). Gulf of mexico oil spill update information current as of Wednesday, August 11, 2010. Available at http://www.fda.gov/Food/FoodSafety/Product-SpecificInformation/Seafood/ucm210970.htm. Accessed August 12, 2010.

USFDA (2010b). Aquaculture. Available at http://www.fda.gov/AnimalVeterinary/DevelopmentApprovalProcess/Aquaculture/default.htm. Accessed August 12, 2010.

USFDA (2010c). Approved drugs. Available at http://www.fda.gov/AnimalVeterinary/DevelopmentApprovalProcess/Aquaculture/ucm132954.htm. Accessed August 12, 2010.

USFDA (2010d). Minor use/minor species. Available at http://www.fda.gov/AnimalVeterinary/DevelopmentApprovalProcess/MinorUseMinorSpecies/default.htm. Accessed August 12, 2010.

USFDA (2010e). The judicious use of medically important antimicrobial drugs in food-producing animals, draft guidance 209. Available at http://www.fda.gov/downloads/AnimalVeterinary/GuidanceComplianceEnforcement/GuidanceforIndustry/UCM216936.pdf. Accessed August 11, 2010.

USFDA (2010f). The melamine story. Available at http://www.fda.gov/AnimalVeterinary/ResourcesforYou/AnimalHealthLiteracy/ucm215253.htm. Accessed August 12, 2010.

USFDA (2011). Phish-pharm listing. Available at http://www.fda.gov/AnimalVeterinary/ScienceResearch/ToolsResources/Phish-Pharm/default.htm. Accessed March 20, 2013.

Usydus Z, Szlinder-Richert J, Polak-Juszczak L, Komar K, Adamczyk M, Malesa-Ciecwierz M, Ruczynska W (2009). Fish products available in Polish market—Assessment of the nutritive value and human exposure to dioxins and other contaminants. Chemosphere 74(11):1420–1428.

Van Donkersgoed J, Sit D, Gibbons N, Ramogida C, Hendrick S (2010). Drug residues recovered in feed after various feedlot mixer truck cleanout procedures. J Food Prot 73(1):75–80.

Van Leeuwen SP, van Velzen MJ, Swart CP, van der Veen I, Traag WA, de Boer J (2009). Halogenated contaminants in farmed salmon, trout, tilapia, pangasius, and shrimp. Environ Sci Technol 43(11):4009–4015.

Varanasi U, Stein JE (1991). Disposition of xenobiotic chemicals and metabolites in marine organisms. Environ Health Perspect 90:93–100.

Viesti G, Cinausero M, Cufaro-Petroni N, D'Erasmo G, Fabris D, Fioretto E, Fioretto E, Fonte R, Lunardon M, Lazzizzera I, Nardulli G, Nebbia G, Palomba, M, Pantaeo A, Pappalardo L, Pesente S, Prati P, Prete G, Reito S, Sartori A, Tecchiolli G, Zavatarelli S, Filippini V (1999). The EXPLODET project: Advanced nuclear techniques for humanitarian demining. Nucl Instrum Methods Phys Res A 422:918–921.

Viesti G, Pesente S, Nebbia G, Lunardon M, Sudac D, Nad K, Blagus S, Valkovic V (2005). Detection of hidden explosives by using tagged neutron beams: Status and perspectives. Nucl Instrum Methods Phys Res B 241:748–752.

Vives I, Grimalt JO, Catalan J, Rosseland BO, Battarbee RW (2004). Influence of altitude and age in the accumulation of organochlorine compounds in fish from high mountain lakes. Environ Sci Technol 38(3):690–698.

VRC (2005). Annual report on surveillance for veterinary residues in food in the UK 2005, Veterinary Medicines Directorate. Surrey: The Veterinary Residues Committee. Available at http://www.vmd.gov.uk/vrc/reports/vrcar2005.pdf. Accessed August 12, 2010.

VRC (2009). Annual report on surveillance for veterinary residues in food in the UK, 2008, Veterinary Medicines Directorate. Surrey: Veterinary Residues Committee. Available at http://www.vmd.gov.uk/vrc/Reports/annual.htm. Accessed August 12, 2010.

Wang L, Ying GG, Zhao JL, Yang XB, Chen F, Tao R, Liu S, Zhou LJV (2010a). Occurrence and risk assessment of acidic pharmaceuticals in the Yellow River, Hai River and Liao River of north China. Sci Total Environ 408(16):3139–3147.

Wang Y, Chen P, Cui R, Si W, Zhang Y, Ji W (2010b). Heavy metal concentrations in water, sediment, and tissues of two fish species (*Triplohysa pappenheimi, Gobio hwanghensis*) from the Lanzhou section of the Yellow River, China. Environ Monit Assess 165(1–4):97–102.

WHO (1999). Food safety issues associated with products from aquaculture. Report of a Joint FAO/NACA(WHO Study Group), WHO Technical Report Series 883. Geneva: World Health Organization. Available at http://www.helid.desastres.net/en/d/Jwho73e/. Accessed August 12, 2010.

Xian Q, Ramu K, Isobe T, Sudaryanto A, Liu X, Gao Z, Takahashi S, Yu H, Tanabe S (2008). Levels and body distribution of polybrominated diphenyl ethers (PBDEs) and hexabromocyclododecanes (HBCDs) in freshwater fishes from the Yangtze River, China. Chemosphere 71(2):268–276.

Xin, H, Stone R (2008). Chinese probe unmasks high-tech adulteration with melamine. Science 322(5906):1310–1311.

Yamprayoon J, Sukhumparnich K (2010). Thai aquaculture: Achieving quality and safety through management and sustainability. J World Aquac Soc 41(2):274–280.

Yess NJ (1991). FDA monitoring program. Residues in foods—1990. J Assoc Off Anal Chem. 74(5):121A–141A.

Yhee JY, Brown CA, Yu CH, Kim JH, Poppenga R, Sur JH (2009). Retrospective study of melamine/cyanuric acid-induced renal failure in dogs in Korea between 2003 and 2004. Vet Pathol 46(2):348–354.

Zhao G, Xu Y, Han G, Ling B (2006). Biotransfer of persistent organic pollutants from a large site in China used for the disassembly of electronic and electrical waste. Environ Geochem Health 28(4):341–351.

Zhou Q, Yue Y (2010). Effect of replacing soybean meal with canola meal on growth, feed utilization and haematological indices of juvenile hybrid tilapia, *Oreochromis niloticus x Oreochromis aureus*. Aquac Res 41(7):982–990.

Zhou SN, Oakes KD, Servos MR, Pawliszyn J (2008). Application of solid-phase microextraction for in vivo laboratory and field sampling of pharmaceuticals in fish. Environ Sci Technol 42(16):6073–6079.

11

RESIDUE AVOIDANCE IN SMALL RUMINANT PRODUCTION SYSTEMS

Kevin Anderson and Reha Azizoglu

Department of Population Health and Pathobiology, College of Veterinary Medicine, North Carolina State University, Raleigh, NC, USA

11.1 INTRODUCTION

Drug use and regulation in sheep and goats, or small ruminants, varies considerably among countries and is influenced by the number and importance of small ruminants in the respective countries. In the United States, for example, sheep and goats are considered "minor" species (Haley, 2006; US Department of Health & Human Services, Food and Drug Administration, 2010b). "Minor species," in the United States, refers to animals other than cattle, horses, swine, chickens, turkeys, dogs, and cats (Fajt, 2001; US Department of Health & Human Services, Food and Drug Administration, 2010b). As of September 5, 2001, sheep have been considered minor species in the United States for collection of data for new animal drug approvals (Fajt, 2001; US Department of Health & Human Services, Food and Drug Administration, 2010b). Although the European Union (EU) does not have a "legislative" definition, the EU considers "sheep (meat animals)" a major species and goats, by default, a minor species (European Medicines Agency, 2010). Australia and New Zealand have not usually used the terminology of

Strategies for Reducing Drug and Chemical Residues in Food Animals: International Approaches to Residue Avoidance, Management, and Testing, First Edition.
Edited by Ronald E. Baynes and Jim E. Riviere.

"major" and "minor" species. However, Australia and New Zealand do have large sheep populations, and sheep are important livestock in these countries. For instance, populations of about 100 and 27 million sheep have been reported in Australia and New Zealand, respectively (Adams, 2005; Ashley-Jones, 2008). Based upon numbers, sheep could be considered a more "major" species in Australia and New Zealand, whereas goats may better fit a designation of a "minor" species.

For other countries, designation of "major" and "minor" could be based upon the populations of the animals in a country, with countries with large populations considering the species as major. The world leader in both sheep and goat numbers is China, with reported populations of 171 million sheep and 196 million goats (Field and Taylor, 2008). Other countries with large sheep populations (more than about 10 million) include developed countries such as Italy, United Kingdom, Russia, Spain, Turkey, Brazil, Argentina, South Africa, Australia, and New Zealand and developing countries such as Algeria, Morocco, Nigeria, Sudan, Iran, Syria, and Pakistan (McKellar, 2006). Obviously, sheep would be a more important species in these countries.

A significant problem in drug use for sheep and goats can be the lack of "approved" or "authorized" or "licensed" or "tested" drugs for use in these species. There are fewer "labeled" (or "registered" or "authorized" or "approved," depending upon the language used in the country of concern) drugs available for sheep and goats, as compared to cattle. This is largely a result of financial considerations, as the overall economic significance of the sheep and goat industries can be much less when compared to cattle, and drug development, approval, and marketing costs are quite high. This leaves the situation where much of the drug use in small ruminants is, of necessity, "extralabel" or "off-label." Specifically, in the United States, "extralabel" refers to the use of a drug other than exactly as indicated "per label." This includes everything, from a species or class of animals other than as stated on the label to a route, dose, frequency or use in a manner other than as exactly stated on the label (Fajt, 2001; US Department of Health & Human Services, Food and Drug Administration, 2010d). The user, then, must assure that the drug is administered without resulting in residues of meat and milk. It may be common, in "extra-" or "off-label" use, to administer drugs licensed or labeled for other food animal species or to use drugs labeled for sheep in goats. These so-called "extralabel" uses impose specific requirements and conditions on the user. In the United Kingdom, as in the United States and Canada (Lanthier, 2010), a "veterinarian–client relationship" is a prerequisite for such "off-label use," with the requirement that the animals be under the clinical care of a veterinary surgeon (i.e., veterinarian) before drugs are prescribed for such "off-label use" (Matthews, 2009). Also, in the United

Kingdom, as in the United States, the regulatory authorities consider all goats to be "food-producing" animals, not pets (Matthews, 2009). When used extralabel, the veterinarian is required to keep appropriate records and to specify adequate meat and milk withholding times in order to prevent any residues. Interestingly, in the United Kingdom, all drugs that are not specifically licensed for goats reportedly are used with "a mandatory 7-day withholding time for milk and 28-day withholding time for meat," although it has been questioned whether, in some cases, these times may be adequate (Matthews, 2009).

In the EU, the "cascade system (off-label use)" is the name for the process that allows "off-label" use of drugs in food animals. The cascade limits a veterinarian treating food-producing animals to prescribing medicines that contain only substances in products "authorized" for use in food-producing species (Matthews, 2009; Subasinghe, 2002). Of course, this use must be accompanied by observing adequate meat/milk withdrawal periods by the veterinarian. There are several restrictions in the system. Regardless of species, products can only be used by the same route for which they are authorized (Subasinghe, 2002). Additionally, off-label use is restricted to use on a "small number" of animals, although it may be difficult to define what a "small number" is (Subasinghe, 2002).

Gastrointestinal (GI) parasitism is a major problem for both sheep and goats and accounts for a significant proportion of drug use in these species. Especially in humid areas, the most important challenge for the goat industry comes from internal parasites, and this is a problem largely due to anthelmintic resistance of parasites, particularly *Haemonchus contortus* (Sahlu et al., 2009). It has been reported that over 60% of U.S. sheep producers consider stomach/intestinal nematodes as an important concern and that 75% of southeast sheep producers considered stomach/intestinal nematodes as a major concern (Miller, 2010). Parasites (or "worms" as they were described) accounted for 90% of deaths in goats submitted to Kentucky diagnostic laboratories (Miller, 2010). A study on anthelmintic resistance on sheep and goat farms in the southeastern United States showed that anthelmintic resistance overall and multiple anthelmintic resistance of *H. contortus* were widespread in the southeast (Howell et al., 2008). It is clear that parasitism is a significant problem for most small ruminant producers, especially in humid areas.

The Food Animal Residue Avoidance Databank (FARAD) calls for residue avoidance information provide an indication of the types of drugs used extralabel in sheep and goats in the United States. Data in Table 11.1 from the FARAD for the 2-year period from January 1, 2005, to January 1, 2007, indicated a total of 339 calls relative to sheep and goat drug use (Fitzgerald and FARAD, personal communication). Calls relative to antiparasitics were most common for goats, with the queries including 45 on ivermectins and

TABLE 11.1 FARAD calls on sheep and goats (January 1, 2005 to January 1, 2007)

	Goats		Sheep	
	Number	Percent	Number	Percent
Antiparasitic	99	43.2	31	28.2
Antimicrobials	79	34.5	35	31.8
Other drugs/chemicals	15	6.6	18	16.4
Anti-inflammatory agents and pain relief	13	5.7	14	12.7
Vitamins, minerals, nutrients	11	4.8	7	6.4
General information	9	3.9	2	1.8
Insecticide–pesticide–pyrethroid	3	1.3	3	2.7
Total	229	100	110	100

Fitzgerald and FARAD, personal communication.

moxidectin, 31 on benzimidazoles, 8 on drugs in the levamisole–morantel–pyrantel group, 12 on anticoccidial drugs, and 3 on other antiparasitic drugs. Queries were received on a wide variety of antimicrobials, with questions on β-lactams (15), sulfonamides (16), oxytetracyclines (10), fluoroquinolones (7), and several other antimicrobials (Fitzgerald and FARAD, personal communication). Similar patterns were observed for calls on drugs used in sheep.

This chapter considers drug use in sheep and goats in the United States, EU, and Australasia. A major concern in drug use in small ruminants is avoiding meat and milk residues following use, especially extralabel use, of drugs in small ruminants. It is important to note that sheep meat is the "most internationally traded" of meat products, so meeting standards in various countries is important (McKellar, 2006). A major focus of the chapter will be on strategies intended to reduce drug and chemical residues in food products from sheep and goats.

11.2 PROPHYLACTIC USE OF MAJOR DRUG CLASSES (E.G., ANTIBIOTICS, ANTIPARASITICS) IN GOAT AND SHEEP PRODUCTION SYSTEMS IN THE EU, UNITED STATES, AND AUSTRALASIA

Overall, it can be difficult to get good estimates on drug use for prophylactic versus therapeutic purposes in animal agriculture (Sarmah et al., 2006). For the United States, values have ranged from one estimate of 14% of antibiotic use attributed to efforts to improve feed efficiency and enhance growth to another estimate indicating that 70% of antibiotic use was for nontherapeutic uses (Sarmah et al., 2006). Another published (Avery et al., 2008) reference states that the American Animal Health Institute reported that 4.5% of

TABLE 11.2 "Commonly used drugs" with an indication of use for "prevention" or "control"

Drug	Species	Use
Ammonium chloride	Sheep	Prevent urinary calculi/urolithiasis
Ammonium molybdate	Sheep	Prevent copper toxicosis
Amprolium	Sheep, goat	Prevent coccidiosis
Chlortetracycline	Sheep	Reduce incidence of Campylobacter fetus abortion
Decoquinate	Sheep, goat	Prevent/control coccidiosis and toxoplasmosis (sheep)
Lasalocid	Sheep, goat	Prevent coccidiosis
Monensin	Sheep, goat	Prevent coccidiosis and toxoplasmosis (sheep)
Neomycin	Sheep, goat	Control colibacillosis
Niacin	Goat	Prevent ketosis
Oxytetracycline (feed or injectable)	Sheep, goat	Prevent chlamydial, campylobacter abortion
Poloxaline	Sheep	Prevent bloat
Salinomycin	Sheep, goat	Prevent coccidiosis

Fajt VR (2002). Appendix 1. In: Pugh DG, editor. Sheep & Goat Medicine. Philadelphia: Saunders; p 435–445.

antimicrobials were sold for increasing feed efficiency, promoting growth, and maintaining animal health. For the EU, it was estimated that 30% of the total antibiotic use of 5 million kg was "in feed" for growth promotion (Sarmah et al., 2006). It is, likewise, difficult to obtain precise information on the quantity of drugs used in small ruminants for prophylactic use, as compared to therapeutic uses. Drugs commonly used for sheep and goats will vary considerably from country to country. As an example, "commonly used drugs" in the United States for sheep and goats, as listed in a U.S. sheep and goat medicine text (Fajt, 2002), are given in Table 11.2. Of the >120 drugs, only 10 are for "prevention" or "control." One of the more common uses is for prevention and control of coccidiosis, a very common disease of young small ruminants; this accounts for 5 of the 10 approved drugs with a label for prevention or control of disease in small ruminants. Two of the drugs are tetracycline-type drugs approved for oral use in preventing or reducing bacterial abortion (Fajt, 2002).

A case could be made that use of anthelmintics could be considered "prophylactic" in many cases. In the past, it was common for producers to administer deworming anthelmintic treatments on a regular scheduled basis, which could be considered (prophylactic). In the case of *H. contortus*, a major parasite of small ruminants, the FAMACHA© system has been advocated and commonly used. In this system, to be discussed in Section 4, "Strategies for Reducing Antimicrobial and Antiparasitic Drug Use," deworming for *H. contortus* is performed selectively, with only those animals showing some degree of anemia being dewormed. The end result can be a significant reduction in the use of dewormers. One study, from India, found that 67% of drug costs in sheep were for prophylactic purposes, with treatment accounting for the other 33% (Kumar and Iyue, 2004). In the Indian study, anthelmintic use was considered prophylactic, rather than for treatment.

Some lambs in the United States and elsewhere are fed in feedlots. A feedlot, for USDA purposes, has been defined as "...any operation with ... inventory of 500 or more market lambs or sheep that identified themselves as a feedlot and fed a high-energy diet for the purpose of getting their animals to an acceptable slaughter weight" (US Department of Agriculture, APHIS-VS, 2002). In a USDA study, 5.1% of U.S. sheep operations were classified as feedlots, and 12.5% of all sheep and lambs were reported as being in feedlots (US Department of Agriculture, APHIS-VS, 2002). This USDA report provided information from 32 feedlots from 11 states, representing a sample of at least 70% of sheep in the United States. Results of that study provide an indication of the level of prophylactic use of drugs at the time the study was conducted. In the study, 30% of feedlots reported using antibiotics in water as part of arrival processing. A total of 84.4% of feedlots used antibiotics in feed or water during the period of the study, with 78.1% of the 84.4% using feed antibiotics and "Aureomycin" premix or tetracycline used in all cases where feed antibiotics were given. Of the feedlots using antibiotics in feed or water, 51.9% used them for disease treatment, 88.9% used the antibiotics for prevention, and 40.7% used them for growth promotion. A total of 84.4% of feedlots also used coccidiostats either in feed or water. As of early 2010, both oxytetracycline and chlortetracycline were U.S. FDA approved and reported as "presently marketed" for use in feed in sheep for increased rate of weight gain and improved feed efficiency (NRSP-7, 2010).

The issue of utilization of antimicrobials in feed or water for prophylactic purposes has been debated for some time. In mid-2010, the U.S. FDA issued a "Draft Guidance: The Judicious Use of Medically Important Antimicrobial Drugs in Food-Producing Animals; Availability" (US Department of Health & Human Services, Food and Drug Administration, 2010c). This "draft" guidance proposed changes that would curtail the use of antimicrobials in feed for production purposes and restrict the use of medically important antimicrobial drugs in food-producing animals, which would only be used with veterinary oversight or consultation. There will obviously be considerable ongoing debate about this issue. However, it seems likely that in the future there will be reduced use of antimicrobials for growth promotion in animal agriculture as a whole. In that regard, the Animal Health Institute reported a 12% decline in spending on feed additives in 2008 (Animal Health Institute, 2009).

11.3 THERAPEUTIC USE OF MAJOR DRUG CLASSES (E.G., ANTIBIOTICS, ANTIPARASITICS) IN GOAT AND SHEEP PRODUCTION SYSTEMS

The two most common uses of drugs for sheep and goats include the administration of antimicrobials for treatment of various infections (e.g., pneumonia, enteritis) and the administration of anthelmintics for treatment of parasitic disease (e.g., primarily endoparasites but also for ectoparasites and other organisms such as coccidia). Data from the United States via Sheep 2001 (US Department of Agriculture, APHIS-VS, 2002) indicated that 16.6% of ewes died or were culled during 2000 for all U.S. operations. Predators accounted for the largest portion of all adult sheep mortality, accounting for 23.5% of all adult losses. The next most common causes of loss were digestive problems (bloat, scours, parasites, enterotoxemia, acidosis, etc.) at 6.7% of all adult losses and respiratory problems (pneumonia, shipping fever, etc.) at 7.0% of all adult losses. The same trends were found for lamb losses. The percentages of operations reporting that the following diseases (among others) were suspected or confirmed in the prior 3 years were (in order of occurrence) as follows: stomach or intestinal worms in 74%, sore mouth in 40%, enterotoxemia in 39%, foot rot in 35%, coccidiosis in 30%, and caseous lymphadenitis in 20%. More than 50% of operations reported using coccidiostats in feed or water (US Department of Agriculture, APHIS-VS, 2002). Overall, 23.6% of operations reported using antibiotics in feed (19.6%) or water (4.0%) for disease treatment, and 14.5% of operations used antibiotics or ionophores in feed (14.1%) or water (0.4%) for growth promotion.

In the United States, there are limited FDA-approved antibiotics for therapeutic use in sheep and goats. Some antibiotics are approved for sheep, but not for goats. As stated before, because of the limited number of approved drugs for small ruminants, most use of drugs is "extralabel."

A Canadian study provides a perspective on use of drugs in small ruminants in North America. In a study of 212 sheep producers in Canada, antimicrobial use in the sheep industry was described (Avery et al., 2008). By far, the most common route of administration was by injection, followed by in-feed, oral (pills, liquid, bolus), and via water medication. For injection, the most commonly used drugs were penicillin and tetracyclines. Adult sheep were the most commonly treated (injected) sheep. Common reasons for treatment (in descending order) were mastitis, respiratory problems, ewes at postlambing, lameness, and scours (Avery et al., 2008). Overall, 93% of the antimicrobial use was "extralabel use" (*The Flock*, 2009).

As an example of one European country, an EU candidate country, Turkey, in 2009 listed populations of 22 million sheep and 5 million goats. In Turkey, there are 792 approved drug products for sheep and 478 approved drug products for goats (this includes biologicals and pesticides). A considerable number of the approved drugs were antibacterials, with 318 approved for sheep and 127 for goats (Turkish Statistical Institute, 2010). Obviously, there are major variations among countries. However, the large sheep populations and the increased number of drug approvals would appear to be related.

One major study divided strategies to reduce therapeutic antibiotic use into two major categories as follows (Committee on Drug Use in Food Animals, 1999):

A. **Efforts for prevention of disease and infection, using measures such as the following:** (i) providing controls on hygiene, population dynamics, feed quality, and environmental conditions to minimize stress; (ii) eradicating specific diseases; (iii) optimizing nutrition in order to enhance natural immunity or using nutrient feeding regimens to lessen the impact of changes or stress; and (iv) breeding for genetically disease-resistant animals.

   For item 1, efforts to provide controls on hygiene, population dynamics, feed quality, and environmental conditions to minimize stress were suggested to include (Committee on Drug Use in Food Animals, 1999):

   1. Controlling ambient temperature and heat stress
   2. Avoiding overcrowding and behavioral stress
   3. Practicing vaccination
   4. Use of beneficial microbial products, such as probiotics and competitive-exclusion products
   5. Practicing biosecurity
   6. Adhering to fly control
   7. Focusing on moisture, mud, and manure management
   8. Enhancing natural modulators of immune function
   9. Using killed bacterial adjuvants—biomodulation

B. **Documented diagnosis of the presence of a pathogen and selection of an antibiotic that is effective and thorough in eliminating infection:** Two overriding major considerations have major impact on therapeutic use of antimicrobials in sheep and goats. On the one hand, illness and suffering are to be avoided and animal welfare maximized. On the other hand, any use must assure that no harm comes to the consuming public via illegal residues or transfer of antibiotic-resistant organisms.

11.4 PREVALENCE OF DRUG RESIDUES IN SHEEP AND GOAT MEAT AND MILK

Worldwide, animal industries and governments are establishing measures to monitor the presence of residues in food products from food animals and to reduce their occurrence in food. Quality assurance programs have been developed for key commodities for both export and domestic markets. The small ruminant industry in the United States is not as large as that for cattle, swine, or poultry (e.g., animals slaughtered in the United States in 2008 included 3,040,559 sheep, lambs, and goats vs. 34,770,197 cattle and 9,395,830,635 poultry) (Food Safety and Inspection Service [FSIS], 2009). Worldwide, there is comparatively much less screening for residues in foods produced from small ruminants as compared to that for cattle, swine, or poultry. In this section, the available occurrence data on drug and chemical residues from various small ruminant producer countries are given.

In the United States, the Food Safety and Inspection Service (FSIS) of the USDA issues National Residue Program (NRP) data (FSIS NRP data October 2009, also known as the "Red Book"), in which chemical residue sampling plans and annual testing results are reported. The NRP's chemical residue sampling plans include import and domestic sampling plans and the guidelines for sampling are presented. The domestic sampling plan consists of two parts: scheduled sampling and inspector-generated sampling. The scheduled sampling plan is the random sampling of tissues of food animals at slaughter. The inspector-generated sampling plan is pursued when an in-plant public health veterinarian suspects that a food animal might have chemical residues at violative levels. Signs of animal disease and producer history may lead to inspector-generated samplings. The imported plan consists of sampling of animal products (meat, poultry, and egg products) at U.S. ports and includes three levels of chemical residue inspections: normal sampling, increased sampling, and intensified sampling.

As a percentage of total meat consumption, the consumption of foods from small ruminants in the United States is very low compared to foods from cattle, swine, or poultry. According to the Red Book data in 2009, 197,054,452 lbs of ovine meat was consumed out of 112,397,340,148 lbs of total meat consumption, which amounts to 0.175%, in the United States. Cattle, swine, and poultry meat consumptions were 23.5, 20.6, and 52.3% of that total, respectively (Fig. 11.1). In 2008, a total of 658,199 goats, 2,263,191 lambs, and 119,169 sheep were reportedly slaughtered in the United States (FSIS, 2008). Among the slaughtered goats, lambs, and mature sheep, a total of 980 (0.15%) goats, 814 (0.04%) lambs, and 472 (0.40%) mature sheep were tested for veterinary drug, food additive, and environmental contaminant residues under the scheduled sampling plan (Table 11.3). In addition,

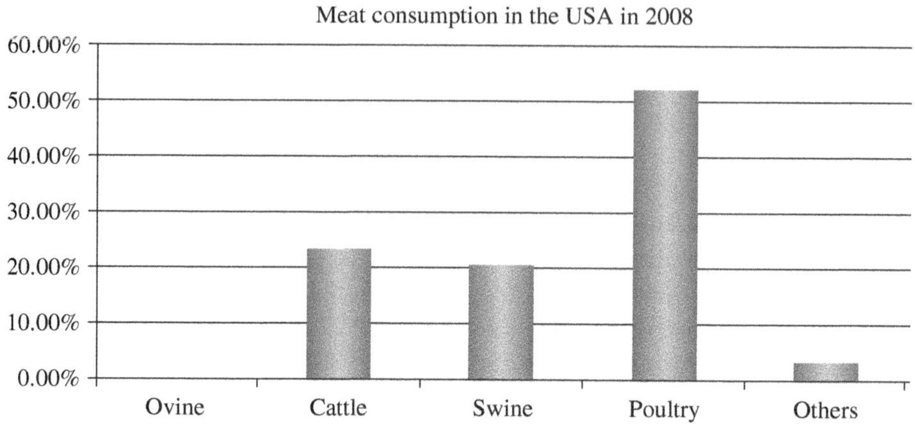

FIGURE 11.1 Meat consumption in the United States in 2008. The data in the graph represents the percentage of the total meat production based on total dressed weight of the slaughtered animals (Red Book, 2009, Food Safety and Inspection Service, 2008 National Residue Program data).

TABLE 11.3 Total number of samples tested in 2008 by the FSIS in the United States for veterinary drug, food additive, and environmental contaminant residues, as well as kidney samples tested for antibiotic residues

Species	Scheduled testing	Inspector-generated testing	Kidney samples for antibiotic testing
Goats	980	180	85
Lamb	814	374	251
Mature sheep	472	137	62

Red Book, 2009, Food Safety and Inspection Service, 2008 National Residue Program data.

among slaughtered goats, lambs, and mature sheep, a total of 180 (0.03%), 374 (0.02%), and 137 (0.11%) were sampled under the inspector-generated sampling plan in 2008 (Table 11.3). Kidney samples from 85 (0.01%) goats, 251 (0.01%) lambs, and 62 (0.05%) mature sheep were tested for antibiotics (Table 11.3).

Test results for 2008 indicated that no lamb or mature sheep sample exceeded the U.S. violative limit for the tested antibiotics. On the other hand, one goat sample had oxytetracycline at levels between 2.51 and 5.00 ppm (4.66 ppm), where the violative limit of oxytetracycline according to 21CFR556.500 is 12 ppm in the kidney (US Department of Health & Human Services, Food and Drug Administration, 2002). As shown in Table 11.4a, there were 227 goat, 287 lamb, and 213 mature sheep liver samples tested for avermectins, and no residue was detected at violative levels. Avermectin was present at detectable but nonviolative levels only in one lamb liver sample.

TABLE 11.4 Results of scheduled sampling in (a) 2008 and (b) 2007 in the United States

Species	Sample tested	Tested for	Number sampled	Number of samples with chemical at detectable levels	Number of samples with chemical at violative levels
(a) 2008 sampling					
Goats	Liver	Avermectins and milbemycins	227	0	0
Lamb	Liver	Avermectins and milbemycins	287	1 avermectin	0
Mature sheep	Liver	Avermectins and milbemycins	213	0	0
Goats	Fat	Chlorinated hydrocarbons and chlorinated organophosphates	214	5 chlorinated hydrocarbons	0
Lamb	Fat	Chlorinated hydrocarbons and chlorinated organophosphates	276	2 chlorinated hydrocarbons	0
Mature sheep	Fat	Chlorinated hydrocarbons and chlorinated organophosphates	197	4 chlorinated hydrocarbons	0
Goats	Liver	β-agonist	221	0	0
Goats	Liver	Sulfonamide	233	0	0
(b) 2007 sampling					
Goats	Liver	Sulfonamides and ivermectin	317	2 moxidectin	0
Lamb	Liver	Sulfonamides and ivermectin	342	0	0
Mature sheep	Liver	Sulfonamides and ivermectin	283	1 moxidectin	0
Goats	Fat	Veterinary drug, food additive, and environmental contaminants	264	1 chlordane (1.01–2.5 ppm)	0
Lamb	Fat	Veterinary drug, food additive, and environmental contaminants	246	3 DDT methoxychlor (1) (0.21–0.3 ppm)	0
Mature sheep	Fat	Veterinary drug, food additive, and environmental contaminants	240	6 DDT	0

In addition, 214 goat, 276 lamb, and 197 mature sheep fat samples were tested for chlorinated hydrocarbon and chlorinated organophosphate residues (Table 11.4a). Detectable but nonviolative levels of chlorinated hydrocarbons were found in five goats, two lambs, and four mature sheep samples (Table 11.4a). A total of 221 and 233 goats liver samples were tested for residues of β-agonists and sulfonamides, respectively, and no residues were detected. The inspector-generated samples in 2008 revealed only one lamb sample violating the residue limits among 180 goats, 374 lambs, and 137 mature sheep samples tested (Table 11.3). In the one violative lamb sample, sulfadimethoxine was detected.

No chemical residues were detected in small ruminants imported into the United States in 2008. In one out of the 68 tested lamb samples imported from Australia, chlorinated hydrocarbon residue was detected but at levels below violative limits. Data from 2007 in the United States indicated the number of samples tested, and results were similar to the results from 2008 (FSIS, 2008). Similarly, no violative levels of chemicals were found in the tested samples. However, in two goat and one mature sheep liver samples, moxidectin was found at nonviolative levels (Table 11.4b). Although, numbers of samples tested in these years were small, it can be concluded that the precautionary measures currently in place appeared successful in the effort of avoiding chemical residues from small ruminant products in the United States.

In the EU, the Commission of the European Communities issued a report that summarizes the actions taken in the member states during 2007, in the implementation of Council Directive 96/23/EC, and the measures to monitor certain substances and residues in live animals and animal products (Commission of the European Communities, 2007a, b). The member states in the EU are required to adopt and implement a national residue monitoring plan for specific residues. Moreover, member states are required to assign a central public department or body for coordinating the implementation of the controls. This body is also responsible for collecting the data for various residues and notifying the Commission annually about the results of the surveys. The Commission provides member states a questionnaire to collect information on actions taken as a result of noncompliant results. The Commission grouped these actions with respect to noncompliant results into three categories:

1. Sampling based upon suspicion of noncompliance, including noncompliant results, suspicion of an illegal treatment, or suspicion of *improper withdrawal period of veterinary drugs.*
2. Modifications of the national plans for the following year: Noncompliant results for a specific substance or a specific food commodity result in

intensified controls for this substance or food commodity in the plan for the following year.

3. Other actions:
   a. To carry out investigations in the farm of origin, such as verification of records and additional sampling
   b. To hold animals on the farm as a consequence of positive findings
   c. To slaughter animals in case of confirmation of illegal treatment and to send them to a high-risk processing plant
   d. To intensify the controls in the farms where noncompliant results were found
   e. To impound carcasses at the slaughterhouse when noncompliant results have been found
   f. To declare the carcasses or products of animal origin unfit for human consumption

The Commission of the European Communities published National Residue Monitoring Plans for the member states in 2007 (Commission of the European Communities, 2007b). In 2007, a total of 40,935,665 sheep and goats were slaughtered in the EU, and 26,599 (or 0.06%) were tested for the presence of chemical residues such as veterinary drugs and hormones. Anthelmintic residues were found in two samples (one abamectin (avermectin) and one ivermectin). Sheep and goat samples from different countries in Europe were found to be noncompliant with MRLs for chloramphenicol (one sample), amoxicillin (one sample), chlortetracycline (three samples), doxycycline (one sample), inhibitors (four samples), oxytetracycline (four samples), sulfadiazine (18 samples), sulfamethoxypyridazine (one sample), and tetracycline (one sample). These results suggest that the measures taken to minimize the chemical residue levels are somewhat successful as only 34 of the 26,599 tested samples were noncompliant with MRLs (Commission of the European Communities, 2007b).

In Oceania, the Food Standards Australian and New Zealand Food Authority (FSANZ) is the combined body that sets and develops food standards for Australia and New Zealand. This includes requirements for foods such as additives, food safety, labeling, and genetically modified foods. The implementation of these standards is the responsibility of State/Territory Health Departments of New Zealand and Australia. In Australia, the Health Department of Western Australian is responsible for adopting and enforcing these standards (Goodchild and Casey, 2001). In addition, the Australian Pesticide and Veterinary Medicines Authority (APVMA) is the national regulator of the use of pesticides and veterinary medicines in Australia. A National Residue Survey (NRS) is conducted yearly to monitor the residues of agricultural and veterinary chemicals and environmental contaminants. The NRS is a unit of the Food and Product Safety and Integrity Branch of the Product Integrity, Animal and Plant Health Division within the Australian Government Department of Agriculture, Fisheries and Forestry. The NRS is largely funded by industry.

In New Zealand, the New Zealand Food Safety Authority (NZFSA) is responsible for providing the food regulatory program for the food produced and consumed in New Zealand as well as import and export of food products. The NZFSA is a semiautonomous body connected to the Ministry of Agriculture and Forestry (MAF) (New Zealand Food Safety Authority, 2009a). For meat products, the MAF conducts monitoring of meat products during the year to determine the level of residues in these products. Fat samples are required for the regular surveillance for pesticides, and the procedure for the preparation of the samples are clearly defined by the NZFSA. Sampling for lamb and sheep from every export slaughterhouse and packing house is done periodically. Kidney, muscle, or urine samples are required for surveillance of residues of heavy metals, antibiotics, hormones, and drugs (New Zealand Food Safety Authority, 2009b).

In Australia, an NRS was established to monitor agricultural and veterinary chemicals and environmental contaminants in foods. This program is largely funded by industry and residue reports are published annually. During the 2007–2008 period, among goat (105) and sheep (338) fat samples tested for anthelmintics, only moxidectin was detected in 47 sheep fat samples but was below the Australian residue standards (<0.5 mg/kg) (Table 11.5). For this time period, no benzimidazoles were found in 326 sheep liver samples tested for benzimidazoles. In four of the 332 sheep liver samples tested for

TABLE 11.5 Results of sheep and goat residue samplings in Australia for 2007–2008

Species	Sample tested	Number of sample	Number of residue found	Violations
Goat	Fat	105	0	0
Sheep	Fat	338	47 moxidectin	0
Sheep	Liver	332	4 closantel	0
Goats	Fat	101	1 cypermethrin	0
Sheep	Fat	801	6 organochlorines, 5 organophosphates, 14 synthetic pyrethroids	0
Goats	Liver	50	20 cadmium, 10 lead	N/A
Sheep	Liver	325	268 cadmium, 176 lead, 23 mercury	15 cadmium (mostly in old sheep), 5 lead (old batteries, access to lead-based paint)
Sheep	Kidney	339	1 dicyclanil	0

salicylanilides, closantel was detected at low concentrations (<5 mg/kg) (Table 11.5). No pesticide residue above Australian residue standards were found in goat (101) and sheep (801) fat samples tested for pesticides. In one goat sample, cypermethrin was detected but at levels below the Australian residue standards (<0.5 mg/kg). In sheep fat samples, six organochlorines, five organophosphates, and 14 synthetic pyrethroids were detected but at concentrations below the Australian residue standards (<0.5 mg/kg). A total of 50 goat and 325 sheep liver samples were tested for selected heavy metals. Among these, residues below the Australian residue standards were found for cadmium in 20 goat and 268 sheep samples, for lead in 10 goat and 176 sheep samples, and for mercury in 23 sheep samples. No mercury residue was detected in goat liver samples. Kidney samples from 339 sheep were tested for amidines (cyromazine, dicyclanil, and melamine), and in one of these samples, dicyclanil was detected. Also, 412 sheep kidney samples were tested for aminoglycosides, β-lactams, cephalosporins, lincosamides, macrolides, sulfonamides, and tetracyclines, and 342 sheep muscle samples were tested for antimicrobials such as chloramphenicol, florfenicol, and thiamphenicol. None of these tested samples contained any of the tested antibiotics (Australian Government, Department of Agriculture, Fisheries and Forestry, 2009).

In the last decade, livestock production in China has increased dramatically. During the same time, there have also been dramatic changes in veterinary services in the country. The majority of the veterinary services in China are provided by the government; however, the number of private veterinarians is increasing as a result of an increase in the number of larger production farms (Bedard and Hunt, 2004). Improvements in food safety in China have been initiated as a result of international trade requirements. Additionally, domestic awareness about food safety is increasing and food safety has become a major concern of Chinese consumers (Wang et al., 2008). A survey on consumer awareness and demands and the implications on food safety has demonstrated that consumers in China are willing to pay more for milk products manufactured using the HACCP system (Wang et al., 2008). Although China has the largest number of goats of any country worldwide (FAOSTAT, 2007), one author reports that, as of 2009, there was no developed screening program for the prevalence of drug residues or national performance record keeping in small ruminants (Schoenian, 2009). In China, herbal medicines are commonly used for treatment purposes, but it is not clear whether withdrawal times have been established (Schoenian, 2009). It is reported that most goats are milked by hand and washing the udders and the use of germicidal teat dip are not common practices (Schoenian, 2009). Currently, the Chinese government and milk processing plants, along with the support of the World Bank, are establishing a dairy improvement project in order to increase the efficiency and safety of the milk production (Schoenian, 2009).

Yamaki et al. (2004) reported on the occurrence of antibiotic residues in 2686 raw goat milks from 490 farms from the Manchega ewes grown in Castilla–La Mancha region of Spain. They tested milk for the presence of β-lactams and sulfonamides in both raw and heat-treated (82°C, 10 min) bulk tank milks by using a microbial inhibitor-based test (Delvo SP test, DSM Food Specialties, Delft, the Netherlands), with further testing to identify specific drug classes. This test is used as a standard tool for industry and detects β-lactam antibiotics, penicillins, sulfonamides, tetracyclines, and macrolides (Matthew, 2009: p 225). From the 2686 raw milk samples, 1.7% were positive for β-lactams and sulfonamides and 2.1% showed doubtful results. When the milk samples were heat treated to inactivate natural inhibitors, the occurrence of the β-lactams and sulfonamides decreased to 1.3%, and the doubtful positive samples decreased to 0.4%. They reported that a majority of the doubtful samples may have had antimicrobial agents other than β-lactams and sulfonamides, possibility from contamination of milk with chlorine from the milking equipment sanitation.

11.5 APPROACHES TO MINIMIZE ANTIMICROBIAL USE AND COST OF ELIMINATING SUBTHERAPEUTIC USE

A variety of approaches can and have been used to minimize antimicrobial use in small ruminants. Most of these approaches focus on either management effort to reduce animal illness and, therefore, drug use or regulatory efforts to monitor for presence of residues. Some of these strategies include the following:

1. **Health management and health promotion: A major key to reducing antimicrobial and anthelmintic use is to practice herd/flock health programs and health management, including:**

   Appropriate use of vaccines to prevent disease occurrence, thereby reducing drug use for therapeutic purposes (Lupton, 2008; Sahlu et al., 2009). An example would be the use of clostridial bacterin–toxoids to reduce the occurrence of enterotoxemia.

   Use of the FAMACHA system for management of the parasite *H. contortus*. This system and its application have been shown to save money by decreasing anthelmintic drug use and aid in avoiding development of further resistance of the parasite to chemical dewormers (Di Loria et al., 2009; Lupton, 2008; Sahlu et al., 2009). Kaplan et al. (2004) evaluated the effectiveness of this method in dealing with *H. contortus* by testing it on 847 sheep and 537 goats from various breeds and ages in the United States. They reported that the FAMACHA method is much more effective than conventional dosing practices in decreasing anthelmintic use in sheep and goats. Burke et al. (2007) tested 552 mature sheep and 676 goats in southern United States and Puerto Rico and stated that FAMACHA is an effective tool for identifying anemic sheep and goats for targeted deworming. In another study conducted in South Africa, it was reported that following the use of FAMACHA treatment costs at commercial sheep farms significantly dropped (58% drop was reported) (Bath and van Wyk, 2001).

Use of alternative management strategies, such as rotational grazing to reduce exposure to endoparasites or decreased stocking density to reduce parasite exposure and to reduce the possibility of disease transmission (Lupton, 2008; Sahlu et al., 2009).

Food restriction around the time of treatment (Lupton, 2008) has been shown to enhance effectiveness of the anthelmintic ivermectin (Lupton, 2008). Natural products such as tannins have been found to have an inhibitory effect on some GI parasites and are being investigated for the potential value in control (Lupton, 2008).

Selecting resistant animals, where, for instance, Kiko and Spanish breeds were reportedly more resistant than Boer goats (Lupton, 2008; Sahlu et al., 2009).

Good husbandry and management. Supplementary protein around the time of deworming has been shown to enhance resilience and resistance (Lupton, 2008). Reduction in stocking density has been shown to have a beneficial effect on the occurrence of parasitism and disease (Lupton, 2008).

Preferred Production Practices in "Herd Health" from Merkel (2005) include establishing and using a herd health program, establishing a valid veterinarian–client–patient relationship and using extralabel drugs as directed, storing/administering drugs properly and using proper withholding periods, using proper injection techniques, and providing training on proper drug use to all individuals treating animals (Merkel, 2005).

2. **Legislative efforts**

In the EU, there are highly regulated controls on the use of veterinary medicines, especially for food animals. Current terminology refers to a required "market authorization," which is needed prior to marketing of a veterinary medicinal product. The term "marketing authorization" replaces the previously used "product license." Criteria for approval include quality, safety, and efficacy. Safety refers to safety to the consumer, the user, the target species, and the environment. Also a requirement is the restriction that the compound is below a predetermined "safe" level prior to marketing. This is the "maximum residue level" or "MRL" in Europe and the "tolerance" in the United States. (Subasinghe, 2002). In the EU, biologicals (vaccines) are currently included with veterinary medicinal products.

In the United States, the U.S. FDA is responsible for "approvals" of veterinary drugs. Biologics fall under the auspices of the USDA. Pesticides are regulated by the Environmental Protection Administration (EPA). Most residue monitoring programs for food animals and food animal species are under the FSIS of the USDA. However, monitoring for veterinary residues for minor species falls under the responsibility of the U.S. FDA. The approval process requires demonstrating similar aspects as under the EU system.

3. **Mandated residue monitoring**

   EU directives require that member states monitor animal products for residues of veterinary medicines to ensure compliance. Each member state produces a plan and gives results of sampling to demonstrate compliance. This refers to a minimum program for the member state. In the United States, the FSIS of the USDA is responsible for most residue detection method approval and monitoring.

4. **Regulation of "extralabel" or "off-label" use**

   In the EU-defined "cascade" system, a prescribing cascade is described for "off-label" use. The effect is to limit veterinarians who treat food-producing animals to prescribing medicines composed only of substances authorized for use in food-producing species. Very specific rules limit how the prescribing cascade is carried out.

   In the United States, "extralabel" use is defined as "actual use or intended use of a drug in an animal in a manner that is not in accordance with the purpose approved on the label" (Subasinghe, 2002). Modifications have been addressed in the Animal Medicinal Drug Use Clarification Act of 1994 (AMDUCA) (US Department of Health & Human Services, Food and Drug Administration, 2010d) and the Minor Use and Minor Animal Species Health Enhancement Act of 2001 (US Department of Health & Human Services, Food and Drug Administration, 2010e).

5. **Producer education, including quality assurance programs or schemes, as described hereafter**

   All of the producer groups rely on quality assurance programs with a heavy focus on producer education.

11.6 QUALITY ASSURANCE PROGRAMS WITH SPECIAL EMPHASIS ON MANAGEMENT OF DRUG RESIDUES IN GOAT AND SHEEP PRODUCTION SYSTEMS

Small ruminant quality assurance programs are developed based on a wide variety of factors. As the focus of this chapter is drug and chemical residues, we will emphasize management of drug and chemical residues in goat and sheep production systems. The first step in establishing a valid program is to understand the possible causes of chemical residues. The Canadian Food Inspection Agency provided a list of possible causes for drug and chemical residues in food animals, which is quite comprehensive and included the following:

- Using veterinary drugs at other than label directions or dosage
- Not following the recommended withdrawal times
- Administering volumes of drugs at more than the recommended amount at a single injection site
- Mishandling and sanitation of the drug-contaminated equipment
- Personal errors in dosing, measuring, or mixing of drugs
- Improper animal husbandry, allowing animals to get in contact with chemical spills or medicated feeds
- Possible chemical interactions between drugs
- Variation between animals, such as age, pregnancy, allergies, etc.
- Environmental contamination
- Misuse of agricultural chemicals, such as pesticides

A major cause of residues in sheep and goats is the improper use of veterinary drugs in food animals. Worldwide, governments are addressing this issue in order to control residues in food animals. In the United States, extralabel use of drugs is allowed, with very strict requirements, in minor species, including small ruminants, defined by Code of Federal Regulations 21 CFR 514.1 (d)(1)(ii)(US Department of Health & Human Services, Food and Drug Administration, 2010a). It is stated that a drug can be used in an extralabel fashion only when the animal is suffering or death is the possible outcome unless treated. In other cases, such as for improving weight and feed efficiency, use of extralabel drugs is not allowed. Similarly, in the EU, extralabel prescription of drugs is also allowed, as long as the product from the treated animal enters the food chain only following full withdrawal of the drug (Pengov and Kirbis, 2009). In the EU, the prescribing veterinarian is responsible for determining the withdrawal time of the drug prescribed in extralabel fashion (Pengov and Kirbis, 2009).

All of the major food-producing industries in the United States have developed "quality assurance programs" specific to the industry. Examples would include the programs of the swine industry, the beef industry, and that of the Milk and Dairy Beef Quality Assurance Program. The programs all have similarities in approaches and goals.

The U.S. sheep industry began development of an industry-wide quality assurance program in 1991. This has culminated in the "Sheep Safety and Quality Assurance Program (SSQA)," with a stated purpose of maximizing consumer confidence and consumer acceptance of sheep products. The program involves the use of research and education to enhance management during production of high quality and safer sheep products (ASI-Colorado State University, 2009).

Efforts to develop quality assurance programs for goats have been less centralized. In North Carolina, a Meat Goat Quality Assurance and training program was developed starting in 2001 (Anderson, 2008; Anderson et al., 2004). In the early phases of the program, producers were required to attend a training session lasting several hours and covering the fundamentals of proper animal drug use, residue avoidance, and health management. Then, producers were required to pass an examination prior to becoming "certified." In an evaluation of the effectiveness of the training program, attendees were given written examinations before and after the training. A significant increase in test scores was found when attendees were tested after the training, suggesting increased knowledge of the participants (Anderson et al., 2004).

In another more widely applicable program, funded by the USDA FSIS, the American Institute for Goat Research of Langston University has led a group of institutions in the development of meat goat quality assurance materials (Merkel, 2005). An excellent Web-based training and certification program for meat goat producers is available, with a large quantity of excellent materials (Langston University Goat & Research Extension, 2010). Practices are referred to as "Preferred Production Practices," and these are intended to guide the producer in the production of healthy, productive animals. The benefit of educational programs in decreasing antibiotic residues was reported by Gonzalo et al. (2010). They screened bulk tank milk samples from 209 dairy ewe flocks in Spain over 5 years from 2004 to 2008. They found that the yearly occurrence of antibiotic residues progressively decreased from 2004 (1.36% occurrence) to 2008 (0.30% occurrence) (Gonzalo et al., 2010). The authors reported that improvements in occurrence of residues (less residues) were associated with education and training programs to increase milk safety (Gonzalo et al., 2010). Specific recommendations for avoiding milk residues have been presented, as shown, in Table 11.6 (Matthews, 2009).

A European example of quality assurance efforts is the "farm assurance scheme" referred to as RUMA, the Responsible Use of Medicines in Agriculture Alliance, used in the British livestock industry (RUMA, 2010). It involves organizations at all stages of the food chain, in order to produce a

TABLE 11.6 Recommendations for avoiding drug residues in goat milk

Avoiding drug residues in goat milk
Use veterinary drugs labeled or licensed for farm animals
Label all dispensed drug with withholding
Assure that farmer is aware that drug is used off-label and that company is not liable
Correctly identify all treated animals
Separate treated animals and milk last
Be extremely careful with IMM treatment
Discard milk from both halves
Observe correct withdrawal
Use antimicrobial screening tests

Adapted from Matthews J (2009). Diseases of the Goat. 3rd ed. Oxford: Blackwell Publishing Ltd; p 227, Table 13.8.

coordinated and integrated approach to "best practices" in the use of medicines. RUMA produces guidelines focused on "guiding principles" for use of sheep producers in management of flocks. Some of the guidelines for farmers include (RUMA, 2010):

1. Commitment to producing safe food
2. Duty/responsibility to safeguard the health and welfare of animals
3. Flock health plan, developed and reviewed with a veterinarian
4. Appropriate use of various medicines
5. Use of appropriate withdrawal times and practices to prevent any drug residues
6. Keeping a medicine record book and accurate information on the identity and treatments given to animals
7. Adequate training and recording for preventing and identifying health problems

Multiple pressures are being focused on the small ruminant industries. There are changes in animal numbers, an increased push for efficiency, increased costs for drug companies to develop and market products, and greater concerns of the public toward animal welfare and public health safety (McKellar, 2006). Anthelmintic resistance is recognized as a global problem, with frequent resistance and even multiresistance (McKellar, 2006). Among the strategies used to deal with anthelmintic resistance are:

1. Use of anthelmintic combinations to delay resistance (McKellar, 2006)
2. Augmented strategies for quarantine, along with combination or sequential treatment with drugs from different anthelmintic classes, followed by posttreatment egg per gram counting (McKellar, 2006)

3. Approaches to improve activity of available anthelmintics, like use of the cytochrome P450 inhibitor piperonyl butoxide (McKellar, 2006) to increase residence time of anthelmintics
4. Food restriction at or around the time of deworming has worked for avermectins (McKellar, 2006)
5. Development of new anthelmintic products: cyclodepsipeptides and the semisynthetic derivative of paraherquamide, 2-desoxo paraherquamide A (McKellar, 2006).
6. Alternative controls (Sahlu et al., 2009):
   a. Immunity enhancement via nutrition
   b. Vaccination
   c. Pasture management such as cograzing with cattle
   d. Genetic resistance (Sahlu et al., 2009)

In the United States, there are no intramammary antibiotic infusion products specifically approved for use in small ruminants. Therefore, in the treatment of mastitis in small ruminants, antibiotics approved for use in bovine mammary glands are used. As these drugs are designed for animals that have different metabolism than small ruminants and the "volume" of the udder in small ruminants is much smaller, the withdrawal time of the drugs vary. Therefore, this extralabel use of drugs in small ruminants accounts for the uncertainty of withdrawal times of the drugs and the possible presence of drug residues in the final product. Pengov and Kirbis (2009) reported that the milk of the dairy ewes that were treated with antibiotics, penicillin, and amoxicillin, designed for bovine, has drug residues over the maximum residue limits (MRL) following the proposed withdrawal times. They tested the withdrawal time of two commercial lactating cow products from ewe's milk following mastitis treatment. Although the milk withdrawal time for penicillin in lactating cows was 36 h, the investigators were able to detect penicillin residues above MRL values after 192 h in lactating ewes. Pengov and Kirbis (2009) concluded that the withdrawal of antibiotics shows significantly longer withdrawal time in the ovine mammary glands compared to bovine mammary glands. This study emphasizes the importance of taking extra precautions when using drugs in extralabel fashion.

As the number of drugs approved for use in small ruminants is limited and extralabel use is common, it is especially crucial for small ruminant farmers to establish valid measures to control the drug residues. For management of drug residues in goat and sheep production systems, the FDA requires the following measures in treated animals:

- Identifying and tracking animals that were treated with drugs
- Keeping a medication/treatment records that identifies the treated animals, treatment date, drugs administered, amounts administered, and

withdrawal times of the drugs administered and the person who administered the drugs

- Appropriate storing, labeling, and accounting of all drug products and medicated feeds
- Obtaining and using veterinary prescription drugs only through a licensed veterinarian based on a valid veterinarian–client–patient relationship
- Educating all employees and family members involved in treating, hauling, and selling the animals on proper administration techniques, observance of withdrawal times, and methods to avoid marketing adulterated products for human food (US Department of Health & Human Services, Food and Drug Administration, 2009)

In addition, the FDA-CVM established a policy to control the extralabel usage of veterinary drugs. According to this policy, the primary step is careful diagnosis of the animals by the attending veterinarian. The next step is to determine if there is no drug on the market or labeled dosage is clinically ineffective for treating the condition diagnosed. The identity of the animals that are treated by using drugs in extralabel fashion should be maintained carefully, and significantly extended period should be assigned for the withdrawal of the drug. In making the decision about the withdrawal times of the drugs that are used extralabel, the veterinarian can benefit from using the FARAD (2010).

11.7 THE FUTURE

As always, it is difficult to "look into the crystal ball" and predict the future. However, it seems likely that current trends will continue, including more sustainable production systems, reducing antimicrobial and chemical use, and a growing demand for safe, high-quality food products. With growth of technology and deeper media penetration, it seems likely that production practices will be watched ever more closely. It seems likely that producers will be asked to grow food and produce animal products using less antibiotics and artificial inputs. Increased testing of products and product verification will be more important, and producer efforts in quality assurance programs will be more intense.

REFERENCES

Adams D. Livestock trucking—an Australian perspective. AATA 31st Annual Conference; 2005; Calgary. Available at http://www.aata-animaltransport.org/conference/2005_calgary/Presentations/AdamsAATA2005_TtruckingAust.pdf. Accessed March 15, 2010.

Anderson KL (2008). Quality assurance for meat goats. In: Haskell SRR, editor. Blackwell's Five-Minute Veterinary Consult: Ruminant. Oxford: Wiley-Blackwell; p 329–330.

Anderson KL, Brownie C, Luginbuhl J-M, Mobley M (2004). Drug use survey and evaluation of quality assurance training for meat goat producers. Intern J Appl Res Vet Med 2(4):261–268.

Animal Health Institute (2009). New Release, November 30, 2009, Spending on animal health care products slips in 2008. Washington, DC: Animal Health Institute.

Ashley-Jones C (2008). Agriculture production statistics in New Zealand as of June 30, 2008. Available at http://search.stats.govt.nz/nav/ct2/industrysectors_agriculture/ct1/industrysectors/0. Accessed March 15, 2010.

ASI-Colorado State University (2009). Sheep safety and QA program. Available at http://www.colostate.edu/programs/SSQA/. Accessed November 9, 2009.

Australian Government, Department of Agriculture, Fisheries and Forestry (2009). National residue survey. Available at http://www.daffa.gov.au/agriculture-food/nrs. Accessed October 22, 2009.

Avery BP, Rajić A, McFall M, Reid-Smith RJ, Deckert AE, Irwin RJ, McEwen SA (2008). Antimicrobial use in the Alberta sheep industry. Can J Vet Res 72: 137–142.

Bath GF, Van Wyk JA (2001). Using the FAMACHA© system on commercial sheep farms in South Africa. Proceedings of the 5th International Sheep Veterinary Congress, January 22–25, 2001, Stellenbosch, South Africa.

Bedard BG, Hunt T (2004). The emerging animal health delivery system in the People's Republic of China. Rev Sci Tech Off Int Epiz 23(1):297–304.

Burke JM, Kaplan RM, Miller JE, Terrill TH, Getz WR, Mobini S, Valencia E, Williams MJ, Williamson LH, Vatta AF (2007). Accuracy of the FAMACHA© system for on-farm use by sheep and goat producers in the southeastern United States. Vet Parasitol 147(1–2):89–95.

Commission of the European Communities (2007a). Commission staff working document on the implementation of national residue monitoring plans in the member states in 2007. Available at http://ec.europa.eu/food/food/chemicalsafety/residues/workdoc_2007_en.pdf. Accessed October 23, 2009.

Commission of the European Communities (2007b). National residue monitoring plans for the member states in 2007 (Council Directive 96/23/EC, 2007). Available at http://ec.europa.eu/food/food/chemicalsafety/residues/workdoc_2007_en.pdf. Accessed October 23, 2009.

Committee on Drug Use in Food Animals. Board on Agriculture, National Research Council (1999). The Use of Drugs in Food Animals: Benefits and Risks. Washington, DC: National Academy Press; p 188–234.

Di Loria A, Veneziano V, Piantedosi D, Rinaldi L, Cortese L, Mezzino L, Cringoli G, Ciaramella P (2009). Evaluation of the FAMACHA© system for detecting the severity of anaemia in sheep from southern Italy. Vet Parasitol 161(1–2):53–59.

European Medicines Agency (2010). Policy for classification and incentives for veterinary medicinal products indicated for minor use minor species (MUMS)/limited markets (EMEA/429080/2009). Available at www.emea.europa.eu. Accessed April 6, 2010.

Fajt VR (2001). Label and extralabel drug use in small ruminants. Vet Clin North Am Food Anim Pract 17(2):403–419.

Fajt VR (2002). Appendix 1. In: Pugh DG, editor. Sheep & Goat Medicine. Philadelphia: Saunders; p 435–445.

FAOSTAT (2007). Food and agricultural commodities production. Available at http://faostat.fao.org/site/339/default.aspx. Accessed July 7, 2010.

Field TG, Taylor RE, editors. (2008). Scientific Farm Animal Production. 9th ed. Columbus, OH: Pearson Prentice Hall; p 43–44.

The Flock (August 2009). Canadian Sheep Federation. Newsletter 36(8):1–2.

Food Animal Residue Avoidance Databank [FARAD] (2010). Available at http://www.farad.org. Accessed May 21, 2010.

Food Safety and Inspection Service [FSIS] (October 2008). 2007 FSIS National Residue Program Data. Washington, DC: United States Department of Agriculture Food Safety and Inspection Service Office of Public Health Science.

Food Safety and Inspection Service [FSIS] (October 2009). 2008 FSIS National Residue Program Data. Washington, DC: United States Department of Agriculture Food Safety and Inspection Service Office of Public Health Science.

Gonzalo C, Carriedo JA, Garcia-Jimeno MC, Perez-Bilbao M, de la Fuente LF (2010). Factors influencing variation of bulk milk antibiotic residue occurrence, somatic cell count, and total bacterial count in dairy sheep flocks. J Dairy Sci 93:1587–1595.

Goodchild S, Casey R (June 2001). Sheep and goats strategic assessment-food safety and residues. Available at http://www.agric.wa.gov.au/objtwr/imported_assets/content/pw/ah/residues.pdf. Accessed November 11, 2009.

Haley CJ (2006). The minor use and minor species animal health act: past, present and future. Food Drug Law J 61(1):13–43.

Howell SB, Burke JM, Miller JE, Terrill TH, Valencia E, Williams MJ, Williamson LH, Zajac AM, Kaplan RM (2008). Prevalence of anthelmintic resistance on sheep and goat farms in the southeastern United States. J Am Vet Med Assoc 233(12):1913–1919.

Kaplan RM, Burke JM, Terrill TH, Miller JE, Getz WR, Mobini S, Valencia E, Williams MJ, Williamson LH, Larsen M, Vatta AF (2004). Validation of the FAMACHA© eye color chart for detecting clinical anemia in sheep and goats on farms in the southern United States. Vet Parasitol 123:105–120.

Kumar RA, Iyue M (2004). Cost of treatment and prophylaxis in sheep farms. Indian Vet J 81:908–911.

Langston University Goat & Research Extension (2010). Goat research. Available at http://www2.luresext.edu/goats/training/QAtoc.html. Accessed February 24, 2010.

Lanthier F (2010). Prudent use of antimicrobials in small ruminants, Canadian Sheep Federation. Available at http://www.ontariosheep.org. Accessed January 1, 2010.

Lupton CJ (2008). ASAS centennial paper: Impacts of animal science research on United States sheep production and prediction for the future. J Anim Sci 86: 3252–3274.

Matthews J (2009). Diseases of the Goat. 3rd ed. Oxford: Blackwell Publishing Ltd; p 225–227, 402–411.

McKellar QA (2006). The health of the sheep industry and the medicines to maintain it. Small Rumin Res 62(1–2):7–12.

Merkel R (2005). A meat goat quality assurance program. Preferred Production Practices in "Herd Health". Proceedings of the 20th Annual Goat Field Day, Langston University, Langston, OK; p 34–42.

Miller J (2010). Internal & external parasites of goats. Available at http://www.luresext.edu/GOATS/training/parasites.pdf. Accessed May 21, 2010.

New Zealand Food Safety Authority (2009a). Meat, game and game meat monitoring services. Available at http://www.nzfsa.govt.nz/animalproducts/meat/meatman/manual-17/man17a.htm. Accessed November 20, 2009.

New Zealand Food Safety Authority (2009b). About the New Zealand Food Safety Authority. Available at http://www.nzfsa.govt.nz/about-us. Accessed November 20, 2009.

NRSP-7 (2010). NRSP-7 On-line drug database for minor species (MUMSRx). Available at http://www.nrsp-7.org/mumsrx/. Accessed June 30, 2010.

Pengov A, Kirbis A (2009). Risks of antibiotic residues in milk following intramammary and intramuscular treatments in dairy sheep. Anal Chim Acta 637:13–17.

RUMA (2010). The responsible use of medicines in agriculture alliance used in the British livestock industry. Available at http://www.ruma.org.uk. Accessed April 7, 2010.

Sahlu T, Dawson LJ, Gipson TA, Hart SP, Merkel RC, Puchala R, Wang Z, Zeng S, Goetsch AL (2009). ASAS centennial paper: Impact of animal science research on United States goat production and predictions for the future. J Anim Sci 87:400–418.

Sarmah A, Meyer MT, Moxall ABA (2006). A global perspective on the use, sales, exposure pathways, occurrence, fate and effects of veterinary antibiotics (Vas) in the environment. Chemosphere 65(5):725–759.

Schoenian S (2009). Dairy goat production in China. Available at http://www.sheepandgoat.com/articles/China.htm. Accessed November 11, 2009.

Subasinghe R (2002). Basic overview of the regulatory procedures for authorization of veterinary medicines with emphasis on residues in food animal species. FAO Fisheries and Aquaculture Department Final March 17 2002. Available at ftp://ftp.fao.org/docrep/fao/004/AC343E/AC343E00.pdf. Accessed June 8, 2010.

Turkish Statistical Institute (2010). Hayvansal uretim, 2009. Report issued on May 18, 2010. Available at http://www.tuik.gov.tr/PreHaberBultenleri.do?id=6250. Accessed June 6, 2010.

US Department of Agriculture, APHIS-VS, National Animal Health Monitoring System (July, 2002). Sheep 2001. Part I: Reference of Sheep Management in the United States, 2001. Fort Collins: US Department of Agriculture, APHIS-VS, National Animal Health Monitoring System; p 88.

US Department of Health & Human Services, Food and Drug Administration (2002). Code of Federal Regulations Title 21, Volume 1 Revised April 1, 2002; p 346–347.

US Department of Health & Human Services, Food and Drug Administration (2009). CPG Sec. 615.200 proper drug use and residue avoidance by non-veterinarians. Available at http://www.fda.gov/ICECI/ComplianceManuals/CompliancePolicyGuidanceManual/ucm074660.htm. Accessed October 23, 2009.

US Department of Health & Human Services, Food and Drug Administration (2010a). Code of Federal Regulation Title 21 Volume 6 Part 514 (21CFR514.1). Available at http://www.accessdata.fda.gov/scripts/cdrh/cfdocs/cfcfr/CFRSearch.cfm?fr=514.1. Accessed July 12, 2010.

US Department of Health & Human Services, Food and Drug Administration (2010b). Animal & Veterinary; Minor Use/Minor Species. Available at http://www.fda.gov/AnimalVeterinary/DevelopmentApprovalProcess/MinorUseMinorSpecies/default.htm. Accessed June 30, 2010.

US Department of Health & Human Services, Food and Drug Administration (2010c). Draft guidance: The judicious use of medically important antimicrobial drugs in food-producing animals; availability. Fed Regist 75(124):37450–37451.

US Department of Health & Human Services, Food and Drug Administration (2010d). Animal Medicinal Drug Use Clarification Act (AMDUCA) of 1994. Available at http://www.fda.gov/RegulatoryInformation/Legislation/FederalFoodDrugandCosmeticActFDCAct/SignificantAmendmentstotheFDCAct/AnimalMedicinalDrugUseClarificationActAMDUCAof1994/default.htm. Accessed July 11, 2010.

US Department of Health & Human Services, Food and Drug Administration (2010e). Minor Use and Minor Species Animal Health Act of 2004. Available at http://www.fda.gov/RegulatoryInformation/Legislation/FederalFoodDrugandCosmeticActFDCAct/SignificantAmendmentstotheFDCAct/MinorUseandMinor SpeciesAnimalHealthActof2004/default.htm. Accessed July 11, 2010.

Wang Z, Mao Y, Gale F (2008). Chinese consumer demand for food safety attributes in milk products. J Food Protec 33(1):27–36.

Yamaki M, Berruga MI, Althaus RL, Molina MP, Molina A (2004). Occurrence of antibiotic residues in milk from Manchega ewe dairy farms. J Dairy Sci 87: 3132–3137.

12

RESIDUE AVOIDANCE IN SWINE PRODUCTION SYSTEMS

Ronald E. Baynes and Glen Almond

Department of Population Health and Pathobiology, College of Veterinary Medicine, North Carolina State University, Raleigh, NC, USA

12.1 INTRODUCTION

Most of the drugs approved for use in swine herds are antimicrobials followed by antiparasitics and one anti-inflammatory drug, flunixin meglumine. Many drugs are administered as feed additives and water additives with several drugs approved for oral, intramuscular (IM), or subcutaneous (SQ) administration. Residue violations used to be a major concern to the swine industry, but the industry through its various organizations such as the National Pork Board (NPB) and National Pork Producers Council (NPPC) and state councils have made every effort to reduce residue violations. This chapter highlights prophylactic and therapeutic uses of the major drug classes used in the swine industry in the United States and the EU with examples of why there may be periodic residue violations for some prophylactic and therapeutic regimens.

Strategies for Reducing Drug and Chemical Residues in Food Animals: International Approaches to Residue Avoidance, Management, and Testing, First Edition.
Edited by Ronald E. Baynes and Jim E. Riviere.

12.2 PROPHYLACTIC USE OF DRUGS IN SWINE

In 2006, the EU prohibited the use of antimicrobials for growth promotion based on the precautionary principle, while the United States still approves the use of these drugs in swine herds for growth promotion and feed efficiency, although this has been changing as described below. Denmark prohibited the use of antimicrobials as growth promotants in finishing pigs in 1998 and nursery pigs in 1999. The ban was based on the concerns from political and some in the scientific community that human health was at risk from emergence of antimicrobial resistance. It should be noted that antimicrobial use in feed and water to control and treat disease was not associated with this ban. Recent reports indicate that antimicrobials are being used more frequently for treatment and control than prior to the ban in use of antimicrobials as growth promotants. Denmark only allows veterinarians to prescribe antimicrobials but cannot sell to producers. This may be associated with Denmark using less antimicrobials per kg meat, that is, 0.05 g/kg meat for all species versus 0.1–0.3 g/kg meat in other EU countries and the United States.

The concern of emerging antimicrobial resistance associated with antimicrobial use in livestock production has triggered numerous studies and consequential policy guidance documents in many regulatory organizations across the globe. More recently, the U.S. FDA-CVM implemented a plan in a new guidance for industry document (GFI #213, 2013) to phase out the use of medically important antimicrobials in livestock production primarily for production purposes such as increase weight gain with less feed. This is consistent with their guidance document entitled "The Judicious Use of Medically Important Antimicrobial Drugs in Food Animals" (GFI #209, 2012). This GFI #213 plan does not apply to therapeutic use of these antimicrobials that includes treatment, control, and prevention of specific diseases. Its aim is to provide sponsors with specific recommendations on how to voluntarily modify the use conditions of medically important antimicrobial drugs to align with the principles outlines in GIF #209. The plan will include changing the over-the-counter (OTC) status to a marketing status that requires veterinary oversight and sponsors given the opportunity to voluntary withdraw current approved production uses of these drugs and their combinations. That is a change from OTC to Veterinary Feed Directive (VDF) status for medicated feed additives and from OTC to Rx status for medicated drinking water additives. None of these changes should be surprising as this shift started in 1993 when all new antimicrobial drugs for livestock production since then have been labeled with Rx or VDF marketing status with exceptions given to generic copies of existing OTC drugs and their combinations.

It should however be noted that according to the U.K. Five-Year Antimicrobial Resistance Strategy 2013–2018 (Departments of Human Health and Agriculture), stated for the first time, "Increasing scientific evidence suggests that the clinical issues with antimicrobial resistance that we face in human medicine are primarily the result of antibiotic use in people, rather than the use of antibiotics in animals."

There are many feed additives and water additives approved for use in swine for growth promotion. For example, Aureomycin Type A medicated feed is approved for *growing swine to be used for growth promotion and improved feed efficiency and fed at levels of 10 g to 50 g/ton. For swine in general, it is approved* for the reduction of the incidence of cervical lymphadenitis (jowl abscesses) caused by *Group E streptococci* susceptible to chlortetracycline, feed at levels of 50 g to 100 g/ton. It can be used for the control of porcine proliferative enteropathies (ileitis) caused by *Lawsonia intracellularis* susceptible to chlortetracycline and for the treatment of bacterial enteritis caused by *Escherichia coli* and *Salmonella choleraesuis* and bacterial pneumonia caused by caused *Pasteurella multocida* susceptible to chlortetracycline. It should be fed approximately 400 g/t, varying with bwt and feed consumption to provide 10 mg/lb/day and should not be fed for not more than 14 days. The meat withdrawal time (WDT) for this feed additive is zero days. However, there are other chlortetracycline feed additives, for example, Chloratet 50 Type A medicated feed, which have longer WDTs such as 5 days at levels similar to the aforementioned formulation that provides 10 mg/lb bwt/day for not more than 14 days. This example highlights the point that not all chlortetracycline feed additives have the same WDT and could be attributable to tetracycline residue violations if one assumes that they all have a zero meat WDT. Finally, there are several "Aureo" formulations such as Aureomycin 500 that contains a sulfonamide, which means that the WDT should be extended to at least 15 days.

A more controversial feed additive is the use of a β-agonist for growth promotion. A good example of this is ractopamine hydrochloride (e.g., Engain 9/ Engain 45 by Zoetis), which continues to be vigorously debated among scientists, regulatory agencies, and media (CFS, 2013). This drug is approved for use in the United States and several other countries for finishing swine, weighing over 149 lbs: to increase rate of weight gain, improve feed efficiency, and increase carcass leanness in finishing swine, weighing not less than 150 lbs, fed a complete ration containing at least 16% crude protein for the last 45–90 lbs of gain prior to slaughter. The drug is used at a dietary concentration of 4.5–9.0 g ractopamine HCl/ton of complete feed, although no increased benefit has been shown when ractopamine concentration in the diet is greater than 4.5 g/ton. A withdrawal period of zero days has been established for ractopamine hydrochloride in swine. This drug was approved in the United States in 1999, and the U.S. tolerance is 0.15 part per million (ppm) for the liver and 0.05 ppm for meat. In 2012, the Codex Alimentarius set maximum residue limits (MRLs) for ractopamine. The European Food Safety Authority believes there is insufficient data upon which to assign an MRL for ractopamine as there are potential risks to human health. This drug may also be formulated with another drug such as tylosin (Paylean®, Tylan®). As of spring 2013, about 25 EU countries, Russia, and China, among many other countries, have banned or restricted the use of ractopamine in livestock.

12.3 THERAPEUTIC USE OF DRUGS IN SWINE

Table 12.1 and Table 12.2 provide examples of therapeutic drugs from various drug classes that are approved for use in swine production facilities in the United States. The reader should be reminded that the water additives

TABLE 12.1 Examples of therapeutic antimicrobials approved for use in swine from 10 drug classes

Trade name	Active ingredient	Route	Rx/ OTC	WDT (days)
Terramycin/Liquamycin Solution	**Oxytetracycline** hydrochloride	IM	OTC or Rx	22
Sulmet Drinking Water Solution 12.5%t	Sulfamethazine	Oral	OTC	15
Tylan Injection 50; 200	**Tylosin**	IM	OTC	2
Spectam Scour-Halt	**Spectinomycin** dihydrochloride pentahydrate	Oral	OTC	21
Prinzone Powder; Pyradan Powder; Vetisulid® Powder	**Sulfachlorpyridazine**	Oral	OTC	4
Lincocin Sterile Solution; Lincomix Injectable	**Lincomycin** hydrochloride	IM	OTC	0
Princillin Injection 200 mg	**Ampicillin** trihydrate	IM	Rx	15
Amoxicillin Oral Suspension	**Amoxicillin** trihydrate	Oral	Rx	15
AQUA-CILLIN; Norocillin	**Penicillin G** procaine	IM	OTC	7
Dihydrostreptomycin	**Dihydrostreptomycin** sulfate	IM	Rx	30
Tet-Sol 10; Tet-Sol 324™; Duramycin-10	**Tetracycline**	Water additive	OTC	4
Tetracycline (*)	Tetracycline	IM	OTC	0
Garacin Piglet Injection	**Gentamicin** sulfate	IM	OTC	40
Naxcel®; Naxcel® Sterile Powder	**Ceftiofur** sodium	IM	Rx	4
Excenel® RTU; Excenel Sterile Suspension	Ceftiofur hydrochloride	IM	Rx	4
Baytril® 100 Injectable Solution	**Enrofloxacin**	SQ	Rx	5
Excede™ for Swine	Ceftiofur crystalline free acid	IM	Rx	14
Draxxin Injectable Solution	**Tulathromycin**	IM	Rx	5

TABLE 12.2 Examples of antiparasitic drugs approved for use in all swine classes

Trade name	Active ingredient	Route	Rx/OTC	WDT (days)
Thibenzole Pig Wormer Paste	Thiabendazole	Oral	OTC	30
Purina Worm-A-Rest Litter Pak	Fenbendazole	Oral	OTC	0
Atgard® Swine Wormer	Dichlorvos	Oral	OTC	0
Purina® Ban Worm for Pigs	Pyrantel tartrate	Oral	OTC	1
Tramisol Gel 11.5%	Levamisole hydrochloride	Oral	OTC	11
Ivomec® Injection for Cattle and Swine	Ivermectin	SQ	OTC or Rx	18
SKYCIS 100 Type A medicated article	Narasin	Oral	OTC	0

OTC, over the counter and do not require a prescription from a veterinarian.

and not feed additives can be used in an extralabel manner according the AMDUCA legislation in the United States. Below are four examples of drugs with label therapeutic indications and approved WDTs and tolerance levels.

12.3.1 Aminoglycosides

In the United States, gentamicin is the only aminoglycoside approved for use in swine, and it is only approved for use in neonatal pigs. Gentamicin sulfate (e.g., Gentamicin Piglet Injection) is approved for use in piglets up to 3 days old for the treatment of porcine colibacillosis caused by strains of *E. coli* sensitive to gentamicin. The label directions require administering 5 mg gentamicin (1 ml) as a single IM dose. The oral formulation, Gentocin® (Garacin) Pig Pump Oral Solution, is also approved for neonatal pigs 1–3 days old with a single 5 mg dose as with the IM formulation. The IM formulation has a meat WDT of 40 days, while the oral formulation is 14 days. The shorter WDT for the oral formulation is indicative of significantly less systemic oral bioavailability compared to the IM route. A tolerance of 0.1 ppm in the muscle, 0.3 ppm in the liver, and 0.4 ppm in the fat and kidney is established for gentamicin residues in the uncooked edible tissues of swine. Aminoglycosides accumulate in the kidneys of all livestock species, which is the slowest depleting tissue and therefore the target tissue for residue analysis. Because of this, FARAD recommends an 18-month WDT for extralabel use of this drug.

12.3.2 Sulfonamides

Sulfamethazine (e.g., Sulmet Drinking Water Solution 12.5%) is labeled for the treatment of porcine colibacillosis (bacterial scours) caused by *E. coli* and bacterial pneumonia caused by *Pasteurella* spp. The labeled dose is 112.5 mg/lb bwt on the first day and 56.25 mg/lb bwt/day on the second, third, and fourth days of administration as a drench or in drinking water. Treatment should continue 24–48 h beyond remission of symptoms but should not exceed 5 days.

The meat WDT is 15 days, although previous work in our laboratory in the last 8 years suggests that the WDT should be extended to as long as 21 days (Buur et al., 2006; Mason et al., 2008). The tolerance and MRL values for all sulfonamides are 0.1 ppm and are harmonized across all countries and regulatory agencies.

12.3.3 Tetracyclines

There are at least 69 tetracycline drugs approved for use in swine with most them being chlortetracycline used as a feed additive. Most of the tetracyclines (oxytetracycline and tetracycline) used therapeutically in swine herds are administered as IM, oral, or as a water additive. One water additive product, Tetrasol Soluble Powder, is approved in the United States for the control and treatment of bacterial enteritis (scours) caused by *E. coli* and bacterial pneumonia associated with *Actinobacillus pleuropneumoniae*, *Pasteurella* spp., and *Klebsiella* spp. sensitive to tetracycline HCl. This drug is administered in a medicated drinking water to provide a dose of 10 mg/lb bwt/day for 3–5 days.

There are several IM oxytetracycline formulations such as Liquamycin LA-200, which have long WDTs of 28 days for a dose administered IM at 3–5 mg/lb bwt/day. The U.S.-labeled meat WDT for many of the tetracycline water additives is 4 days, but there is only one IM product, and it has a zero-day meat WDT. The U.S. tolerance for tetracycline varies considerably from MRL values (see Table 12.2).

12.3.4 Nonsteroidal Anti-inflammatory Drugs

Flunixin meglumine (Banamine) is the only approved nonsteroidal anti-inflammatory drug (NSAID) for use in swine at 2.2 mg/kg bwt by IM route of administration as a single dose for control of pyrexia with swine respiratory disease. The meat WDT is 12 days. It should be known that flunixin is approved in one other food animal species and it is approved for intravenous route only at doses of 1.1–2.2 mg/kg for up to 3 days but has a meat WDT of 4 days. Recent research has suggested that the emergence of flunixin drug residues in meat from cattle may be associated with extralabel use in beef and dairy cattle (Kissell et al., 2012) and disease (Wu et al., 2012). There is no data to suggest that extralabel use or disease may be associated with flunixin residues in swine. Recent pharmacogenomic studies have also suggested that there may be breed and gender differences clearance of flunixin in swine (Howard et al., 2013).

Meloxicam (Metacam®, Boehringer Ingelheim) was approved in 2010 by the European Medicines Agency (EMA) for use in the EU as an oral suspension in a concentration of 15 mg/ml for use in pigs, and there is an IM formulation (2% meloxicam) with a label single dose of 0.4 mg/kg. This drug is approved in the EU to reduce symptoms of lameness and inflammation in animals suffering from noninfectious locomotor disorders and acts as adjunctive therapy in the treatment of puerperal septicemia and toxemia (mastitis–metritis–agalactia (MMA) syndrome) in sows. It should be noted that as of December 2013, this drug is not approved for use in pigs in the United States although Metacam® was first launched for pigs in 2003 and is now licensed in over 40 countries around the world. Prior to its approval in the EU, this drug was approved for use in cattle as a single dose IM or SQ (0.5 mg/kg) for the treatment of acute respiratory infection in combination with appropriate antimicrobial therapy. The EU MRLs for this drug in both cattle and swine tissues are 65 ppb for either the liver or the kidney and 20 ppb for the muscle as it accumulates in the liver and kidneys of pigs. The EMA has recommended a meat WDT of 5 days for the approved label IM dose of 0.4 mg/kg in pigs although the cattle dose (0.5 mg/kg) by the same routes requires a 15-day WDT.

While it may be tempting for swine producers and veterinarians in the United States to use meloxicam and other NSAIDs because only one NSAID is approved for use in the United States, the AMDUCA law in the United States requires that the veterinarian use the approved drug, flunixin meglumine, before considering other NSAIDs.

12.4 PREVALENCE OF DRUG RESIDUES

According to the U.S. FSIS (2012) data for calendar year 2010, the main drug residues are antibiotics and sulfonamides in grower and slaughter weight pigs. Observations described earlier may be associated with these violations. In recent years, there have been concerns about penicillin and ampicillin residues in cull sows. In 2010, FSIS laboratories conducted analyses for clenbuterol, salbutamol, ractopamine, and cimaterol (β-agonists) on 14 selected market show hogs, but there were no violations. Carbadox, which is frequently used to prevent and treat enteritis and improve feed efficiency, was not detected in any of the 220 market weight pigs and 242 roaster pigs sampled in 2010 by the U.S. FSIS. Table 12.3 summarizes the results from the U.S. FSIS 2010 monitoring of market weight pigs, roaster pigs, and sows, which are the traditionally targeted swine class for drug and chemical residue violations. Residue violations in 2010 were very low and ranged from 0.06 to 0.13% for the three swine production classes.

TABLE 12.3 Summary of U.S. FSIS drug residue monitoring in 2010 of three swine classes

Swine class	Number of samples			Number of nonviolatives			Number of violatives		
	Sulfas	AB	β-agonist	Sulfa	AB	β-agonist	Sulfas	AB	β-agonist
Market hogs	221	278	1	0	1	0	2	0	0
Roaster pigs	136	292		0	4		0	1	
Sows	250	300		0	5		0	0	

AB, antibiotics; nonviolatives, residue levels detected but below tolerance level; sulfas, sulfonamides.

The following drugs and chemicals were identified in these very few residue violations: sulfamethazine, gentamicin sulfate, and polybrominated diphenyl ether. In addition to the aforementioned drug classes, the U.S. FSIS also sample for the following drug classes in pigs: nitrofurans, pesticides, and chloramphenicol and thyreostats in sows only.

The reader should be aware that sampling plans by the U.S. FSIS consist of the random sampling of tissue from healthy-appearing food animals. The number of samples for each drug class and pig production class as listed in Table 12.3 is dependent on many factors, which include determination of chemicals/drugs that are of a food safety concern and pairing these compounds with appropriate pig production classes. The U.S. FSIS has established a scheduled sample size of about 300 per drug class (see Table 12.3) as a public health standard, ensuring that with 95% probability, residues are detected if they exist at a rate greater than or equal to 1%. This domestic scheduled sampling plan also includes an inspector-generated sampling, and in 2010, there were 13,080 such samples tested in market pigs although 1,509 market pigs were scheduled for sampling. There are also import sampling plans for pork and pork products. In Denmark, more than 20,000 samples are analyzed for the antibiotic residues in pigs, which represent 0.1% of the entire pig population slaughtered per year. However, the prevalence of antibiotic residues in Denmark pork industry is low (~0.2 and 0.01% in sows and slaughter pigs, respectively), which is very similar to drug residue prevalence in the United States. Scientists in Denmark have recently proposed reduction in sample size that would be more cost-effective than current practices while increasing and maintaining probability of detection (Baptista et al., 2012). This is another example as cited elsewhere in this book of how more updated statistical and mathematical approaches could be utilized by regulators to manage drug residue in livestock.

Readers should be aware that USDA FSIS prioritized drugs as it did below for its 2013 sampling plans (FSIS, 2013) (Table 12.4). This includes

TABLE 12.4 FSIS domestic scheduled sampling for 2013

	Production	Class
Methods	Market hogs	Sows
Multiclass	600	600
Aminoglycoside	600	600
Pesticides	260	260
Metals	100	100
β-agonists	0	0
Avermectins	300	300
Carbadox	300	0
Nitrofurans	200	200
Arsenic	300	300

Multiclass can include analgesics/anti-inflammatory, β-agonists, β-lactams/cephalosporins, fluoroquinolones, hormones, macrolides/lincosamides, phenicols, sulfonamides, tetracyclines, and general drugs.

several drug classes that are in the multiresidue method such as **analgesics/anti-inflammatory** (oxyphenbutazone, flunixin, phenylbutazone, dexamethasone), **β-agonists** (salbutamol, cimaterol, ractopamine), **β-lactams/cephalosporins** (amoxicillin, cefazolin, desfuroylceftiofur cysteine disulfide (DCCD), ampicillin, penicillin G, oxacillin, cloxacillin, nafcillin, dicloxacillin), **fluoroquinolones** (desethylene ciprofloxacin, norfloxacin, ciproflox acin, danofloxacin, enrofloxacin, sarafloxacin, and difloxacin), **hormones** (prednisone, melengestrol acetate, zeranol), **macrolides/lincosamides** (lincomycin, pirlimycin, clindamycin, gamithromycin, tilmicosin, erythromycin, tylosin, tulathromycin), **phenicols** (florfenicol and chloramphenicol), **sulfonamides** (sulfadiazine, sulfathiazole, sulfapyridine, sulfamerazine, sulfamethizole, sulfamethazine, sulfamethoxypyridazine, sulfachloropyridazine, sulfadoxine, sulfamethoxazole, sulfaethoxypyridazine, sulfadimethoxine, sulfaquinoxaline, sulfanitran), **tetracyclines** (oxytetracycline, tetracycline, chlortetracycline), and **general drugs** such as 2-quinoxalinecarboxylic acid (QCA). There are a number of pesticides that are also listed in the high priority list of compounds and for which they have multiresidue methods.

12.5 MINIMIZING RESIDUES IN SWINE

The pork industry is concerned about consumer confidence and maintaining export markets for its pork products, and various organizations have provided guidance to producers and veterinarians to minimize the presence of violative residues in national and export markets. For the most part, this simply involves following quality assurance programs on the farm regarding prudent drug and chemical use on farm and stresses the involvement of a veterinarian in any drug or chemical use on the farm. It is also important as demonstrated in previous studies (Houpert et al., 1993) that IM injections are correctly administered, that is, injections into the neck perpendicular to the skin surface just behind the base of the ear. There are some regions in the United States that do not allow sulfonamides to be used on the swine farm because of the very long WDTs and the long history of residue violations in swine being predominantly associated with this drug class.

There is also the USDA Export Verification Program that aims to verify compliance with MRLs in export markets. For example, the Russian Product Verification Program (PVP) provided specific product requirements for marketing U.S. pork and pork products to the Russian Federation. This required a 14-day withdrawal period for all soluble and feed-grade tetracycline products to satisfy Russia's tetracycline MRL. If the pigs are for the domestic U.S. market, the label WDT of 4 days will apply for these specific products. Regarding the recent concerns about penicillin residues in cull sows, the NPB (National Hog Farmer, 2011), and other organizations have been recommending a 14-day WDT for procaine penicillin at labeled doses and 50-day WDT for benzathine penicillin. The latter is a long-acting penicillin salt that is not approved for use in swine but can result in persistent residues. In 2013, the USDA also launched a new certification program for livestock producers that permit them to market their products with a special "Never Fed Beta Agonists" label. The aim here is to open up previously closed or restricted markets in Russia, South Korea, China, Taiwan, and others to U.S. meat products.

REFERENCES

Baptista FM, Alban L, Olsen AM, Petersen JV, Toft N (2012). Evaluation of the antibacterial residue surveillance programme in Danish pigs using Bayesian methods. Prev Vet Med 106:308–314.

Buur JL, Baynes RE, Smith G, Riviere JE (2006). Use of probabilistic modeling within a physiologic based pharmacokinetic model to predict sulfamethazine residue withdrawal times in edible tissues in swine. Antimicrob Agents Chemother 50(7):2344–2352.

CFS (2013). Ractopamine Fact Sheet. Food Safety Fact Sheet. Washington, DC: Center for Food Safety.

Danish Agriculture and Food Council (2012). Danish Pig Production in a European Context. Available at www.lf.dk/~/media/lf/Aktuelt/…/LFEUBenchUK110318.ashx. Accessed April 6, 2014.

FSIS (2012). United States National Residue Program for Meat, Poultry, and Egg Products 2010 Residue Sample Results. Washington, DC: United States Department of Agriculture Food Safety and Inspection Service Office of Public Health Science.

FSIS (2013). United States National Residue Program for Meat, Poultry, and Egg Products 2013 Residue Sampling Plans. Washington, DC: United States Department of Agriculture Food Safety and Inspection Service Office of Public Health Science.

GFI #209 (2012). Guidance for Industry #209. The Judicious Use of Medically Important Antimicrobial Drugs in Food-Producing Animals. Washington, DC: Department of Health and Human Services, Food and Drug Administration, Center for Veterinary Medicine.

GFI #213 (2013). Guidance for Industry #213. New Animal Drugs and New Animal Drug Combination Products Administered in or on Medicated Feed or Drinking Water of Food-Producing Animals: Recommendations for Drug Sponsors for Voluntarily Aligning Product Use Conditions with GFI #20. Washington, DC: Department of Health and Human Services, Food and Drug Administration, Center for Veterinary Medicine.

Houpert P, Combrisson H, Nain SL, Autefage A, Toutain PL (1993). Intra- vs intermuscular injections in swine. Vet Res 24:278–285.

Howard JT, Baynes RE, Brooks JD, Yeatts JL, Bellis B, Ashwell MS, Routh P, O'Nan AT, Maltecca C (2014). The effect of breed and sex on sulfamethazine, enrofloxacin, fenbendazole and flunixin meglumin pharmacokinetic parameters in swine. J Vet Pharmacol Therap. DOI: 10.1111/jvp.12128 [Epub ahead of print].

Kissell LW, Smith GW, Leavens TL, Baynes RE, Wu H, Riviere JE (2012). Plasma pharmacokinetics and milk residues of flunixin and 5-hydroxy flunixin following different routes of administration in dairy cattle. J Dairy Sci 95:7151–7157.

Mason SE, Baynes RE, Buur JL, Riviere JE, Almond GW (2008). Sulfamethazine water medication pharmacokinetics and contamination in a commercial pig production unit. J Food Prot 71(3):584–589.

National Hog Farmer (2011). USDA reports violative penicillin residues in sows. Minneapolis, MN: National Hog Farmer. Available at nationalhoffarmer.com. Accessed April 6, 2014.

Wu H, Baynes RE, Tell LA, Riviere JE (2013). Prediction of flunixin tissue residue concentrations in livers from diseased cattle. Food Chem Toxicol 62:876–879.

13

CONFIRMATORY METHODS FOR VETERINARY DRUGS AND CHEMICAL CONTAMINANTS IN LIVESTOCK COMMODITIES

Hui Li

Division of Residue Chemistry, Office of Research, Center for Veterinary Medicine, FDA, Laurel, MD, USA

13.1 INTRODUCTION AND ESSENTIAL CONCEPTS

Foods from animal origin, such as meat (terrestrial and aquatic), milk and dairy products, eggs, honey, and so forth, comprise a significant nutrition source for the human diet, especially protein. With the continued economic growth around the globe, the demand for high-protein food is ever increasing. In modern times, the mass production of food animals, such as beef cattle, milk cows, hogs, poultry, and aquacultured seafood, relies on proper use of veterinary drugs to prevent and treat diseases. In addition to treating individual sick animals, these drugs can also help maintain herd and flock health, promote growth, improve meat quality, reduce production costs, and drive profit. In the United States, only approved veterinary drugs can be used in compliance with the prescribed indication, administration route, dose, and period (FDA, 2013a) unless used in accordance with the extralabel drug use regulations (FDA, 2013b). In addition, there is usually a withdrawal period

Strategies for Reducing Drug and Chemical Residues in Food Animals: International Approaches to Residue Avoidance, Management, and Testing, First Edition.
Edited by Ronald E. Baynes and Jim E. Riviere.

requirement (or discard period requirement for milk) associated with a predetermined tolerance level (FDA, 2013c), during which the treated food animal should not be sent for slaughter or the milk from the treated animal should be discarded. This measure ensures that drug residues in edible products decrease below levels of health concern before the animal-derived products enter the market.

Nevertheless, improper use of drugs in food animal farming is a leading cause of drug residue presence in human food above the tolerance limit; approved drugs are being applied contra product label and qualified veterinarian instructions (extralabel use is allowed for some drugs under restrictive conditions). In addition, harmful, unapproved chemicals may enter the human food chain illegally because of the low cost and immediate efficacy of using the unapproved drugs for disease control and growth promotion. Such drugs include malachite green and nitrofurans. Environmental pollution (e.g., dioxin), deliberate economic adulteration (e.g., melamine and Sudan Reds), natural toxin production (e.g., mycotoxins and shellfish poisons), or agricultural activities (e.g., pesticide application) are other ways toxic substances are delivered to food and feedstuff. Issues causing public health concern and even serious consequences, such as melamine addition to pet food that led to animal kidney failure and death, have continuously arisen over the past decade. To address these issues, the United States passed a series of laws laying the foundation for regulation of food safety and veterinary drug use: Food, Drug, and Cosmetic Act of 1938 and amendments thereafter (FD&CA), Animal Medicinal Drug Use Clarification Act of 1994 (AMDUCA), and Food Safety Modernization Act of 2011 (FSMA). The Food and Drug Administration (FDA) and U.S. Department of Agriculture (USDA) control the use of veterinary drugs by approving safe uses based on scientific data and implementing routine inspection and monitoring programs to test for unsafe residues in both domestic and imported products. To effectively enforce these laws, analytical methods are needed to unambiguously identify, confirm, and accurately quantify trace concentrations of harmful chemicals in various food samples. These methods are specifically called regulatory methods because they directly serve government agency decision-making, litigation, and public health surveillance.

For many veterinary drugs, specific residue levels (termed "tolerance" levels in the United States or "maximum residue limit (MRL)" in the European Union [EU]) are explicitly stipulated in the pertinent regulation or directive, based on toxicology and risk analysis. Approved veterinary drugs typically have an approved nonzero tolerance (associated with a specific type of biological matrix), or the tolerance has not been established (the drug is conditionally approved), while banned substances have a zero tolerance (such as a "group A substance," defined in 96/23/EC [Council Directive 96/23/EC,

1996]). Substances are banned mainly due to concerns about side effects such as endocrine disruption, carcinogenicity, or neoplastic anemia, or the likelihood of generating antimicrobial resistance to the critically important human antibiotics. Commonly encountered tolerance levels are often part per million (ppm), mg/kg, or part per billion (ppb), μg/kg, for chemical residues in food samples (matrix). When the statutory level is "zero" or is not explicitly expressed, the target level is often set at a practical and technically achievable limit ("action level" or "safe level" in the United States, "minimum required performance limit [MRPL]" in the EU). It is important for regulatory methods (both quantitative and qualitative) to establish a concentration range to bracket the tolerance concentration and give reliable assay results throughout the entire range.

Common biological matrices that are relevant to monitoring veterinary drug use in food animals are the kidneys and livers of many terrestrial animals because these organs are often the metabolizing and excreting organs where drug residues concentrate or persist the longest. For aquatic species, meat (muscle) is usually the target matrix. Other common matrices include milk, eggs, and honey. It is typical to use a bench procedure to extract target analyte(s) out of the sample matrix before instrumental analysis. The extract is a very complex mixture, analysis of which is often metaphorically dubbed as "finding a needle in a haystack." In the United States, USDA routinely inspects commodities such as beef, pork, poultry, egg, and lamb by implementing the National Residue Program (FSIS, 2011), which includes not only veterinary drugs but also pesticides and other organic and inorganic contaminants. The FDA has its own residue monitoring program covering aquaculture, milk, and honey, for both domestic and imported goods. Alternative biological matrices such as plasma, urine, hair, tears, or saliva might be of interest to analysts, because banned substances (or metabolites) could be present and persist in these matrices (Gratacós-Cubarsí et al., 2006).

Depending on the question to be answered, there are four basic types of regulatory methods (Table 13.1). Among them, screening, confirmation, and identification methods are considered qualitative analysis. A confirmatory method differs from the other three categories in that it is the only type of method that must be conducted for regulatory action against violative goods. Screening is not always needed before confirmatory or quantitative analysis if intelligence from other avenues suggests the possibility of a specific substance's presence. For unapproved drugs or banned substances, quantitation is not as critical as confirmation because any level of confirmed analyte constitutes adulteration. In practice, many regulatory methods can answer more than just one type of question listed in Table 13.1. For example, there are methods that can both quantify and confirm the identity of the sought-after analyte. Some others can serve multiple purposes such as screening, confirmation, and

TABLE 13.1 Four types of regulatory methods for residue analysis

Terminology	Plain language interpretation[a]	Information type	Confidence metrics	Example of applications
Screening (indication)	Is it possibly present in there?	Logical (yes/no)	FP rate (usually allows a low FN rate)	On-farm or in-factory residue testing
Determination (quantitation)	How much is it in there?	Numeric	Accuracy and RSD	Laboratory residue analysis of suspected samples or surveillance samples
Identification (semitargeted; nontargeted)	What is it (that could have generated the specific signal)?	Unique chemical ID (molecular formula or structure)	Similarity score; ranking (among possible hits)	Emergency response to find out what has caused certain adverse effects in animals; metabolite identification in drug discovery or toxicology studies
Confirmation	Is it truly what it is thought to be?	Logical (yes/no)	FN rate (must have very low FP rate)	Laboratory residue analysis of suspected samples

[a]"It" refers to a chemical substance of interest; "there" refers to a sample or target commodity.

semiquantitation at the same time. In a slightly different system, four types of regulatory method are defined in 2002/657/EC (qualitative screening, quantitative screening, qualitative confirmation, and quantitative confirmation; Commission Decision 2002/657/EC, 2002), and three types are classified by the Codex Alimentarius for residues of veterinary drugs in foods (screening, quantitation, and confirmation, along with quantitation if necessary; FAO/IAEA, 1998). Confirmatory analysis is an indispensable part of routine analysis in regulatory laboratories around the world for surveillance and food safety testing. It is the critical piece of evidence to support the finding of specific residues in a sample.

One common way to evaluate the performance characteristics of a qualitative method (including confirmation) is to undertake a predetermined number of repeated assays with both known negative and positive controls. The result can be presented in the format as shown in Table 13.2. Four metrics, that is, false-positive (FP) rate, false-negative (FN) rate, specificity rate, and selectivity rate, are often used to characterize the method's performance. It should be noted that there is a subtle difference between the terms "positive" and "noncompliance." In this chapter, "positive" refers to definitive confirmation of identity (a technical term), while "noncompliance" means the sample is in violation of applicable law because of the residue (a regulatory

TABLE 13.2 Parameters to evaluate performance of qualitative methods

	Negative result (=tn + fn)	Positive result (=fp + tp)	Total counts = tn + fn + tp + fp
Negative sample (=tn + fp)	tn = number of true negative outcome	fp = number of FP outcome	FP rate = fp/(tn + fp) Specificity rate = tn/(tn + fp)
Positive sample (=fn + tp)	fn = number of FN outcome	tp = number of true-positive outcome	FN rate = fn/(fn + tp) Sensitivity rate = tp/(tp + fn)

FN is failing to declare what is present (β-error); FP is to assert something that is absent (α-error); sensitivity rate = 1 − FN rate; specificity rate = 1 − FP rate. That is, the lower the FP rate, the more specific a method is; the lower the FN rate, the more sensitive a method is.
Both "specificity" and "sensitivity" are narrowly defined terms used here to characterize a qualitative method, not to be confused with their more general meaning in other places in this chapter.
There is no direct correlation between specificity rate and sensitivity rate.

term). Not only must the identity of the residue be confirmed, its concentration must exceed the pertinent tolerance or safe level. On the other hand, while a "negative" finding (not confirmable) is sufficient to conclude "compliance" (though only for the analytes being monitored), the reverse may not necessarily be true. A compliant sample may contain a drug residue at a level that can be confirmed by a particular method yet is still below tolerance. Sometimes, depending on need, quantitative (or semiquantitative) results can be converted to qualitative ones by comparing individual values to a preset threshold, yielding "positive" or "negative" readings. Information content will decrease, however, when numeric data are turned into binary data. The "performance characteristic curve" is another way to descriptively evaluate a qualitative method, by plotting the frequency/probability of "being confirmed" (from actual assay outcomes) against analyte concentrations (Trullols et al., 2004).

The term "confirmation" is defined in the Oxford online dictionary as "(to) establish the truth or correctness of (something previously believed, suspected, or feared to be the case)" (Oxford Dictionaries, 2013). Within the residue analysis community, it is widely accepted that confirmation (generally described as "analyte 'X' in matrix 'M'") of chemical identity can be established if a group of preset measurable characteristics that is unique to this analyte matches the values of the contemporaneously analyzed reference standard (the same compound). In theory, every molecule can be considered unique from all other chemicals, because every compound has a distinctive structure (restricting our discussion to organic compounds). Due to the limited number of physical parameters that can be measured by real-world instruments and their finite resolving power, there is no guarantee that every compound in the "chemical space" gets a unique combination of defined physical parameters that can distinguish it from everything else. For example, LC or GC can separate analytes according to their interaction with both the stationary phase and mobile phase, and the resulting retention time (RT) is the measurable value that can discriminate analytes. Mass spectrometry (MS) can distinguish analytes based on their molecular weight and ion formation characteristics, where the mass-to-charge ratio (m/z) value is the physically measurable parameter. If the original (precursor) ion is broken up on the flight path and the fragments (product ions) are monitored, additional differentiation power is supplied because structurally distinct gas-phase ions tend to disintegrate upon excitation in characteristic ways. In contemporary residue analysis, MS interfaced with GC or LC is deemed the de facto gold standard for confirming existence of compounds of relatively low molecular weight (roughly < 1000 Da) in complex matrices, although use of more than one distinctive non-MS methods is sometimes recognized. The most significant advantage of MS is its combined selectivity, sensitivity, and applicability to chemical substances. In most applications, these techniques provide enough total resolving power so the chance for one analyte to be indistinguishable from a different compound is minimal. The trustworthiness of a confirmation finding is usually better if two fundamentally different methods are used (Heller et al., 2010), or even two analysts (Lehotay et al., 2008) perform analysis independently on the same sample lot. This is more expensive, however, and the intended purpose may not always warrant the effort. In this context, the meaning of "confirmatory method" is restricted to qualitative, single-lab analysis, while the contemporaneous use of an authentic, high purity chemical standard is a prerequisite.

There is a subtle difference between "confirmation" and "identification," although sometimes they are used interchangeably. Confirmation of identity proves with high confidence that a (found) substance is what it is claimed to be, whether it is pure matter or is mixed with other material. Residue analysis is ideal for the latter situation when the concentration of a target compound in the matrix is extremely low. In Center for Veterinary Medicine's (CVM) Guidance for Industry (GFI)-118, confirmation is specifically defined as "… unambiguous identification of a compound's presence by comparison to a reference standard" (FDA, 2003). Conversely, identification is to reveal the identity (e.g., chemical structure) of a previously unknown substance or one that is not yet known to the analyst. Confirmation is more often a requirement to establish a solid basis for regulatory action, while identification is more likely to be part of early investigation or scientific research, such as new drug discovery, toxicology, or not-for-cause surveillance. With an authentic chemical reference standard, confirmation can verify a "known unknown"; however, an "unknown unknown," for which no reference standard is available, cannot be confirmed. Identification can be applied in both situations, even in the absence of a chemical reference, if adequate knowledge exists (such as searchable chemical databases) about the nature of the detection technique and the specificity of the signals obtained. Under this circumstance, the level of confidence in the assignment of identity is lower. In practice, identification effort is initiated due to a suspicious finding in screening, clinical signs, or even a nonscientific cause. If the analyst knows what particular analyte(s) to look for before analysis, it is called "targeted analysis," and confirmation can be done along with the concurrently analyzed chemicals for the exact analytes. At other times, the analyst does not know exactly what to look for, such as at the beginning of a criminal investigation. The work is generally referred to as "non-/untargeted analysis" or "semitargeted analysis." For example, if it is known only that "novel" or "wonder" drugs were illegally designed and used, then the analyst has no knowledge of the drugs exact chemical identities until identification is made based upon various analytical methodologies. If the legal or social stakes are high enough, the suspected compound may be synthesized and analyzed along with the "unknown unknown." Then confirmation is possible. That being said, there is considerable overlap in the techniques to either confirm or identify a chemical substance.

13.2 INSTRUMENTATION AND TECHNIQUES

Today, a wide variety of instruments are available for trace analysis of organic substances, such as GC coupled with electron capture detector (ECD), nitrogen–phosphorus detector (NPD), flame ionization detector (FID), or EI-MS, and HPLC coupled with UV/Vis, diode array detector (DAD), fluorescence, IR, or various types of mass spectrometers. Some of these instruments, however, are considered inadequate for confirming a chemical's identity. For example, HPLC-UV (single wavelength) is customarily used for quantitation, yet it cannot differentiate coeluted analytes that have absorbance at the set wavelength, which is quite common. In contrast, signals generated by liquid chromatography–tandem mass spectrometry (LC-MS/MS) have enough selectivity that the probability is quite low that another chemical that coelutes with the target analyte will have the same molecular weight and fragmentation pattern upon collision-induced disassociation (CID). In essence, specificity, deemed as having enough selectivity to confidently distinguish the analyte from any other molecule, is at the heart of all confirmatory methods. Even though, in practice, confirmation is achieved by successfully matching the signal's profile between the unknown sample (usually a fraction isolated by chromatography) and a reference standard, it is critical to understand that the essence of confirmation is the exclusivity and uniqueness of the signal–analyte relationship rather than the signal's similarity to that of a known reference. During the past decades, MS has improved significantly, and it is widely accepted as the gold standard for confirmation. For example, in Appendix A of USDA's Red Book (2009) reporting on the National Residue

Program (FSIS, 2011), three types of methods are listed for screening, quantitation, and confirmation. While screening and determinative methods use both bioassay and instrumental methods, confirmatory methods are almost exclusively MS based (except atomic absorption spectroscopy [AAS] for arsenics).

Many different types of mass spectrometers are based on the principle and design of the analyzer, the core of a mass spectrometer. These include single quadrupole (Q1), triple quadrupole (QqQ), quadrupole ion trap (QIT) (3-D or linear trap), time of flight (TOF), Fourier transform ion cyclotron resonance (FT-ICR), Orbitrap, and hybrids of the above (Q-TOF, Q-Orbitrap, ion trap–Orbitrap, ion trap–FT-ICR, TOF–TOF, and so forth). Each of these MS technologies is based on different hardware constructions and electrical field dynamics, and all are capable of manipulating ion movement to obtain signals at the *m/z* of interest. Magnetic field MS was among the earliest of MS technologies, but its use in regulatory laboratories is relatively limited today. QqQ is able to perform a variety of scan modes such as selected ion monitoring (mode) (SIM), selected reaction monitoring (mode) (SRM) (or multiple reaction monitoring [mode] [MRM]), neutral loss scan, product ion scan, and precursor ion scan. When used in the SRM (MRM) mode, it is generally considered more sensitive than other types of MS because the duty time is spent on selected transitions only, though it is not exactly fair to compare a point-scan type of analyzer to a full-range-scan one. QIT MS has a unique feature to conduct ion fragmentation to the *n*th level (precursor ion to product ion to subproduct ion, and so forth), which provides an in-depth view of the chemical structure of the ion. The most significant advantage of TOF, FT-ICR, and Orbitrap MS is the ability to acquire high-resolution *m/z*—with resolving power from 50,000 to 500,000 to distinguish isobaric ions—providing a basis for calculating a precise chemical formula, which is invaluable for unknown identification.

The underlying principle for the quadrupole type of MS is the Mathieu Equation. It describes movement of an ion in an alternating electromagnetic field and the "parameter space," where ions of specific *m/z* can travel in stable and confined trajectory, while other ions will sooner or later fly out of boundary in the flight path or ion trap and be unable to reach the detector. The use of single-stage quadrupole in confirmation was mostly coupled with GC and electron impact (ionization) (EI), and four ions are typically monitored for one analyte. If "soft" ionization is used, the cone voltage (or equivalent) must be tuned so that in-source CID generates meaningful fragments; then the analysis can satisfy the pertinent confirmation criteria. It is worth noting that although the fragment ions were probably generated from the precursor ion, there is no intrinsic mechanism like the QqQ MS to ensure there is a causal relationship between the precursor ion and the hypothetical product ions. The latter could be merely a coincidental interference.

QqQ MS is probably the most widely used instrumentation for not only confirmatory methods but also multipurpose methods that combine screening, confirmation, or quantitation in one analysis. QqQ has three consecutive quadrupole stages, where the first and third quadrupoles can be precisely controlled in concert. Thus, the precursor and product ions (after fragmentation) can be selected simultaneously, drastically enhancing the instrument's selectivity over Q1-MS. QIT is another type of single-stage MS, which can physically hold the "ion cloud" within the trap by applying alternating current/direct current (AC/DC) voltage on the opposite electrodes. It is usually used in full-scan mode to generate mass spectra with many more data points, and consequent information content, than the MRM data. Moreover, it has the advantage of generating higher stages of fragmentation (power ≥ 3) sequentially to provide additional structural information. This is in contrast to a single CID process, which possibly produces more than just primary product ions, with the causal relationship among all the different fragments unapparent in a full spectrum. Additionally, helium is the typical collision gas, so QIT sometimes generates a different mass spectrum than QqQ or Q-Orbitrap MS, which use nitrogen or argon as the collision gas. For method developers, the choice between QqQ and QIT for confirmatory analysis will eventually depend on the fragmentation characteristic of the analyte ion. If the precursor ion yields an adequate number of structurally informative product ions (≥ 2) with stable relative intensity (ion ratio), MS/MS is deemed specific enough. High-stage fragmentation is often unnecessary because the relatively slow scanning speed for QIT MS overrides its advantage. On the other hand, if a precursor ion does not generate characteristic product ions (such as merely losing a universal H_2O or CO_2 molecule) after the first CID process or it generates a number of nontrivial fragments but their signal intensity is too low or unstable over time, QIT is worth considering.

Another technique, single-stage high-resolution MS (HRMS), is quickly gaining popularity in residue analysis, including confirmation. This instrument can provide resolving power from about 10,000 ($R = M/\Delta M$ at full width half maximum [FWHM]) to over a million. In contrast to the unit resolution quadrupole MS models, HRMS has superior mass accuracy to enable sufficient selectivity to unequivocally elucidate an ion's elemental composition up to 300 *m/z*, in combination with isotope abundance ratios (Grange and Sovocool, 2008), or to differentiate molecular formulas with very subtle differences in mass, such as one subunit of CO, C_2H_2, or N_2, at 70,000 FWHM resolution (valid for singly-charged ion at *m/z* 400 or lower) (Nielen et al., 2007). This makes meaningful database search and match possible with only the molecular ion's accurate mass as information. The TOF analyzer uses a pulsed electric field to accelerate all the ionized species and then measure the time the species take to reach the detector. For ions with the same charge, the imposed kinetic energies will be identical, and their acquired velocities will depend only on their masses. Thus, lighter ions will have higher velocity and reach the detector sooner (shorter flight time) given the same initial push and the same flight path length. TOF instruments have exceedingly high acquisition speeds (e.g., 20,000 full spectra per second) supported by both ultrafast analog-to-digital converter (ADC) and microchannel-plate-multitime-to-digital converter (TDC) technologies. Current commercial models, however, still have lower resolution (up to 40,000 FWHM) than Orbitrap MS or FT-ICR-MS.

Although Orbitrap MS includes a component that can temporarily store ions, its signal detection principle is completely different from QIT's. The core of the integrated mass analyzer and detector is a spindle-shaped electrode around which ions are trapped by electrostatic force. The ions are both orbiting around the central electrode (x- and y-axis) and oscillating back and forth along the electrode's long (z-)axis. Oscillation of each ion generates an image current of specific frequency, and the resulting composite signal is detected by the outer electrode and converted to full-scan mass spectra via Fourier transformation. The Orbitrap MS not only has much higher mass resolution but also has comparable, or even better, sensitivity and dynamic range than QqQ MS (Hu et al., 2005). The high resolution of Orbitrap MS, however, is accompanied by a relatively slow acquisition speed, typically only a few full spectra per second. Although this aspect could be a concern for quantitative applications, in general, it does not affect its use in confirmation because the minimum number of scans across the liquid chromatographic peak for a determinative method (e.g., 15) is not usually required. Lastly, FT-ICR-MS is a type of MS using a fixed magnetic field to trap ions and measure them via excitation-induced current. It can provide higher resolving power than other types of MS because the superconducting magnet provides a much more stable and consistent field than an alternating radio frequency field. Commonalities between Orbitrap and FT-ICR-MS are as follows: (i) The mass analyzer and detecting circuit are spatially inseparable; (ii) all ions are detected simultaneously without being time or spatially resolved first; (iii) ions are monitored via the image current from their movement, rather than having to physically hit the photomultiplier to induce a signal; and (iv) the same mathematical approach (Fourier transfer) is used to process the raw "signal package." The apparent disadvantage for FT-ICR-MS is its high cost, which prevents it from being widely used in food safety monitoring. For small MW compounds that comprise the bulk of chemicals of regulatory concern, its ultrahigh mass resolution is probably unnecessary.

Moreover, dual-stage HRMS, such as Q-TOF and Q-Orbitrap, can provide higher overall selectivity for confirmation purposes. With this instrumentation, the first-stage analyzer is usually a unit mass resolution filter to select out the targeted ion (along with other isobaric ions). After CID or high-energy collision disassociation (HCD) in a second quadrupole, all resulting ions are analyzed in the second-stage, high-resolution analyzer. Although the acquired full spectrum contains more than just fragments of the target ion, high mass resolution provides a way to differentiate them from interfering ions. For example, if the elemental composition of both the precursor and product ions can be narrowed down to small numbers, some obviously impossible combinations can be ruled out, and thus, the causational relationship between the right ions can be established. This is a useful feature for identification of "known unknown" or even "unknown unknown" compounds and for structural elucidation when a reference standard is unavailable. Full capacity is best used in investigative works rather than routine analysis. Dual-stage HRMS provides the flexibility for using the second-stage HR analyzer only (as a single-stage HRMS) for combined screening, confirmation, and/or quantitation. A significant benefit of acquiring full-scan MS data over a MS/MS experiment is that the data file can be revisited to look for something previously unchecked, without redoing the whole sample analysis procedure (assuming the "old" sample is still available).

Presently, LC or GC is virtually an indispensable part when using MS for confirmation applications. In the chromatographic process, a mixture of compounds is either dissolved in the mobile phase or vaporized in a carrier gas, passing through a solid adsorbent material packed in a column. Each analyte in the loaded (injected) sample aliquot interacts differently with the adsorbent material, depending on its specific molecular structure, functional groups, hydro-/lipophilicity, size, ionization form (in the case of LC), and solubility in the mobile phase. Based upon the fundamental mechanisms of analyte–stationary phase interaction, liquid chromatography can be classified into normal phase, reverse phase, ion exchange, or size exclusion, but it is not unusual for one type of packing material to interact with analytes via more than one mechanism. Because all compounds constantly adsorb onto and desorb from the stationary phase, the carrier gas or liquid flow drives each "packet" of analyte downstream to the other end of the column. In general, the stronger the interaction with the solid phases, the longer the elution time. By separating each analyte from the other molecules in the liquid or gas phase, chromatography also provides significant resolving power for confirmation of identity. Note that the "analyte-specific" RT is more than just a function of the intrinsic molecular properties. It also depends on chromatograph conditions such as column type, mobile phase, and pH. This is in contrast to MS analysis where molecular weight is the sole factor to differentiate chemical species, although in MS/MS experiments, CID condition also significantly affects the relative abundance of the characteristic fragments. In theory, GC and LC's resolving potential can be considered unlimited, because if the column is infinitely long, there will be enough resolving capacity for all analytes. In real applications, however, the column is always of limited length. The amount of resolution used in a chromatographic analysis is a balance between the need to separate the analyte from other compounds in the mixture and the practical need to achieve a reasonable sample throughput in the testing laboratory. In recent years, small-size LC column packing material with sub-2-micron particles has become popular. Compared to previous generations of columns that use larger particles (3–5 microns), resolution may be increased several times. A specialized LC system such as ultrahigh performance liquid chromatography (UHPLC) must be used, however, to achieve very high pressure (e.g., 120 MPa) needed to reach the optimum flow rate correlating with the highest column efficiency. The chromatographic peak width under such conditions can be as narrow as a few seconds. While the resolving power of LC is increased in this way, the coupled MS needs to have adequate scanning speed to generate a statistically sufficient number of data points. Other techniques to enhance chromatographic selectivity are to use GC×GC or two-dimensional (2-D) LC to achieve the extra resolution supplied by the two columns, usually with orthogonal retention mechanisms. FPs caused by inadequate resolution from LC have been discussed in several publications (Berendsen et al., 2013a; Croley et al., 2012).

The commonly used soft ionization sources such as electrospray ionization (ESI) (Ionspray, TurboIonSpray), atmospheric pressure chemical ionization (APCI) (Heated Nebulizer), and atmospheric pressure photoionization (APPI) (other, less popular, ionization techniques are used as well) are not considered a significant contributor of selectivity for confirmatory purposes. When choosing among these ionization mechanisms for various applications, ionization efficiency is usually the main factor to consider rather than whether the source can differentiate the analyte with the coeluting interference. An ionization mechanism sometimes plays a significant role, however, in a confirmatory method's functioning. For example, it has been reported that when either ESI or APCI source was used, malachite green and gentian violet's mass spectra were acquired with high background (Li et al., 2006). However, when the voltage (or charge current) was turned off, the signal intensity was much higher and the background reduced significantly to become very clean. In the same paper, APCI negative mode was set for toltrazuril sulfone to produce an easily fragmenting precursor ion at 456 *m/z* (probably a radical anion). If negative ESI mode was used, only the $[M\text{-}H]^-$ ion was generated, which did not produce any significant fragment ion no matter how high the collision voltage is. The *m/z* 456 anion (or anionic radical) was deemed to result from electron capture processes at the source.

Sample preparation should not be overlooked as another source of selectivity for confirmatory methods, although in most cases its role is rather limited compared to MS and chromatography. The primary goal of sample treatment is to separate, as much as possible, the analytes of interest from the bulk part of matrices, such as protein, fat, polysaccharide, and inorganic salts. Depending on the scope of the intended use, some methods use very generic extraction schemes, that is, as much as possible, collecting into the final extract many analytes of very different chemical–physical properties at the expense of coextracting a large amount and variety of potentially interfering or signal-suppressing substances. Examples include direct buffer dilution ("dilute and shoot") of urine and plasma and protein precipitation/hexane defatting of whole milk. Other commonly used extraction and cleanup techniques, such as liquid–liquid extraction (LLE), solid-phase extraction (SPE), supercritical fluidic extraction, microwave-assisted extraction, MW-cutoff membranes, and so forth, are all possible choices for prospective methods. Versatile formats exist for convenient use of SPE, such as cartridge, 96-well plate, disc, solid-phase microextraction, matrix solid-phase dispersion (MSPD), dispersive solid-phase extraction (dSPE), in-tube capillary column, and pipette-tip SPE. In 2003, a new extraction with the acronym "QuEChERS" (quick, easy, cheap, efficient, rugged, safe) was introduced and has been widely adopted (Michelangelo et al., 2003). Briefly, the sample is first extracted with an organic solvent (such as acetonitrile [ACN], ethyl acetate [EtOAc], or acetone), with salts added (usually NaCl and $MgSO_4$) to induce phase separation. After shake-mixing and centrifugation, a defined quantity of sorbent (e.g., C18, PSA) was added to the organic phase (or a portion of it) to absorb excess "background" material for a cleaner extract while the analytes remain dissolved in solution. Hexane wash is sometimes performed as necessary. In contrast, rather complex extraction procedures exist, with parallel or sequential extraction steps followed by multistage SPE cleanup, to achieve adequate recovery for analytes of various compound classes in a relatively clean extract before injecting into LC (Hammel et al., 2008).

The opposite of generic extraction is compound/class-specific extraction, which takes advantage of unique molecular structure moiety common to a compound class, such as the phenylamino group of sulfonamides, β-carbonyl carboxylic acid/amide group of fluoroquinolones and tetracyclines, cyclic polyether structure of ionophores, and multiple carbohydrate units of aminoglycosides. The most powerful sample preparation techniques in terms of selectivity are immunoaffinity cartridges or molecularly imprinted polymers to retain and purify targeted analytes. The selectivity is usually very high, yielding extracts with low levels of soluble solids, and serves as supplemental evidence by itself in confirmation of identity. The downside of using highly substrate-specific SPE is that it applies only to a narrow class of compounds and may not be compatible with use in multiclass methods. There are a few excellent reviews on sample preparation of food matrices for analysis of veterinary drugs, albeit not solely for confirmatory methods (Berendsen et al., 2013b; Marazuela and Bogialli, 2009; Moreno-Bondi et al., 2009). Lastly, some of the cleanup work traditionally done in offline sample preparation is now performed by chromatography. A variety of LC-related technologies, such as restricted-access material, core–shell (a thin layer of a porous stationary phase supported by a solid silica core), and Turboflow (large particle size) are all available for online cleanup. These stationary phase materials can add an orthogonal function of separating small molecules from large biomolecules via size exclusion per se to the analytical column's resolving capacity toward small molecules.

As with other analytical techniques, the LC (or GC)-MS^n measured signal contains both random error and nonrandom bias. Many factors contribute to the magnitude of the overall error, such as volume/weight-measuring devices, instrument and laboratory environment stability, lot-to-lot quality variability of reagents and other lab consumables, and sample nonhomogeneity. Moreover, ubiquitous existence of coanalyzed compounds in the matrix may generate a signal that overlaps with that of the analyte (a nonrandom interference with its own random error). The relative effect becomes larger when the analyte's concentration (or analyte-to-matrix ratio) is lower. Particularly for qualitative methods, the combined signal distortion from all errors may exceed a certain threshold, causing false readings. Therefore, a method's specificity is defined as its power to "distinguish between the analyte being measured and other substances. This characteristic is predominantly a function of the measuring technique described, but can vary according to class of compound or matrix" (Commission Decision 2002/657/EC, 2002). Specificity depends not only on the method's own intrinsic merit but also the complexity of the sample matrix.

When developing a new LC-MS-based confirmatory method with a firm scientific basis and sufficient ruggedness, all the aspects discussed in this section deserve consideration. On the other hand, it is also important to remember the "fit-for-purpose" concept to avoid overexploiting the potential of each resolution-contributing factor while unnecessarily increasing complexity, raising costs, "chasing zero," and decreasing practicality and adoptability (Bethem et al., 2003).

13.3 METHOD DEVELOPMENT, VALIDATION, AND OFFICIAL GUIDELINES

Regulatory methods are used in laboratories for analysis of samples that are of commercial or legal relevance. Therefore, it is critical that these methods are based on sound science, are precisely written, and, preferably, are not too technically demanding or laborious to be routinely implemented. These

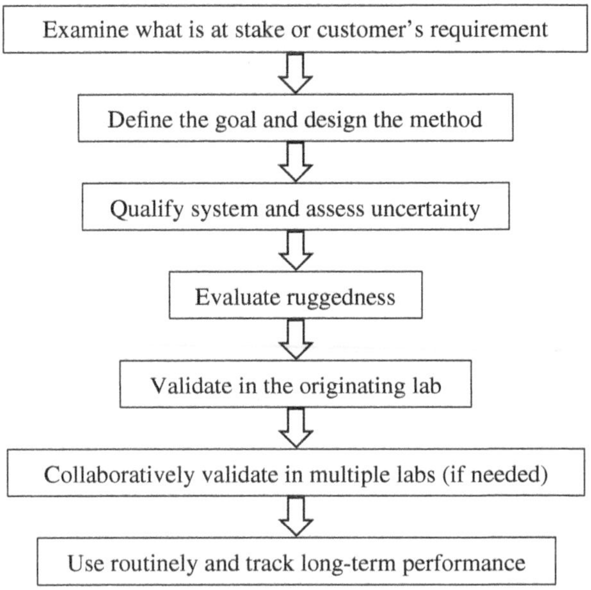

FIGURE 13.1 Typical steps in regulatory method development, validation, and routine use.

methods are often prescribed as Standard Operating Procedures (SOPs). There should be components that define the scope, scientific principle, reagents and apparatus, standard solution preparation, sample treatment (homogenization, extraction, cleanup, and so forth), instrument parameters, expected performance (system suitability and quality control [QC]), data acceptance criteria, calculations, what/how to report, and other issues. Regulatory laboratories are often run under good laboratory practices (GLP) or International Standards Organization/International Electrotechnical Commission (ISO/IEC) 17025 (General Requirements for the Competence of Testing and Calibration Laboratories) accreditation. Figure 13.1 shows the typical steps in regulatory method development, from the design stage to routine use.

The most fundamental elements for confirming the identity of an analyte in a sample are based on chromatographs and/or mass spectra "extracted" from the raw data file. When raw data are presented as reconstituted ion chromatograph (RIC) for a certain ion or transition, the RT, peak height or area, signal-to-noise ratio (S/N), and occasionally peak width and shape characteristics can be used for the basis of (i) comparison to an arbitrary threshold (e.g., $S/N > 3$) or (ii) comparison with a reference standard (e.g., $\Delta RT/RT < 5\%$ or peak area ratio within ±10% of mean value). For data acquired in SRM or SIM mode, it is common to select more than one transition or ion as the identification feature for each analyte; therefore, consistency among the corresponding RICs (RT, peak shape) is vital for confirmation. When data are presented as full-range mass spectra (if acquired in full-spectrum mode), the existence of diagnostic ions and their relative abundance, predominance of the base peak, mass accuracy and isotope ratio (for HRMS), and sometimes overall cleanness of the spectrum are examined either against a reference standard or arbitrary values to determine whether the analyte can be confirmed with an acceptable level of confidence in the sample. Correspondingly, suitable confirmation criteria for a method will depend on instrument type, acquisition mode, and intended use. The minimal (least stringent) acceptable criteria are usually prescribed in official guidance documents, some of which are discussed later.

Validation of a regulatory method is required for its formal use as part of an overall quality management plan. There are established protocols or guidance on how to validate a confirmatory method or a method that has a confirmatory aspect. Validation should address all the essential items such as what elements and criteria constitute confirmation of chemical identity, what kinds of samples and how many repeats are to be assayed, and at what concentration level(s) fortification occurs. Three types of samples are relevant to validation: negative (blank) control, positive control (spiked directly onto the sample), and incurred specimen (obtained from dosed animals). Although incurred samples are preferred, these may not always be readily available, especially for multiresidue methods; therefore, they are not required in every circumstance. Typically, a set (batch) of samples, including at least negative and positive controls, are processed manually or robotically over the same time span and then analyzed by instrument virtually continuously. Control samples spiked at a pivotal concentration, such as tolerance, MRL, or limit of confirmation (LOC), should be analyzed during the method development stage to establish that the analyte can be reliably confirmed. Usually, there is no requirement regarding the recovery of analytes if a method will not be used for quantitation purposes. System suitability and ruggedness testing before validation is highly recommended. The former measure can limit the extent of error originating from the instrument. Carefully designed ruggedness testing may reveal critical or sensitive parameters that affect the method's performance, so that a different user can be aware of potential pitfalls, as minor operational deviations are almost unavoidable in the laboratory. Because of the high cost of operation of collaborative studies, a method should be thoroughly validated by the originating laboratory or analyst before being distributed for further validation. In the method validation stage, results from multiple batches (done on different days according to the same SOP) are compiled and evaluated. If all preset acceptance requirements are met, including multiple-lab check as appropriate, the method is deemed as fully validated. One way to understand the essential elements of confirmatory methods and their evaluation is illustrated in Figure 13.2.

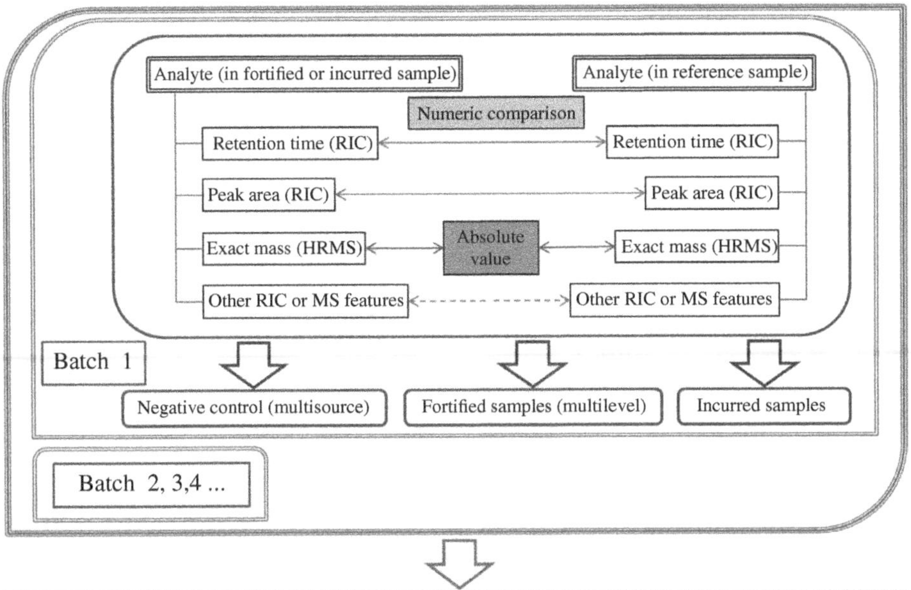

	Validation requirements		Batch QC requirements	
Method validation	FP rate < threshold	**Negative controls**	No FP	Method routine use
	FN rate < threshold	**Fortified samples**	No FN	
	FN rate at tolerance < threshold	**Incurred samples**	*To be determined*	

FIGURE 13.2 Essential elements of confirmatory methods.

The method SOP should be written to describe the experiment procedures clearly. Modifications are allowed to the SOP following the validation to address ambiguities, add needed details, and provide ancillary information to make the SOP reflect what is actually done in the laboratory. If significant changes occur as a result of validation, revalidation of the method is needed. In routine use, the laboratory (including the analyst) should monitor each batch's QC samples and conduct proficiency testing periodically. Over time, the accumulated data will indicate how reliably the method performs by revealing long-term trends for those features relevant to confirmation (the "microscopic" level) and the accumulated number of FP and FN that actually occurred (the "macroscopic" level). Ideally, a laboratory will be able to use a method over a long period without any trouble. This may not always be the case, however, because many factors other than the scientific merit of the method can be responsible, resulting in failed batches. Possible causes include operator error, contamination of automated systems, drifting of MS response, decay of instrument modules, or other causes not necessarily of a scientific nature. The reasons for rejecting data as unacceptable need to be carefully documented. Otherwise, a researcher can be accused of "cherry picking" the data to achieve the desired conclusion.

In 2003, CVM published a GFI titled "Mass Spectrometry for Confirmation of the Identity of Animal Drug Residues" (GFI-118) (FDA, 2003). This document intended to "facilitate and expedite coordination between CVM and its stakeholders so these activities may be carried out in a consistent and timely manner," wherein the "activities" include confirmatory method development, evaluation, collaborative trial, consultation, and review of confirmatory data. The scope concerns new animal drug application, use of unapproved substances, and extralabel use of approved drugs. This guidance applies only when a reference standard is available. Since its publication, many regulatory methods have used the confirmatory criteria listed therein, sometimes with fit-for-purpose modifications.

Section I of GFI-118 describes minimum requirements for the validation package, which include five blank controls, five fortified samples at tolerance/safe level, and 5 or 10 incurred samples depending on whether it is single-laboratory or interlaboratory validation. All analyses must have a zero FP rate, while up to a 10% FN rate at or above the tolerance/safe level is allowed. Data should be acquired over multiple days. Through interference testing, researchers must demonstrate that no animal drugs or endogenous matrix components are causing FPs. The required components of an SOP are listed in Section II. Section III describes confirmation criteria for various types of mass spectrometers and highly recommends the use of chromatography. Minimum S/N threshold for chromatogram peaks is 3:1, and deviation of RT for an analyte should be within 2% (GC) and 5% (LC) relative to that of the reference. MS data may be acquired with different types of instruments in various modes, and the data can be presented in different ways. Confirmatory criteria are set as described in Table 13.3. When all applicable criteria are satisfied, the identity of the targeted analyte is confirmed. For batch-QC samples, negative controls must fail the criteria; positive controls must meet all criteria needed to confirm the compound. Specific criteria for HRMS-generated data are not provided in GFI-118, because, before 2003, use of TOF MS was not widespread in regulatory laboratories, and the Orbitrap MS was not on the market. Section IV spells out QC requirements for routine use of a validated confirmatory method: (i) system suitability, (ii) at least one negative and one positive control per batch, (iii) carryover avoidance, (iv) system suitability as a condition for repeat analysis of the same sample, and (v) confirmation criteria needs to be fixed before validation. Lastly, the guidance has a section recommending a minimal data package for use in investigations and emergency response. This data package for ad hoc situations requires fewer repeat analyses than the validation package for routine use methods. Nevertheless, both negative and spiked positive controls are still necessary. Good quality assurance, staff training, and analyst expertise also contribute to CVM's confidence in this type of validation package.

TABLE 13.3 Summary of confirmation criteria for various MS type and acquisition modes in CVM GFI-118

	Number of structurally specific ions[a]	Ion ratio matching range[b]	Other criteria
MS^1 full scan	≥3, all above a minimum level	No requirement	Visually match reference spectrum; prominent nonrelated ion(s) must be explained
MS^1/SIM; or MS^1 full or partial scan, data treated as SIM	Option 1: 3 (or >3 if not all structurally specific)	Within ± 10%	N/A
	Option 2: ≥4	Within ± 15%	
MS^n full or partial scan	≥3, all above a minimum level	No requirement	Visually match reference spectrum; prominent nonrelated ion(s) must be explained
MS^n/SRM; or MS^n full scan, data treated as SRM	Option 1: 2 (if precursor ion completely disassociates)	Within ± 10%	N/A
	Option 2: ≥3	Within ± 20%	

[a]Usually, loss of common moiety such as water (18 Da) or ammonia (17 Da), quasimolecular ion, or nonmonoisotopic ions is not considered structurally specific relative to the precursor ion.
[b]Arithmetic percentage (e.g., ±10% of 45% is a range of 35–55%).

Recently, the Office of Foods and Veterinary Medicine issued an internal guidance directing the validation of both quantitative and qualitative chemical methods for regulatory use (FDA, 2012). It states that "Confirmation of identity for each analyte must be performed as part of the method validation for regulatory enforcement. Unambiguous confirmation of identity usually requires analytically identifying key features of each analyte in the scope of the new method being validated such as with mass spectral fragmentation patterns or by demonstration of results in agreement with those obtained using an independent orthogonal analysis." According to the extent of validation work, a method can be categorized into one of the four levels (Table 13.4). The higher the number of labs, analysts, and instruments testing a method, the higher degree of reliability of that method.

Another important guidance document was published in 2002 by the European Commission, Commission Decision 2002/657/EC (2002), in order to implement Council Directive 96/23/EC (1996) in a harmonized way by the community and national reference laboratories. Common criteria for analytical method performance and results interpretation are stipulated in this document. Decision limit (CC_α) and detection capacity (CC_β) are defined for key performance indicators, instead of commonly used limit of detection (LOD), limit of quantitation (LOQ), or LOC. The following performance

TABLE 13.4 Foods program key validation parameter requirements for chemical methods

	Level 1: Emergency use	Level 2: Single-laboratory validation	Level 3: Multilaboratory validation	Level 4: Full collaborative study
Number of participating labs	1	1	≥2	8 (quantitative), 10 (qualitative)
Number of matrix sources per matrix	≥1	≥3 recommended	≥3 recommended	≥3 recommended
Number of analyte(s) spike levels per matrix source	≥2 spike levels + 1 matrix blank	≥3 spike levels + 1 matrix blank	≥3 spike levels + 1 matrix blank	≥3 spike levels + 1 matrix blank
Replicates required per matrix source at each level tested per laboratory	≥2 (quantitative), ≥2 (qualitative)	≥2 (quantitative), ≥3 (qualitative)	≥2 (quantitative), ≥3 (qualitative)	≥2 (quantitative), ≥3 (qualitative)
Replicates required at each level tested per laboratory if only one matrix source used	≥4 (quantitative), ≥6 (qualitative)	≥6 (quantitative), ≥9 (qualitative)	≥3 (quantitative), ≥6 (qualitative)	≥2 (quantitative), >6 (qualitative)

characteristics must be determined for a confirmatory method: CC_β, CC_α, selectivity/specificity, and applicability/ruggedness/stability. Additionally, confirmatory methods shall provide information on the chemical structure of the analyte, but trueness/recovery and precision data are unnecessary if the same method is not used for quantitation purposes. Particularly for MS-based confirmatory methods, either an online or an offline chromatographic separation must precede MS analysis. An "identification point" (IP) system is introduced to evaluate whether the techniques used in a method possess enough selectivity for confirmation purposes. The total IP is the sum of all contributing components, such as those listed in Table 13.6. Minimal IP required for a method depends on the category of target analyte defined in Annex I of Directive 96/23/EC: 4 IP for group A substances (zero tolerance) and 3 IP for group B substances (veterinary drugs and contaminants with MRL). Table 13.5, Table 13.6, Table 13.7, and Table 13.8 (Commission Decision 2002/657/EC, 2002) relate to confirmatory method selection, technical criteria for MS signals, and the designation and application of IP. If one single method provides inadequate IP for intended analysis, combining two independent technologies to achieve the required IP is allowed. Therefore, this system provides flexibility in designing a confirmatory method.

With the increasing use of HRMS in compound and metabolite identification, the need for supplementing the official guidelines accordingly becomes apparent. Hernández et al. (2004) pointed out that "high resolution" is defined in 2002/657/EC, while "mass accuracy" is not. They suggested assigning an IP value based on mass accuracy, for example, single ion (SIM or precursor) with errors of (i) higher than 10 ppm, (ii) between 2 and 10 ppm, and (iii) below 2 ppm, being assigned point values of 1.0, 1.5, and 2.0, respectively. Using the same assumptions for product ions, the IP assignment would be 1.5, 2.0, and 2.5 points, respectively. Nielen et al. (2007) indicated that coeluting matrix components or structure analogs in LC-MS and isobaric product ions in LC-MS/MS could increase the uncertainty of mass measurement in veterinary drug residue analysis. Using stanozolol as an example, because it generates two isobaric product ions (*m/z* 161.1073 and 161.1324) on Orbitrap or FT-ICR, but only one undissolved "composite" ion (*m/z* 161.1223) on Q-TOF, a false compliant actually occurred in the latter case, in both screening and confirmation. Kaufmann (2009) commented on several practical problems with direct application of the criteria onto multiclass methods and proposed ways to solve them, such as using alternative techniques of calculating CC_α in the absence of noise (HRMS), designing a simultaneous validation scheme for banned and regulated substances within the same analytical method, and evaluating whether CC_β is really important. On another issue, Wang and Leugn (2007) argued that using UPLC-TOF, two ions (no precursor selection) was adequate for confirmation (≥3 IP for all

TABLE 13.5 Suitable confirmatory methods for organic residues or contaminants ("Table 1" in 2002/657/EC)

Measuring technique	Substances Annex 1 96/23/EC	Limitations
LC or GC with mass spectrometric detection	Groups A and B	Only if following either an online or an offline chromatographic separation Only if full-spectrum techniques are used or using at least three (group B) or four (group A) IPs for techniques that do not record the full mass spectra
LC or GC with IR spectrometric detection	Groups A and B	Specific requirements for absorption in IR spectrometry have to be met
LC full-scan DAD	Group B	Specific requirements for absorption in UV spectrometry have to be met
LC-fluorescence	Group B	Only for molecules that exhibit native fluorescence and to molecules that exhibit fluorescence after either transformation or derivatization
2-D TLC full-scan UV/Vis	Group B	2-D HPTLC and cochromatography are mandatory
GC–electron capture detection	Group B	Only if two columns of different polarity are used
LC-immunogram	Group B	Only if at least two different chromatographic systems or a second independent detection method is used
LC-UV/VIS (single wavelength)	Group B	Only if at least two different chromatographic systems or second independent detection method is used

Disclaimer: Only EU documents published in the Official Journal of the EU are deemed authentic.

TABLE 13.6 Assignment of IP to various types of MS-derived signals ("Table 5" in 2002/657/EC)

MS technique	IP
Low-resolution MS (LRMS)	1.0
LRMS^n precursor ion	1.0
LRMS^n transition products	1.5
HRMS	2.0
HRMS^n precursor ion	2.0
HRMS^n transition products	2.5

Disclaimer: Only EU documents published in the Official Journal of the EU are deemed authentic.

TABLE 13.7 Maximum permitted tolerances for relative ion intensities using a range of mass spectrometric techniques ("Table 4" in 2002/657/EC)

Relative intensity (% of base peak)	EI-GC-MS (relative) (%)	CI-GC-MS, GC-MS^n, LC-MS, LC-MS^n (relative) (%)
≤10%	±50	±50
>10–20%	±20	±30
>20–50%	±15	±25
>50%	±10	±20

Disclaimer: Only EU documents published in the Official Journal of the EU are deemed authentic.

TABLE 13.8 Examples of the number of IPs earned for a range of techniques and combinations thereof ("Table 6" in 2002/657/EC; *N* or *n* is a positive integer)

Technique(s)	Number of ions	IPs
GC-MS (EI or chemical ionization [CI])	*N*	*n*
GC-MS (EI and CI)	2 (EI) + 2 (CI)	4
GC-MS (EI or CI) 2 derivatives	2 (derivative A) + 2 (derivative B)	4
LC-MS	*N*	*n*
GC-MS/MS	1 precursor and 2 daughters	4
LC-MS/MS	1 precursor and 2 daughters	4
GC-MS/MS	2 precursor ions, each with 1 daughter	5
LC-MS/MS	2 precursor ions, each with 1 daughter	5
LC-MS/MS/MS	1 precursor, 1 daughter, and 2 granddaughters	5.5
HRMS	*N*	2*n*
GC-MS and LC-MS	2 + 2	4
GC-MS and HRMS	2 + 1	4

Disclaimer: Only EU documents published in the Official Journal of the EU are deemed authentic.

analytes at 10 ppm mass accuracy), while Stolker et al. (2008) maintained that "confirmatory analysis of the suspected samples has to be done by the use of an MS/MS technique because criteria for confirmation of the identity of drugs by TOF MS are not included yet in the EU guidelines." Technical obstacles must also be overcome, such as excessively large data file sizes, long data processing times, varying centroiding algorithms, and FN recognition by software. Nevertheless, there are significant advantages to using HRMS: (i) requires very limited effort in tuning; (ii) obtains full-range spectra for retrospective evaluation; and, specifically for confirmation, (iii) achieves a higher level of confidence.

Particularly worth mentioning is the recent work published by Stoev et al. (2012) on a mathematical model to estimate the relative resolution capacity between various types of LC-MS. The underlying principle is to estimate the probability of erroneous spectral assignment by computing the maximum number of hypothetical analytes a particular instrument can differentiate, typical resolving power of commonly used instrumentation, *m/z* value for molecular ions, mode of acquisition, and corresponding confirmatory criteria. Some examples for the calculated probability values (exhibiting only the order-of-magnitude approximations) are MS^1 full spectrum at unit resolution, $P \sim 10^{-3}$; MS^1 full spectrum at 10,000 resolution, $P \sim 10^{-6}$; MS/MS at unit mass resolution, $P \sim 10^{-8}$; $HRMS^1$ full spectrum, $P \sim 10^{-7}$; and $HRMS^2$ (one precursor ion selected) at 100,000 resolution, $P \sim 10^{-13}$. Because P represents the probability of erroneous assignment, this approach introduces a quasiquantitative basis for comparing analytical methods performed on completely different MS platforms (or operation modes) within the current regulatory frameworks.

One significant limitation in using MS for confirmatory analysis is that it typically does not provide much information regarding an analyte's three-dimensional structure. Recent advances in the design of the gas-phase ion mobility interface show promise for use in drug analysis (Kanu et al., 2008; Kolakowski and Mester, 2007). On the other hand, traditional affinity-binding-based techniques such as enzyme-linked immunosorbent assay (ELISA) are generally very selective toward the same-class substrates, but not specific enough to be considered stand-alone confirmation. Marchesini et al. (2009) presented a novel example of combined biosensor-based bioanalysis with LC-TOF MS for confirmation of paralytic shellfish poisons in mussels and cockles. After initial immunoaffinity screening using a surface plasmon resonance detector to identify the possible noncompliant samples, the corresponding extracts were injected to a recovery biochip (CM5 chip with MAb GT13A immobilized on the surface). Then the analytes were desorbed and injected into a nanohydrophilic interaction liquid chromatography (HILIC)-TOF MS system for confirmation. This affinity chip/LC-MS combination strategy is likely to provide high total resolving capacity and high confidence in the confirmatory finding.

The guidance documents described earlier provide a common ground for the residue analysis community, especially regulatory laboratories around the world. A significant amount of work has been published adopting the confirmation criteria and validation concepts in these documents, covering all types of conventional instrument platforms and a wide variety of matrices and chemicals. Nonetheless, there are rare occasions that a confirmatory method for a particular drug–matrix combination consistently produces either FP or FN results (Lehotay et al., 2008; Schürmann et al., 2009). The root cause is often attributable to having inadequate resolving capacity built into the methods. Components in coeluting matrices could distort the monitored ion ratio of an unknown sample, possibly causing an FP or FN. Usually in this instance, an experienced mass spectrometrist carefully examines the situation and reoptimizes operational parameters in sample treatment, LC, or MS, focusing on the analyte that suffered from interference. Other times, an FP is the result of an overly sensitive MS. Setting up semiquantitative threshold criteria may weed out these FPs in method validation (not to be confused with the "positive but compliant" scenario) if the corresponding signal is relatively low. However, the risk in routine use of this method is that truly negative samples from a particular source have a high level of this specific interference component, which causes confirmation error. Designing tighter criteria than the minimum required RT or ion ratio threshold or using a narrower extraction window for HRMS data may also lower the FP rate, though this is likely to increase the probability of false compliance. Postvalidation manipulation of the confirmatory criteria to fit existing results is never allowed.

Government agencies, international organizations, or trade associations that need to make decisions based on confirmatory analysis also have established guidance on confirmation criteria for chemical identity in areas beyond that of human food safety, such as forensic toxicology, emergency response, and drug abuse in many arenas, such as horseracing. Table 13.9 summarizes confirmatory criteria published by the USDA (Pesticide Data Program), European Commission's (EC) Directorate-General for Health and Consumers (DG-SANCO), FDA's Office of Regulatory Affairs (ORA), Association of Official Racing Chemists (AORC), World Anti-Doping Agency (WADA), and Society of Forensic Toxicology/American Academy of Forensic Sciences (SOFT/AAFS). Although the confirmation criteria in each of these documents address different situations, the fundamental structures are similar. AOAC International does not recommended specific criteria for confirmatory methods; nonetheless, it prescribes how to design and conduct a collaborative study (for high confidence) (AOAC, 2002) and advocates the use of "probability of detection" (POD) as a statistical model for validation of qualitative methods (Wehling et al., 2011). The CODEX Committee on Residues of Veterinary Drugs in Foods (CODEX Alimentarius Commission [CAC]/CCRVDF), under World Health Organization (WHO)/Food and Agriculture Organization (FAO), is also working on the revision of official document CAC/GL 71-2009, *Guidelines on Analytical Terminology*, to include an appendix on performance criteria for multiresidue analytical methods in veterinary drug residue analyses, which include confirmatory analysis (CCRVDF, 2009, 2013).

Lastly, database (library) matching has not been favored as a confirmatory strategy because of unpredictable matrix interference, lot-to-lot variability

TABLE 13.9 Confirmation criteria for GC-MS- or LC-MS-based methods (for unit resolution MS)

Guidance or laboratory manual	Scope	RT or RRT criteria (unknown vs. reference)	Minimum number of (A) transitions (SRM mode); (B) ions (SIM or full-spectrum mode)	Acceptable limit for relative abundance variation
USDA PDP (USDA, 2013)	Pesticide in food	GC: RT ±0.05 min or RRT ±0.01 min LC: RT ±0.5 min or RRT ±0.1 min	(A) 2; (B) 3	abs ± 20%
SANCO 12495/2011 (SANCO, 2012)	Pesticide in food	GC: RT ±0.5% LC: RT ±2.5%	(A) 2; (B) 3	rel ± 20% (>50% RA) rel ± 25% (>20–50% RA) rel ± 30% (>10–20% RA) rel ± 50% (≤10% RA)
FDA ORA-LAB.010 (FDA, 2009)	Pesticide in food	RT within experimental precision	(A) 2; (B) 3	rel ± 20% for LC; abs ± 10% for GC (>40% RA) rel ± 25% (>10–40% RA) rel ± 50% (≤10% RA)
WADA TD2010IDCR (WADA, 2010)	Illegal drugs in sports	GC and LC: smaller of ±2% or 0.1 min; or RRT ±1%	(A) 2 or 1[a]; (B) 3 or all ions that >10%RA	abs ± 10% (>50% RA) rel ± 20% (>25–50% RA) abs ± 5% (>5–25% RA) rel ± 50% (≤5% RA)
AORC MS criteria 2011 (AORC, 2011)	Drugs in racing animals	GC: RRT ±1%; RT greater of ±1% or 6 s LC: RRT ±2%; RT greater of ±2% or 12 s UHPLC: RT greater of ±50% HHPW or 3 s	(A) Use of SRM is discouraged; (B) 3 or all that >10%RA	MS^1 full spectrum: greater of abs ± 10% or rel ± 30% Product ion full scan: greater of abs ± 20% or rel ± 40% Use of SIM is discouraged
SOFT/AAFS laboratory guidelines 2006 (SOFT/AAFS, 2006)	Forensic toxicology	GC: ±1–2% LC: slightly larger than GC allowance	(A) N/A; (B) 2	GC-MS: rel ± 20% LC-MS: rel ± 25–30%

RRT: relative RT of analyte to an internal standard.
abs: absolute (e.g., abs ± 10% of 70% is 60–80%).
rel: relative (e.g., rel ± 10% of 70% is 63–77%).
RA: relative abundance, which equals the percentage of ion counts to the base peak in a full-range mass spectrum or RIC peak area ratio of two transitions or ions (full-scan or SRM mode).
HHPW: half height, peak width.

of LC column performance (even with the same brand), and instrument-to-instrument differences in the MS^n fragmentation pattern. Even for the same instrumentation set, the RT and relative abundance/ion ratio of the same monitored ions/transitions often vary over time, possibly due to matrix diversity (Kaufmann et al., 2009), instrument drift, laboratory environment fluctuation, or other factors. All these issues make library matching less reliable than directly comparing a contemporaneously analyzed reference standard, QC samples, and regulatory samples in the same batch. Nonetheless, chemical databases are useful in method ruggedness testing to search for possible interferences with target compounds. Several articles carefully examine the efficiency and reliability of such a strategy (Berendsen et al., 2013a; Little et al., 2011, 2012).

13.4 SELECTED RECENT PUBLICATIONS FOR CONFIRMATION OF VETERINARY DRUGS OR ORGANIC CONTAMINANTS IN FOOD ANIMAL PRODUCTS AND FEED

Decades ago, mainstream analytical methods usually monitored only one or a small number of analytes at a time. In recent years, thanks to rapid advances in analytical instrumentation for residue analysis—especially MS—multiresidue (first single class and then multiclass) methods have significantly increased. The main benefit of using multiresidue methods is the greatly enhanced throughput (the number of analytes per sample per unit analyst/instrument time). In Table 13.10, we used the following standards to select representative articles describing LC-MS confirmatory methods for regulatory purposes: (i) published in a peer-reviewed journal after 2002 when the 2002/657/EC and GFI-118 became publicly available; (ii) discussed veterinary drugs or contaminants relevant to food animal or feed residue issues, (iii) described a confirmatory method for regulatory use (may be combined with screening or quantitation), and (iv) used commercially available MS detectors. Most of the publications are from major regulatory analytical institutions or joint work with external collaborators. Due to the vast number of papers on this topic in recent years, our compilation is not all inclusive. It should be noted that private analytical service companies have their own collection of proprietary methods meeting official guidance on confirmation, and major analytical instrument manufacturers have publicized technical notes to demonstrate their LC or MS instrument's capability to perform residue analysis for a variety of food matrices (not covered in this chapter).

Summaries of the selected methods are listed in Table 13.10. Following each method's target matrix and residue (class) is a brief notation of key steps

TABLE 13.10 Selected examples of LC-MS-based confirmatory methods

Biological matrix	Drug class (total number of analytes)	Instrumentation; sample preparation	Confirmation guidance and characteristics	Reference
Raw milk	Macrolides (5)	LC-MS/MS; ACN, then SPE	2002/657/EC; 2 transitions	Wang et al. (2006)
Milk	Ceftiofur, fluoroquinolones, sulfonamides (11)	LC-MS/MS; methanol (MeOH)/H_2O/formic acid, then centrifuge	2002/657/EC; 2 transitions	Toaldo et al. (2012)
Milk	Nitrofurans (4)	LC-MS/MS; hydrolysis and derivatization, then SPE	GFI-118; 3 transitions	Chu and Lopez (2007)
Milk	Multiclass (58)	LC-MS/MS; two routes for different groups: (1) ACN, then filter (2) 5% TCA, then filter	2002/657/EC; 2–3 transitions	Gaugain-Juhel et al. (2009)
Milk	Spiramycin, neospiramycin	LC-MS/MS; ACN, then SPE	2002/657/EC; 2 transitions	Wang and Leung (2009)
Milk, honey	Multiclass (55)	UHPLC-Q-TOF MS; QuEChERS	Accurate mass and isotope pattern	Wang and Leung (2012)
Milk	Benzimidazoles, avermectins, flukicides (38)	LC-MS/MS; QuEChERS with dSPE	2002/657/EC; 2 transitions	Whelan et al. (2010)
Milk	Multiclass (47)	LC-MS/MS; TCA, McIlvaine, filter, then SPE	2002/657/EC; 2 transitions	Bohm et al. (2009)
Milk	Multiclass (26)	LC-MS/MS; ACN, SPE, then MW-cutoff filter	GFI-118; 3 transitions	Clark et al. (2011)
Milk	Avermectins, moxidectin (4)	LC-QIT MS (APCI/APPI); ACN, then SPE/SPE	GFI-118; 1 precursor, 2 product ions	Turnipseed et al. (2005)
Whole milk	Multiclass (25)	LC-MS/MS; ACN, SPE, then MW-cutoff filter	GFI-118; 3 transitions	Turnipseed et al. (2008)

Milk	Aminoglycosides (6)	LC-QIT MS; conc. HCl, CBA-SPE, derivatization with phenylisocyanate	GFI-118; 1 precursor and 4 product ions, visual matching	Turnipseed et al. (2009)
Milk	Corticosteroids (3) and nonsteroidal anti-inflammatory drugs (4)	LC-MS/MS; ACN/NaCl, hexane wash	2002/657/EC; 2 transitions	Malone et al. (2009)
Milk	Glucocorticoids (12)	LC-MS/MS; ACN/NaCl, hexane wash	2002/657/EC; 2 transitions	Malone et al. (2010a)
Milk	Corticosteroids (16), bisphenol A	LC-MS/MS; ACN/NaCl, then dSPE	2002/657/EC; 2 transitions	Malone et al. (2010b)
Milk	Multiclass (25) and metabolites (ID)	LC-Q-TOF MS; ACN, then MW-cutoff filter	Accurate mass for precursor and 2 product ions (3 for SDM)	Turnipseed et al. (2011)
Milk, meat	Anthelmintic (7), phenylbutazone	UHPLC-Orbitrap or UHPLC-MS/MS (core–shell type column); QuEChERS (ACN + ammonium sulfate), then filter	2002/657/EC as guideline; Orbitrap: 2 accurate mass ions (1 for thiabendazole) including quasimolecular ion; MS/MS: 2 transitions	Kaufmann et al. (2011a)
Egg	Sulfonamides (16)	LC-QIT MS; ACN, then SPE	Structurally specific product ions >2% relative abundance; visual matching of product ion full spectrum	Heller et al. (2002)
Egg	Ionophores (4), macrolide (2), novobiocin	LC-QIT MS; sodium succinate buffer, then silica SPE	GFI-118; 1 precursor and 3 product ions	Heller and Nochetto (2004)
Egg	Multiclass (29)	LC-QIT MS; sodium succinate buffer, then SPE	GFI-118; full spectrum at MS^2 or $MS^2 + MS^3$, precursor ion and at least 1 product ion	Heller et al. (2006)

(*Continued*)

TABLE 13.10 (*Continued*)

Biological matrix	Drug class (total number of analytes)	Instrumentation; sample preparation	Confirmation guidance and characteristics	Reference
Egg	Macrolides (5)	LC-MS/MS; ACN, hexane wash, then SPE	2002/657/EC; 2 or 3 transitions	Wang et al. (2005)
Egg	Tetracyclines (7), quinolones (4)	LC-MS/MS; EDTA/McIlvaine, then SPE	2 transitions	Jia et al. (2008)
Egg	Coccidiostats (10)	LC-MS/MS; ACN, then evaporation	2002/657/EC; 2 transitions	Dubreil-Chéneau et al. (2009)
Egg	Nitroimidazole (11)	LC-MS/MS; ACN/NaCl, hexane wash	2002/657/EC; 2 transitions	Cronly et al. (2009)
Meat (beef, pork, poultry), fish (sea bream, trout)	Macrolides (7)	LC-MS/MS; pressurized liquid extraction (PLE) with MeOH	2 transitions (1 for josamycin)	Berrada et al. (2008)
Salmon	Triphenylmethyl dyes (4)	LC-MS/MS; ACN/McIlvaine, then SPE	2002/657/EC; 2 transitions	Dowling et al. (2007)
Grass carp, eel, salmon, shrimp, and shellfish	Triphenylmethyl dyes (4)	LC-QIT MS; ACN/McIlvaine, dSPE (alumina), CH_2Cl_2 partitioning, evaporation, then SPE	2002/657/EC; 1 precursor and 3 product ions	Wu et al. (2007)
Salmon, shrimp	Fluoroquinolones (4, in salmon); phenicols (4, in shrimp)	LC-QIT MS; fluoroquinolones: acetic acid/ethanol, then SPE. Phenicols: basic EtOAc/ACN, defat with hexane, then SPE	Draft GFI (details given)	Turnipseed et al. (2003)
Finfish	Multiclass (38)	LC-QIT MS; ACN, then hexane wash	Generally follow GFI-118	Smith et al. (2009)

Tilapia, rainbow trout, salmon	17α-Methyltestosterone	LC-MS/MS; ACN, LLE (EtOAc and NaCl), SPE (SAX), LLE (hexane), SPE (Florisil)	GFI-118; RT within 2%; 2 transitions	Chu et al. (2006)
Fish/salmon	Multiclass (7)	LC-TOF MS; ACN/NaCl, then dSPE	2002/657/EC; 1 precursor + 1 product from in-source CID	Hernando et al. (2006)
Shrimp, whole milk	Tetracyclines (3)	LC-MS/MS; milk: succinic acid, then SPE. Shrimp: succinic acid/NaCl; dSPE (alumina), then SPE	3 transitions; RT ±5%, ion ratio ±10%, S/N > 10	Andersen et al. (2005)
Fish	Dyes (3)	LC-QIT MS; buffer + ACN, LLE (CH_2Cl_2), 2,3-dichloro-5,6-dicyano-1,4-benzoquinone (DDQ) (oxidation), then SPE	Precursor + 3 (or more) product ions	Andersen et al. (2009)
Fish	Sulfonamides (3), tetracyclines (3)	LC-MS/MS; MeOH/H_2O, then filter	2002/657/EC; 2 transitions	Cháfer-Pericás et al. (2010)
Shrimp	Multiclass (14)	LC-TOF MS; QuEChERS	Mass accuracy ≤ 2 ppm; multiple ions (precursor and in-source CID fragments)	Villar-Pulido et al. (2011)
Shrimp	Multiclass (18)	LC-QIT MS; TCA, then SPE	Generally follow GFI-118	Li et al. (2006)
Shrimp, crab meat	Chloramphenicol	LC-MS/MS; EtOAc (evaporated), hexane wash, evaporation, back extract with EtOAc	2002/657/EC and draft GFI-118; 4 transitions	Hammack et al. (2003)

(*Continued*)

TABLE 13.10 (*Continued*)

Biological matrix	Drug class (total number of analytes)	Instrumentation; sample preparation	Confirmation guidance and characteristics	Reference
Catfish	Fluoroquinolones (4)	LC-QIT MS; acid/ACN, then dSPE/hexane wash	Full spectrum at MS^2 level; 3 major MS^3 ions	Mcmullen et al. (2009)
Multimatrix (turkey, duck, guinea fowl, chicken, quail, swine, shrimp, whole egg, honey)	Nitrofurans (5)	LC-MS/MS; acid hydrolysis/derivatization, LLE (EtOAc), then evaporation	2002/657/EC; 2 transitions	Verdon et al. (2007)
Chicken muscle	Multiclass (39)	UPLC-MS/MS; MeOH/H_2O, then filter	2002/657/EC; 2 transitions	Chico et al. (2008)
Chicken liver	Coccidiostat (12)	LC-MS/MS; ACN, filter with alumina, then SPE	2002/657/EC; 2–3 transitions	Olejnik et al. (2009)
Bovine kidney	β-lactam (10)	LC-MS/MS; ACN/H_2O, then dSPE	2 transitions (ion ratio ±20%)	Fagerquist et al. (2005)
Meat (chicken, pork)	Nitrofurans (4)	LC-MS/MS; derivatization, EtOAc (evaporated), hexane wash, then SPE	2002/657/EC; 3 transitions	Mottier et al. (2005)
Poultry muscle	Antiviral (7)	LC (online SPE)-MS/MS; MeOH, then SPE/SPE	2002/657/EC; 2 transitions	Berendsen et al. (2012)
Muscle (bovine, porcine, ovine, poultry)	Multiclass (19)	UPLC-MS/MS; EDTA/ACN (evaporated), then filter	2002/657/EC; 2 transitions	McDonald et al. (2009)
Beef	Multiclass (16)	LC-MS/MS; PLE or QuEChERS	2002/657/EC; 2 transitions	Blasco et al. (2011)
Poultry, bovine, equine, and swine kidney	Aminoglycoside (10)	LC-MS/MS; phosphate buffer/TCA/EDTA, hexane wash, then SPE	2002/657/EC; 2 transitions	Almeida et al. (2012)

Beef, pork	8 classes (53)	LC-MS/MS; EDTA/McIlvaine, ultrasound, then SPE	2002/657/EC; 2 transitions	Bohm et al. (2011)
Pig muscle, kidney, liver	Tetracyclines (7), quinolones (14)	UPLC-MS/MS; EDTA/McIlvaine, then SPE	2002/657/EC; 2 transitions	Shao et al. (2007)
Beef, poultry, pork muscle	Macrolides (9)	LC-MS/MS; ACN, hexane wash, then dilution	2002/657/EC; 2 transitions	Martos et al. (2008)
Beef	Multiclass (130)	UHPLC-MS/MS; ACN/H_2O, dSPE, hexane wash, then evaporation	GFI-118; 3 transitions	Geis-Asteggiante et al. (2012)
Bovine kidney	Multiclass (62)	UHPLC-MS/MS; ACN/water, then dSPE/hexane wash Other four sample prep procedures	GFI-118; 3 transitions	Lehotay et al. (2012)
Bovine kidney	Multiclass (120)	LC-MS/MS; ACN/H_2O with or without acid, then dSPE/hexane wash	GFI-118; 2 transitions	Schneider et al. (2012)
Bovine kidney	Phenolic and salicylanilide anthelmintics (8)	LC-MS/MS; ACN (basic) or acetone (acidic), then SPE	2002/657/EC; 2 transitions	Caldow et al. (2009)
Beef	β-lactams (8), tetracyclines (3)	UPLC-MS/MS; ACN/H_2O, hexane wash, dSPE	2002/657/EC; 2 transitions	Rezende et al. (2012)
Kidney, liver, muscle, fish, honey	Multiclass (100+)	UPLC-Orbitrap; ACN + succinic/EDTA buffer, then SPE	2002/657/EC	Kaufmann et al. (2011b)
Egg, raw milk, honey	Macrolides (6)	LC-MS/MS and UPLC-Q-TOF (TOF, W mode); ACN or phosphate buffer, then SPE	LC-MS/MS: 2 transitions UPLC-Q-TOF: mass accuracy 2–10 ppm, 2 ions (no precursor selected)	Wang and Leung (2007)

(*Continued*)

TABLE 13.10 *(Continued)*

Biological matrix	Drug class (total number of analytes)	Instrumentation; sample preparation	Confirmation guidance and characteristics	Reference
Kidney, honey	Multiclass (100+)	UPLC-Orbitrap and LC-MS/MS; ACN + succinic/EDTA buffer, then SPE	HRMS: 2 ions (no precursor selection) MS/MS: 2 transitions	Kaufmann et al. (2011c)
Milk, liver	Benzimidazoles, macrolides, and flukicides (38)	LC-MS/MS; QuEChERS with dSPE	2002/657/EC; 2 transitions	Kinsella et al. (2009b)
Liver	Multiclass anthelmintics (38)	UHPLC-MS/MS; QuEChERS	2002/657/EC; 2 transitions	Kinsella et al. (2010)
Honey	Macrolides (5)	LC-MS[1] and LC-MS/MS; phosphate buffer, then SPE	2002/657/EC; 2 or 3 transitions	Wang (2004a)
Honey	Penicillins (6)	LC-MS/MS; phosphate buffer, then SPE	2002/657/EC; 2 or 3 transitions	Wang (2004b)
Honey	Multiclass (17)	LC-MS/MS; H_2O, then either dilution (STREP) or SPE	GFI-118; 2 transitions	Lopez et al. (2008)
Honey	Multiclass (42)	LC-MS/MS; Na_2HPO_4, then four sequential extractions	2002/657/EC; 2 transitions	Hammel et al. (2008)
Honey	Chloramphenicol	LC-MS/MS; EtOAc (evaporated), SPE, EtOAc/ACN/H_2O (evaporated)	GFI-118 and 2002/657/EC; 3 and 4 transitions, respectively	Quon et al. (2006)

Honey	Nitrofurans (4)	LC-MS/MS; dilution, SPE, hydrolysis and derivatization, LLE (EtOAc)	GFI-118; 3 transitions; ratio ±10%	Lopez et al. (2007)
Honey	Multiclass (37)	LC-MS/MS; McIlvaine buffer, then SPE	2002/657/EC; 2 transitions	Bohm et al. (2012a)
Honey	(Dihydro)streptomycin	LC-MS/MS; phosphate buffer, then SPE	2002/657/EC; 2 transitions	Bohm et al. (2012b)
Honey, royal jelly, propolis	Fluoroquinolones (8)	UHPLC-MS/MS; QuEChERS	2002/657/EC; 2 transitions	Lombardo-Agüí et al. (2012)
Urine	Multiclass (100+)	UPLC-TOF MS; direct dilution	Accurate mass of precursor ion, supplemented by evidence of metabolite, isotope ratio, or class-specific fragment	Kaufmann et al. (2007)
Urine	Nandrolone (α and β), trenbolone (α and β)	LC-MS/MS; enzymatic deconjugation, then immune-affinity cartridge	2002/657/EC; 3 transitions	Gasparini et al. (2009)
Feed/dry distillers grains	Multiclass (13)	LC-QIT MS; EDTA/TCA/H_2O, then MeOH; 2-track SPE	Generally follow GFI-118 except tylosin	De Alwis and Heller (2010)
Feeding stuff	Multiclass (33)	LC-MS/MS; MeOH/ACN/McIlvaine/EDTA, then dSPE	2002/657/EC; 2 transitions	Boscher et al. (2010)

in sample preparation, type of instrumentation, and number of diagnostic MS ions or transitions, because these are the most important factors for total resolution and specificity. In many cases, the guidance that the authors followed in method validation is cited, although there are occasions when the actual criteria were modified to fit a specific situation. On the other hand, information regarding LOC, CC_{α}, and CC_{β} is not included because it is highly instrument dependent, and for multiresidue methods, it demands too much space to present. As discussed previously, use of highly specific immunoaffinity cartridges for sample cleanup and a sub-2-micron particle UHPLC column or larger dimension HPLC column contributes to better overall selectivity, not to mention HRMS or multistage MS^n. In method validation, a larger number of blank control samples from various sources enhance the robustness of the method (against FPs), as it is hard to predict an occurrence of endogenous interference due to variation in feed, animal strain, health condition, husbandry environment, and so forth. In some of these publications, the confirmatory results were presented in detail, including the level of spiking, the total number of repeated analyses at each level including the blank control, and the number or percentage of positive hits. A "performance characteristic curve" could be drawn if the data at several levels are individually presented, which enables readers to have a comprehensive view of overall method performance. The other common reporting format is to give an LOC or limit of identification (LOI) figure for each analyte, yet without any detail on the success/failure rate at each tested spiking level. Some papers simply give a generic statement such as "meets guidance (decision) requirements." A few published reviews are cited in a significant number of publications, though not all of them contain a confirmatory component (De Brabander et al., 2009; Kinsella et al., 2009a; Le Bizec et al., 2009; Samanidou and Nisyriou, 2008).

13.4.1 Examples of LC-MS/MS-Based Multiresidue Confirmatory Methods

Today, LC-MS/MS (unit resolution) is still the most commonly used instrumentation for confirmation. It has the advantage of technological maturity, adequate selectivity, high sensitivity, and relatively straightforward data evaluation. Nonetheless, it can be used only for targeted analysis. If the method aims for a large number of analytes, the instrument has difficulty (i) acquiring a reproducible signal (ion count) in a short time interval (as little as a few millisecond [ms]), (ii) fast switching between MS/MS transitions, and (iii) fast switching between opposite polarities.

Bohm et al. (2009) reported validation of a multiclass method that simultaneously confirms and quantifies 47 substances belonging to tetracyclines, quinolones, macrolides, sulfonamides, diaminopyrimidine derivatives, and lincosamides in milk in a single analytical run. The LC-MS/MS was run in SRM (MRM) mode with ESI ionization, using a 3 μm particle size column. Milk samples were first precipitated with trichloroacetic acid (TCA) and followed by SPE cleanup with Oasis HLB cartridges, which can retain both hydrophilic and lipophilic compounds. This method was used for both quantitation and confirmation. Validation considered that not all of the 47 drugs have established MRLs. The analytes with an MRL were spiked at concentration levels of 0, 0.5, 1.0, 1.5, and 2.0× MRL, except sulfonamides, which were fortified at additional low levels for a specified reason. The analytes without MRLs were spiked at 0, 5, 10, 15, and 20 μg/kg. Seven potentially performance-influencing factors, that is, milk type, trader, fat content, storage temperature, operator, Evaporator (brand), and SPE cartridge lot, were selected for ruggedness testing. Customized software for design and processing the factor-comprehensive validation data was used. Quantification was based on matrix-matched calibration curves that resulted in good accuracy (recovery) for all drugs (close to 100%). Confirmation based on matching the ion ratios of the two MS/MS transition responses (IP = 4) and RT was successful in all cases, in compliance with 2002/657/EC. The seemingly generic extraction/cleanup combined with typical LC-MS/MS provided adequate selectivity for all the targeted analytes at the levels of interest.

In 2011, Clark et al. (2011) published a screening/confirmatory multiclass method for 26 veterinary drugs in milk, which was an extension of previous work (Turnipseed et al., 2008). This method employed a newer generation of LC-MS/MS instrumentation to conduct screening and confirmation at the same time, by simultaneously monitoring three transitions for each drug. Among all analytes, penicillin G was monitored with its isomer (penillic acid), while flunixin and enrofloxacin were monitored with their respective major metabolite, flunixin-OH and ciprofloxacin. The number of repeats at 0.5×, 1×, and 2× for each drug's target level was 5, 21, and 5, respectively, and the confirmatory criteria in GFI-118 were applied. The milk sample was first precipitated with ACN, and then the supernatant was cleaned up with an HLB cartridge. A 30 kDa cutoff filter was used to remove residual large biomolecules in the extract. Because most of these residues have established tolerance or safe levels, a minimum response threshold was needed. The approach suggested by an EC Reference Laboratory for the same type of analysis was adopted (Guagain-Juhel et al., 2009), that is, a standard deviation (SD) at 1× each drug's tolerance (or safe level) was calculated to give the lower threshold values for all residues at the 95% confidence level. Yet for convenience of use, the peak area of the 0.5× spiked sample was considered as a semiquantitative threshold for passing compliance samples.

Among the diversified analytes, some met the confirmation criteria at much lower levels than the 0.5× screening threshold. Therefore, the use of peak area thresholds ensured that only residues present near the level of concern rather than barely above zero were sent for further quantitative analysis. On the other hand, a few listed drugs such as ampicillin could not be confirmed at the 1× target level. Nevertheless, the method will be useful in simultaneously and quickly passing all compliant samples and catching most of the noncompliant ones, leaving only a small portion that screen positive, but cannot be confirmed yet.

In 2012, researchers at USDA published a series of regulatory methods for monitoring veterinary drugs in bovine kidney with LC-MS/MS (Lehotay et al., 2012; Schneider et al., 2012). One method included 120 drugs belonging to 11 classes. A simple sample treatment scheme was followed, that is, using 4:1 ACN/H_2O for extraction from homogenized kidney tissue and then conducting an LLE cleanup with hexane. The instrumentation used was an HPLC-MS/MS using an ODS-3 column (150×3.0 mm, 5 μm) with total analysis time of 30 min, including a 10 min column wash cycle to minimize carryover. The confirmatory criteria for LC-MS/MS in GFI-118 were followed, designating one qualifier transition with an ion ratio allowing ±10% (absolute value) deviation and RRT within ±4 SD of the matrix-matched calibrators' mean value. In validation, for a drug to be deemed confirmable at a certain level, at least 14 of 15 (≥93%) repeat analyses of spiked samples must meet the confirmatory criteria for the particular analyte. Overall, 57% of the analytes met the criteria at the 10 ng/g level and 84% did so at the 200 ng/g level. In general, the method worked well for most drugs except nitrofurans and thyreostats. The second method had 62 drugs on the list. A QuEChERS-type extraction procedure combined with UHPLC-MS/MS analysis, using a sub-2-micron particle column (100×2.1 mm, 1.8 μm), was employed. In this work, the term "identification" has the same meaning as "confirmation" in this chapter. The confirmatory criteria in GFI-118 were generally followed with an additional check item, that is, the RT and chromatographic peak shape of all transitions of a given analyte should coincide. It was observed that for some analytes, the ion ratio value could be very different between the reagent-only standard and the matrix-matched one. The validation threshold for an FN was ≤10%, consistent with the guidance. Occurrence of an FP was very rare (cimaterol only) and could be mitigated by other measures, because the laboratory's protocol directed "positively identified samples" to be subjected to further confirmation using one of the applicable, established single-class methods. Overall, 43 drugs passed confirmation at all three spiking levels (0.5×, 1×, and 2× target levels); the other four and two drugs met criteria at the 1× and 2× levels, respectively.

13.4.2 Examples of LC-QIT MS-Based Multiresidue Confirmatory Methods

QIT MS is one of the earliest types of MS invented. Compared to QqQ MS, it has the ability of conducting multiple-level CID to extract additional information on each of the fragment ions to better elucidate molecular structure. Often used in full-scan mode, the mass spectrum contains much more information than that obtained from SRM mode. Also, because many QIT MS use helium as the collision gas, the fragmentation pattern for some ions can be quite different from that occurring in the tandem MS CID process, where the collision gas is typically nitrogen or argon. One other unique feature of many QIT MS models is the wider selection of activation parameters for optimizing CID conditions compared to QqQ MS. Additional features include activation energy (normalized), tunable activation time, wideband activation, and precursor isolation width. The linear quantitation range, however, is usually narrower than MS/MS due to the space-charge effect. Also, in full-scan mode, the acquisition cycle time can be rather long (e.g., several scan cycles per second), and the *m/z* range for product ion scanning is somewhat restricted due to the fundamental physics of this type of instrument.

Heller et al. developed a series of multiclass methods for veterinary drugs (many of the drugs were unapproved for use in laying hens in the United States), hereby referred to as "Method 1, 2, and 3," for the three papers published in 2002, 2004, and 2006, respectively (Heller and Nochetto, 2004; Heller et al., 2002, 2006). The targeted drug classes are 16 sulfonamides (Method 1); 4 ionophores, 2 macrolides, and novobiocin (Method 2); and 15 sulfonamides, 5 tetracyclines, 4 fluoroquinolones, and 5 β-lactams (Method 3). Because of the grossly diverse solubility and chromatographic profiles of these drugs, a variety of solvents were applied each time to extract a drug subset. The sample preparation procedures were as follows: Method 1, ACN extraction followed by SPE (C18); Method 2, initial ACN extraction followed by solvent evaporation, re-extraction with hexane, and finally SPE (silica) cleanup; Method 3, extraction with sodium succinate buffer and then cleanup with SPE (HLB). MS and acquisition mode were as follows:

Method 1: LCQ Classic in time-scheduled mode. The 16 sulfonamides had very reproducible RT on ODS-AQ YMC column (4 × 50 mm, 3 μm).

Method 2: LCQ Classic in a combination of time-scheduled and data-dependent mode. The four ionophore compounds were not well separated by the YMC phenyl column (4 × 50 mm, 3 μm); therefore, they were monitored within one acquisition segment. In the first two segments that had only one or two precursor ions monitored, the time-scheduled mode was set; in the third segment that contained four precursor ions, data-dependent mode was used, mainly because of the slow scan speed of LCQ Classic (isolation time is in the hundreds ms scale) relative to the LC elution peak width (typically 10–30 s). By this design, less acquisition time is wasted if not all analytes are present in the third segment, which is a more likely scenario.

Method 3: LCQ Deca XP Plus. This instrument offered much shorter isolation time (ca. 25 ms) than the LCQ Classic, so many more analytes could be included in one segment without sacrificing signal quality.

At the time the paper for "Method 1" was published, GFI-118 had not been officially released. The following confirmatory criteria were described in the Method 1 paper: (i) RIC peak's S/N >3; (ii) RT within 2% of the standard's; (iii) relative abundance of the structurally specific product ions > 2%; and (iv) sample product's ion mass spectrum that visually matches that of the same batch standard injection, with a general correspondence between relative abundances. A strict numerical criterion was not required because the full-spectrum data usually had hundreds of significant data points for comparison, and the number of diagnostic ions was more than three for many analytes. Abiding by the criteria above, confirmation was achieved for 14 of the 16 sulfonamides, with reported LOC between 5 and 10 ppb.

In "Method 2," relevant confirmatory criteria already published in GFI-118 were followed. Confirmatory evaluation was conducted only if the screening criteria were met first. The designated screening ion should appear at the correct RT (±5%), with S/N >5:1, and a semiquantitative result >0.5 ppb. Additional criteria for confirmation were as follows: (i) the full mass spectrum had to correspond closely to the standard mass spectra acquired the same day, and (ii) the monitored confirmation ions had to be predominant in the mass spectra. Method 2 could thus confirm the selected veterinary drugs at concentrations below 10 ppb.

A sequel to the previous two methods, "Method 3" was also developed for a long-term internal surveillance program covering a wide variety of drug residues in eggs, with a target level of 100 ppb. The main goal of this effort was to combine the detection of many different compounds into the fewest methods necessary. MS data acquisition time was scheduled from 3 to 19 min, split into multiple 1 min segments. Within the capacity of the instrument, numerous target analytes were monitored in segments from 8 to 14 min, as three or more diagnostic ions were assigned to all analytes. In summary, confirmatory limits achieved on the LCQ Deca XP Plus were 10–20 ppb for fluoroquinolones, sulfonamides, and tetracyclines but higher than 50 ppb for β-lactams. LOC for all analytes ranged from 2- to 5-fold higher than the corresponding screening limits.

Around 2002, due to a series of findings regarding banned chemical substances—such as chloramphenicol, nitrofurans (tissue bound), and malachite green—in imported fish, shrimp, and crab meat, efforts were made to develop multiclass methods to address the problem. In the United States, no drug had been approved for shrimp farming, and thus, only qualitative (screening and/or confirmatory) methods were needed. Li et al. (2006) developed and validated a combined screening/confirmatory method that covered five classes of drugs: oxytetracycline, 6 sulfonamides, 6 (fluoro)quinolones, 2 triphenylmethyl dyes (and their respective metabolites in leuco form), and toltrazuril sulfone (a marker residue of toltrazuril). The LCQ Classic ion trap MS with an APCI source was used to obtain full-scan mass spectra for each analyte. One-time extraction with 5% TCA solution was employed followed by SPE (HLB) cleanup. The recoveries varied considerably among the analytes, seemingly in a qualitative correlation with the order of their respective HPLC RT (i.e., earlier elution corresponds to higher recovery). The confirmation criteria in general followed GFI-118, with at least three diagnostic ions monitored per analyte. The validated LOC were as follows: oxytetracycline, 200 ng/g; toltrazuril sulfone, 50 ng/g; sulfaquinoxaline, 20 ng/g; and the other 15 analytes, 2–10 ng/g.

Smith et al. (2009) developed a screening/confirmatory method for four different finfish species: trout, salmon, catfish, and tilapia, with 38 drugs being monitored simultaneously. A rather simple extraction procedure was employed: Homogenized fish muscle was first extracted with ACN, and the crude extract was washed with hexane. No SPE cleanup was used. For most of the analytes, at least three diagnostic ions were monitored. Control fish fortified with all standards each at 1, 0.1, and 0.01 ppm, respectively, and fishes dosed with selected drugs were used in validation and analyzed for at least 10 repeats per sample lot. Probably due to a lack of extensive cleanup in the extraction procedure, the MS signal of most analytes was found to be suppressed by at least 50% using an ESI source. The LOC for each residue was estimated to be the lowest fortified level where 90% or more of the samples at that level were confirmed. All of the quinolones, fluoroquinolones, macrolides, and malachite green, and most of the imidazoles could be confirmed at 0.01 ppm. Florfenicol amine, metronidazole, sulfonamides, tetracyclines, and most of the β-lactams were confirmable at 0.1 ppm. Ivermectin and penicillin G were only detectable (not confirmable) in samples fortified at 1 ppm.

13.4.3 Examples of LC-(Q)-TOF-Based Multiresidue Confirmatory Methods

In 2011, Villar-Pulido et al. (2011) reported an LC-TOF MS method for simultaneous quantitative multiclass determination of residues of 14 antibiotics and other veterinary drugs—benzalkonium chloride, ethoxyquin, leucomalachite green, malachite green, mebendazole, sulfadiazine, sulfadimethoxine, sulfamethazine, sulfamethizole, sulfanilamide, sulfapyridine, sulfathiazole, and trimcthoprim—in shrimp. Several different extraction procedures were tested: ACN extraction followed by dSPE (PSA and $MgSO_4$), TCA extraction followed by MSPD, protein precipitation with sulfuric acid/sodium tungstate/ACN followed by SPE and LLE, and extraction with metaphosphoric acid in ACN followed by SPE (HLB) cleanup. The first method was chosen as the most appropriate one for best overall quantitation performance. LC-TOF MS with a C18 column (50×4.6 mm, 1.8 μm) was used for confirmatory analysis. Confirmation of the target analytes was based on accurate mass measurements of both the precursor ion and fragment ions generated by in-source CID (up to six fragments). In most analyses, the mass accuracy errors were smaller than 2 ppm. LOD for these analytes was determined to be in the range of 0.06–7 μg/kg, at which level presumably the combination of consistent RT and accurate mass could unequivocally confirm their identity.

Turnipseed et al. (2011) developed and validated an LC-Q-TOF MS method to analyze veterinary drug residues in milk. In a simple procedure, an aliquot of pasteurized whole milk was first vortex mixed with ACN, and then the mixture was centrifuged and the supernatant was further spin filtered through a 3 kDa MW-cutoff filter. Control samples with 25 target compounds, including sulfonamides, tetracyclines, β-lactams, and macrolides, were fortified at 0.5×, 1×, or 2× target levels in validation. Residue screening was based on accurate mass (±10 ppm error) and RT (±0.5 min) acquired in TOF-only mode, using Mass Hunter software's "Find by Formula" algorithm to search for hits in an in-house veterinary drug database. Overall, the targeted residues were detected in samples fortified at 1× the tolerance/safe level 97% of the time. Matrix interference was observed for sulfamerazine, due to a component in the blank control having a very close *m/z* and RT to this analyte. It was suggested that higher MS resolution (~14,000) was needed to resolve these two ions. For confirmation, MS/MS data were also generated in Q-TOF mode from the $[M+H]^+$ ions. Mass accuracy of the selected monoisotopic product ions for each compound was compared to theoretical values. In this study, LOI was defined as the lowest fortification level at which the MS/MS full spectra (averaged around each analyte's RT) visually matched a known standard, allowing a ±15 ppm error for the predominant product ions (confirmation per se). It was also found that relative abundance of the selected ions could meet the criterion for unit resolution in GFI-118. Overall, LOI for 20 of the analytes was achieved at or below the respective 1× target levels. Residue-incurred milk samples from cows dosed with either sulfamethazine, flunixin, cephapirin, or enrofloxacin were analyzed with this method. Several metabolites, including a few that were previously unreported, were also detected by evaluating the MS and MS/MS data.

13.4.4 Examples of Orbitrap-Based Multiresidue Confirmatory Methods

Because Orbitrap first became commercially available only in mid-2000, there is only a relatively short history of its application in drug residue analysis. Orbitrap's mass resolution is high enough in the latest model (well above 100,000 FWHM) to unequivocally assign an elemental composition or molecular formula based on an accurate mass and isotope ratio up to a certain *m/z* range (Grange and Sovocool, 2008); this is a huge technical leap in compound identification and confirmation. Consequently, because of this level of selectivity, database searches with *m/z* can be conducted much more effectively. Orbitrap's main drawback is its slow scanning speed, especially at higher resolution settings.

In 2011, a multiresidue method for more than 100 veterinary drugs or pesticides in the kidney, liver, muscle, fish, and honey was published by Kaufmann et al. (2011b) using UPLC-Orbitrap. Both approved drugs (with MRL) and banned substances comprised the target analyte list. Initially, it was found that if the sample preparation procedure in a previously developed TOF MS-based multiresidue method (Kaufmann et al., 2008) was used, severe postinterface signal suppression occurred, due to the space-charge effect in the C-trap. This caused a major problem for both quantitation and sensitivity, as these "heavy" matrices had a high content of soluble and multiply-charged proteins in the raw extract. Several protein precipitation procedures were tested, such as ACN, cationic heavy metal solutions, polymeric tungstate solution, and concentrated ammonium sulfate. The SPE cartridge for subsequent cleanup was also carefully selected. The new protocol reduced the protein content by about 90%, which was evidenced by a much cleaner total ion chromatogram or total ion current (TIC). The single-stage Orbitrap MS in this work was operated at 50,000 FWHM, compared to the previous TOF MS method in which a resolution of 12,000 FWHM was insufficient for all drugs to be resolved from isobaric interference. This method was validated according to 2002/657/EC, with modifications to fit the particular situation this method was facing, that is, (i) not all analytes had an established MRL, and (ii) matrices from multiple species and organ tissues were very diverse. Significantly better performance was achieved for this Orbitrap MS-based method compared to the older TOF method, which was largely attributable to measures taken in sample preparation and use of the more powerful Orbitrap MS.

Also in 2011, a quantitative and confirmatory method for the determination of 9 benzimidazoles, febantel, 2 avermectins, and phenylbutazone residues in milk and muscle was reported (Kaufmann et al., 2011a). A QuEChERS-type extraction procedure was used for sample preparation, and the extracts were analyzed by a UPLC-single-stage Orbitrap system with a core–shell column (C18, 150×2.1 mm, 2.6 μm) for chromatographic separation. The high mass resolution of 50,000 FWHM and a narrow (10 ppm) mass extraction window led to selective and sensitive detection of analytes without need of a separate fragmentation step for the precursor ions (some of them were sodium adducts). Thus, the class of avermectins whose sodium adduct is known to have difficulty generating a meaningful fragment by CID could be confirmed in this way. For both quantitation and confirmation method validation, 2002/657/EC was followed for the UPLC-Orbitrap MS method and a separate UPLC-MS/MS method (two transitions monitored for each analyte). The corresponding IP for each individual drug depended on its regulatory classification (approved or banned). The Orbitrap-based method offered significantly greater sensitivity for avermectins than the MS/MS method.

13.4.5 Comparison between Unit Resolution MS/MS and HRMS[1] on the Same Analysis Targets

In 2007, Wang and Leung (2007) published two multimatrix methods for 6 macrolides in eggs, raw milk, and honey. One method used UPLC-Q-TOF in TOF mode (W-wave; full-spectrum; 15,000 FWHM), and the other used HPLC-MS/MS in MRM mode (two transitions per analyte). A 1.7 μm BEH C18 column (100×2.1 mm) was used for UPLC, and a YMC ODS-AQ cartridge (S-3, 120Å, 50×2 mm) was used for HPLC. Depending on the matrix, samples were extracted by either ACN or 0.1 M phosphate buffer (pH 8) and then cleaned up with Oasis HLB cartridges. Based on the alternative criterion advocated by Hernández et al. (2004), it was proposed that at least 3 IPs be assigned to the UPLC-TOF method (actual *m/z* deviation, 2–10 ppm), adequate to satisfy the IP requirement defined in 2002/657/EC. The results indicated that LC-MS/MS had lower LOD and better repeatability, while UPLC-TOF MS provided high enough resolving power for unequivocal confirmation of positive samples. The two methods provide complementary information in confirmatory analysis of macrolide antibiotics in these foods.

A recent work by Kaufmann et al. (2011c) provided a fundamental insight into the difference between Orbitrap, TOF, and tandem MS on quantitative and confirmatory performance. More than 100 different veterinary drugs were targeted, though only a few dozen of them were analyzed by MS/MS due to limited monitoring capacity of the instrument. To study background interference, matrices with complex contents (kidney and honey) were chosen. Analytes were first extracted with ACN and a succinate/ammonium sulfate/ethylenediaminetetraacetic acid (EDTA) buffer, and then the raw extract was cleaned up with SPE (ABN). The same set of extracts were analyzed with UPLC-Orbitrap (single stage operated at 50,000 FWHM with three discrete fragmentation mechanisms), UPLC-TOF (12,000 FWHM in "W" mode), and UPLC-MS/MS (50 ms dwell time, two transitions per analyte, two injections per sample). Two characteristic exact-mass ions (precursor ion plus one product ion without precursor selection) were chosen for each analyte for confirmatory analysis by Orbitrap and TOF. Validation was conducted according to 2002/657/EC. Orbitrap showed slightly better quantitative performance than QqQ (MS/MS mode) and much better performance than TOF. Sensitivity was higher for MS/MS only if a limited number of transitions were set up in one MS/MS method. The ion ratio as the key confirmatory criterion was first calculated using the peak area ratio between the quantitative and qualitative traces for both MS/MS and Orbitrap methods (data were acquired under all three fragmentation modes separately). The relative standard deviations (RSDs) of the measured ion ratio ($n=4$) for a few dozen analytes were plotted against the (i) ion ratio (%), (ii) fragmentation efficacy (%), and (iii) analyte concentration. Overall, variation of fragmentation ratios was higher for non-precursor-selected experiments (Orbitrap) than MS/MS, albeit to a lesser extent for HCD mode. It was pointed out that for the MS/MS experiments, because each transition was individually optimized, all should have had optimum fragmentation efficacy. Compared to the Orbitrap-based methods where a generic (not custom optimized for each analyte) collision energy was applied, the MS/MS had better sensitivity during confirmation, especially for the weaker qualitative trace, which tends to have a higher RSD.

13.5 CONCLUSION AND FUTURE PERSPECTIVE

The past decade has been an exciting and dynamic period, with mass spectrometric technologies progressing at a fascinating speed. Widespread use of digital computing power greatly enhances an analyst's ability to handle onerous tasks that used to be unthinkable. We will likely see further advancement and ever-expanding adoption of these state-of-the-art instruments, especially HRMS, for use in confirmatory work. Other advancements in the foreseeable future are automated sample preparation and dedicated data handling software to efficiently process large HRMS data files. Meanwhile, regulatory analysts are making efforts to expand the use of existing validated methods to cover more analytes and matrices, adapt them to other instrument platforms, and reach mutual recognition and harmonization of analytical results. There is also need to update guidance to keep pace with new technologies, embracing both targeted and nontargeted analysis. And, hopefully, the mathematical and theoretical ground for confirmatory analysis will be further explored. Residue analysis as a means of securing food safety for the public is entering a new era.

ACKNOWLEDGMENTS

The author thanks Dr. Mary Carson and Dr. Phil Kijak at CVM for their thoughtful comments.

REFERENCES

Almeida MP, Rezende CP, Souza LF, Brito RB (2012). Validation of a quantitative and confirmatory method for residue analysis of aminoglycoside antibiotics in poultry, bovine, equine and swine kidney through liquid chromatography-tandem mass spectrometry. Food Addit Contam 29(4):517–525.

Andersen WC, Roybal JE, Gonzales SA, Turnipseed SB, Pfenning AP, Kuck LR (2005). Determination of tetracycline residues in shrimp and whole milk using liquid chromatography with ultraviolet detection and residue confirmation by mass spectrometry. Anal Chim Acta 529:145–150.

Andersen WC, Turnipseed SB, Karbiwnyk CM, Lee RH, Clark SB, Rowe WD, Madson MR, Miller KE (2009). Multiresidue method for the triphenylmethane dyes in fish: Malachite green, crystal (gentian) violet, and brilliant green. Anal Chim Acta 637:279–289.

AOAC (2002). AOAC official methods of analysis, Appendix D: Guidelines for collaborative study procedures to validate characteristics of a method of analysis. Available at http://www.aoac.org/imis15_prod/AOAC_Docs/StandardsDevelopment/Collaborative_Study_Validation_Guidelines.pdf. Accessed April 21, 2014.

AORC (2011). AORC guidelines for the minimum criteria for identification by chromatography and mass spectrometry (May 2011). Available at http://www.aorc-online.org/documents/aorc-ms-criteria-may-2011/. Accessed July 2, 2013.

Berendsen BJA, Wegh RS, Essers ML, Stolker AAM, Weigel S (2012). Quantitative trace analysis of a broad range of antiviral drugs in poultry muscle using column-switch liquid chromatography coupled to tandem mass spectrometry. Anal Bioanal Chem 402:1611–1623.

Berendsen BJA, Stolker LAM, Nielen MWF (2013a). Selectivity in the sample preparation for the analysis of drug residues in products of animal origin using LC-MS. Trends Analyt Chem 43:229–239.

Berendsen BJA, Stolker LAM, Nielen MWF (2013b). The (un)certainty of selectivity in liquid chromatography tandem mass spectrometry. J Am Soc Mass Spectrom 24:154–163.

Berrada H, Borrull F, Font G, Marcé RM (2008). Determination of macrolide antibiotics in meat and fish using pressurized liquid extraction and liquid chromatography–mass spectrometry. J Chromatogr A 1208:83–89.

Bethem R, Boison J, Gale J, Heller D, Lehotay S, Loo J, Musser S, Phil P, Stein S (2003). Establishing the fitness for purpose of mass spectrometric methods. J Am Soc Mass Spectrom 14:528–541.

Blasco C, Masia A, Morillas FG, Picó Y (2011). Comparison of the effectiveness of recent extraction procedures for antibiotic residues in bovine muscle tissues. J AOAC Int 94(3):991–1003.

Bohm DA, Stachel CS, Gowik P (2009). Multi-method for the determination of antibiotics of different substance groups in milk and validation in accordance with Commission Decision 2002/657/EC. J Chromatogr A 1216:8217–8223.

Bohm DA, Stachel CS, Gowik P (2011). Validated determination of eight antibiotic substance groups in cattle and pig muscle by HPLC/MS/MS. J AOAC Int 94(2):407–419.

Bohm DA, Stachel CS, Gowik P (2012a). Confirmatory method for the determination of streptomycin and dihydrostreptomycin in honey by LC–MS/MS. Food Addit Contam 29(2):189–196.

Bohm DA, Stachel CS, Gowik P (2012b). Validation of a multi-residue method for the determination of several antibiotic groups in honey by LC-MS/MS. Anal Bioanal Chem 403:2943–2953.

Boscher A, Guignard C, Pellet T, Hoffmann L, Bohn T (2010). Development of a multi-class method for the quantification of veterinary drug residues in feedingstuffs by liquid chromatography-tandem mass spectrometry. J Chromatogr A 1217:6394–6404.

Caldow M, Sharman M, Kelly M, Day J, Hird S, Tarbin JA (2009). Multi-residue determination of phenolic and salicylanilide anthelmintics and related compounds in bovine kidney by liquid chromatography–tandem mass spectrometry. J Chromatogr A 1216:8200–8205.

CCRVDF (2009). Guidelines on analytical terminology (CAC/GL 72-2009). Report of the Nineteenth Session of the CCRVDF. Available at http://www.codexalimentarius.org/input/download/standards/11357/cxg_072e.pdf. Accessed July 2, 2013.

CCRVDF (2013). Proposed draft guidelines on performance characteristics for multi-residues methods (Appendix to CAC/GL 71-2009), N01-2011. Available at ftp://ftp.fao.org/codex/meetings/ccrvdf/ccrvdf21/rv21_07e.pdf. Accessed July 2, 2013.

Cháfer-Pericás C, Maquieira Á, Puchades R, Company B, Miralles J, Moreno A (2010). Multiresidue determination of antibiotics in aquaculture fish samples by HPLC–MS/MS. Aquac Res 41:e217–e225.

Chico J, Rúbies A, Centrich F, Companyó R, Prat MD, Granados M (2008). High-throughput multiclass method for antibiotic residue analysis by liquid chromatography–tandem mass spectrometry. J Chromatogr A 1213:189–199.

Chu PS, Lopez MI (2007). Determination of nitrofuran residues in milk of dairy cows using liquid chromatography-tandem mass spectrometry. J Agric Food Chem 55:2129–2135.

Chu PS, Lopez MI, Serfling S, Gieseker C, Reimschuessel R (2006). Determination of 17α-methyltestosterone in muscle tissues of tilapia, rainbow trout, and salmon using liquid chromatography-tandem mass spectrometry. J Agric Food Chem 54:3193–3198.

Clark SB, Storey JM, Turnipseed SB (2011). Optimization and validation of a multiclass screening and confirmation method for drug residues in milk using high-performance liquid chromatography/tandem mass spectrometry. J AOAC Int 94(2):383–393.

Commission Decision 2002/657/EC (2002). Commission Decision of 12 August 2002 implementing Council Directive 96/23/EC concerning the performance of analytical methods and the interpretation of results (2002/657/EC). Off J Eur Commun L221:8–35.

Council Directive 96/23/EC (1996). Council Directive 96/23/EC of 29 April 1996 on measures to monitor certain substances and residues thereof in live animals and animal products and repealing Directives 85/358/EEC and 86/469/EEC and Decisions 89/187/EEC and 91/664/EEC. Off J Eur Commun L125:10–32.

Croley TR, White KD, Callahan JH, Musser SM (2012). The chromatographic role in high resolution mass spectrometry for non-targeted analysis. J Am Soc Mass Spectrom 23:1569–1578.

Cronly M, Behan P, Foley B, Malone E, Regan L (2009). Rapid confirmatory method for the determination of 11 nitroimidazoles in egg using liquid chromatography tandem mass spectrometry. J Chromatogr A 1216:8101–8109.

De Alwis H, Heller DN (2010). Multiclass, multiresidue method for the detection of antibiotic residues in distillers grains by liquid chromatography and ion trap tandem mass spectrometry. J Chromatogr A 1217:3076–3084.

De Brabander HF, Noppe H, Verheyden K, Vanden Bussche J, Wille K, Okerman L, Vanhaecke L, Reybroeck W, Ooghe S, Croubels S (2009). Residue analysis: Future trends from a historical perspective. J Chromatogr A 1216:7964–7976.

Dowling G, Mulder PPJ, Duffy C, Regan L, Smyth MR (2007). Confirmatory analysis of malachite green, leucomalachite green, crystal violet and leucocrystal violet in salmon by liquid chromatography–tandem mass spectrometry. Anal Chim Acta 586:411–419.

Dubreil-Chéneau E, Bessiral M, Roudaut B, Verdon E (2009). Pascal sanders validation of a multi-residue liquid chromatography–tandem mass spectrometry confirmatory method for 10 anticoccidials in eggs according to Commission Decision 2002/657/EC. J Chromatogr A 1216:8149–8157.

Fagerquist CK, Lightfield AR, Lehotay SJ (2005). Confirmatory and quantitative analysis of β-lactam antibiotics in bovine kidney tissue by dispersive solid-phase extraction and liquid chromatography-tandem mass spectrometry. Anal Chem 77:1473–1482.

FAO/IAEA (1998). Validation of analytical methods for food control. A Report of a Joint FAO/IAEA Expert Consultation. Rome: Food and Agriculture Organization of the United Nations. Available at ftp://ftp.fao.org/docrep/fao/007/w8420e/w8420e00.pdf. Accessed July 2, 2013.

FDA (2003). Center for Veterinary Medicine's Guidance for Industry #118: Mass Spectrometry for Confirmation of the Identity of Animal Drug Residues (May 1, 2003). Available at http://www.fda.gov/downloads/AnimalVeterinary/GuidanceComplianceEnforcement/GuidanceforIndustry/ucm052658.pdf. Accessed July 2, 2013.

FDA (2009). Office of regulatory affairs procedure #ORA-LAB.010: Guidance for the analysis and documentation of support regulatory action on pesticide residues (effective 11/06/2009).

FDA (2012). Guidelines for the Validation of Chemical Methods for the FDA Foods Program (February 2, 2012). Available at http://www.fda.gov/downloads/ScienceResearch/FieldScience/UCM298730.pdf. Accessed July 2, 2013.

FDA (2013a). Title 21: Code of Federal Regulations, Part 520: Oral dosage form new animal drugs; Part 524: Ophthalmic and topical dosage form new animal drugs; Part 526: Intramammary dosage form new animal drugs; Part 529: Certain other dosage form new animal drugs. Revised as of April 1, 2013.

FDA (2013b). Title 21: Code of Federal Regulations, Part 530: Extralabel drug use in animals. Revised as of April 1, 2013.

FDA (2013c). Title 21 Code of Federal Regulations, Part 556: Tolerances for residues of new animal drugs in food. Revised as of April 1, 2013.

FSIS (May 2011). 2009 National Residue Program Data - Red Book. Available at http://www.fsis.usda.gov/wps/portal/fsis/topics/data-collection-and-reports/chemistry/red-books/archive/2009-red-book. Accessed June 22, 2013.

Gasparini M, Curatolo M, Assini W, Bozzoni E, Tognoli N, Dusi G (2009). Confirmatory method for the determination of nandrolone and trenbolone in urine samples using immunoaffinity clean-up and liquid chromatography–tandem mass spectrometry. J Chromatogr A 1216:8059–8066.

Gaugain-Juhel M, Delépine B, Gautier S, Fourmond MP, Gaudin V, Hurtaud-Pessel D, Verdon E, Sanders P (2009). Validation of a liquid chromatography-tandem mass spectrometry screening method to monitor 58 antibiotics in milk: A qualitative approach. Food Addit Contam 26(11):1459–1471.

Geis-Asteggiante L, Lehotay SJ, Lightfield AR, Dutko T, Ng C, Bluhm L (2012). Ruggedness testing and validation of a practical analytical method for >100 veterinary drug residues in bovine muscle by ultrahigh performance liquid chromatography–tandem mass spectrometry. J Chromatogr A 1258:43– 54.

Grange AH, Sovocool GW (2008). Automated determination of precursor ion, product ion, and neutral loss compositions and deconvolution of composite mass spectra using ion correlation based on exact masses and relative isotopic abundances. Rapid Commun Mass Spectrom 22:2375–2390.

Gratacós-Cubarsí M, Castellari M, Valero A, García-Regueiro JA (2006). Hair analysis for veterinary drug monitoring in livestock production. J Chromatogr B 834:14–25.

Guagain-Juhel M, Delepine B, Gautier S, Fourmond MP, Guadin V, Hurtaud-Pessel D, Verdon E, Sanders P (2009). Validation of a liquid chromatography-tandem mass spectrometry screening method to monitor 58 antibiotics in milk: A qualitative approach. Food Addit Contam 26:1459–1471.

Hammack W, Carson MC, Neuhaus BK, Hurlbut JA, Nochetto C, Stuart JS, Brown A, Kilpatrick D, Youngs K, Ferbos K, Heller DN (2003). Multilaboratory validation of a method to confirm chloramphenicol in shrimp and crabmeat by liquid chromatography-tandem mass spectrometry. J AOAC Int 86(6):1135–1143.

Hammel Y-A, Mohamed R, Gremaud E, LeBreton M-H, Guy PA (2008). Multi-screening approach to monitor and quantify 42 antibiotic residues in honey by liquid chromatography–tandem mass spectrometry. J Chromatogr A 1177:58–76.

Heller DN, Nochetto CB (2004). Development of multiclass methods for drug residues in eggs: Silica SPE clean-up and LC-MS/MS analysis of ionophore and macrolide residues. J Agric Food Chem 52:6848–6856.

Heller DN, Ngoh MA, Donoghue D, Podhorniak L, Righter H, Thomas MH (2002). Identification of incurred sulfonamide residues in eggs: Methods for confirmation by liquid chromatography–tandem mass spectrometry and quantitation by liquid chromatography with ultraviolet detection. J Chromatogr B 774:39–52.

Heller DN, Nochetto CB, Rummel NG, Thomas MH (2006). Development of multiclass methods for drug residues in eggs: Hydrophilic solid-phase extraction clean-up and liquid chromatography/tandem mass spectrometry analysis of tetracycline, fluoroquinolone, sulfonamide, and β-lactam residues. J Agric Food Chem 54:5267–5278.

Heller DN, Lehotay SJ, Martos PA, Hammack W, Fernández-Alba AR (2010). Issues in mass spectrometry between bench chemists and regulatory laboratory managers: Summary of the roundtable on mass spectrometry held at the 123rd AOAC International Annual Meeting. J AOAC Int 93(5):1625–1632.

Hernández F, Ibáñez M, Sancho JV, Pozo ÓJ (2004). Comparison of different mass spectrometric techniques combined with liquid chromatography for confirmation of pesticides in environmental water based on the use of identification points. Anal Chem 76:4349–4357.

Hernando MD, Mezcuaa M, Suárez-Barcena JM, Fernández-Alba AR (2006). Liquid chromatography with time-of-flight mass spectrometry for simultaneous determination of chemotherapeutant residues in salmon. Anal Chim Acta 562:176–184.

Hu Q, Noll RJ, Li H, Makarov A, Hardman M, Cooks RG (2005). The Orbitrap: A new mass spectrometer. J Mass Spectrom 40(4):430–443.

Jia X, Shao B, Wu Y, Yang Y, Zhang J (2008). Simultaneous determination of tetracylines and quinolones antibiotics in egg by ultra-performance liquid chromatography–electrospray tandem mass spectrometry. J AOAC Int 91(2):461–468.

Kanu AB, Dwivedi P, Tam M, Matz L, Hill HH Jr. (2008). Ion mobility–mass spectrometry. J Mass Spectrom 43:1–22.

Kaufmann A (2009). Validation of multiresidue methods for veterinary drug residues, related problems and possible solutions. Anal Chim Acta 637:144–155.

Kaufmann A, Butcher P, Maden K, Widmer M (2007). Ultra-performance liquid chromatography coupled to time of flight mass spectrometry (UPLC-TOF): A novel tool for multiresidue screening of veterinary drugs in urine. Anal Chim Acta 586:13–21.

Kaufmann A, Butcher P, Maden K, Widmer M (2008). Quantitative multiresidue method for about 100 veterinary drugs in different meat matrices by sub 2-μm particulate high-performance liquid chromatography coupled to time of flight mass spectrometry. J Chromatogr A 1194:66–79.

Kaufmann A, Butcher P, Maiden K, Widmer M, Giles K, Uria D (2009). Are liquid chromatography/electrospray tandem quadrupole fragmentation ratios unequivocal confirmation criteria? Rapid Commun Mass Spectrom 23(7):985–998.

Kaufmann A, Butcher P, Maden K, Walker S, Widmer M (2011a). Development of an improved high resolution mass spectrometry based multi-residue method for veterinary drugs in various food matrices. Anal Chim Acta 700:86–94.

Kaufmann A, Butcher P, Maden K, Walker S, Widmer M (2011b). Quantification of anthelmintic drug residues in milk and muscle tissues by liquid chromatography coupled to orbitrap and liquid chromatography coupled to tandem mass spectrometry. Talanta 85:991–1000.

Kaufmann A, Butcher P, Maden K, Walker S, Widmer M (2011c). Quantitative and confirmative performance of liquid chromatography coupled to high-resolution mass spectrometry compared to tandem mass spectrometry. Rapid Commun Mass Spectrom 25:979–992.

Kinsella B, Lehotay SJ, Mastovska K, Lightfield AR, Furey A, Danaher M (2009a). New method for the analysis of flukicide and other anthelmintic residues in bovine milk and liver using liquid chromatography–tandem mass spectrometry. Anal Chim Acta 637:196–207.

Kinsella B, O'Mahony J, Malone E, Moloney M, Cantwell H, Furey A, Danaher M (2009b). Current trends in sample preparation for growth promoter and veterinary drug residue analysis. J Chromatogr A 1216:7977–8015.

Kinsella B, Whelan M, Cantwell H, McCormack M, Furey A, Lehotay SJ, Danaher M (2010). A dual validation approach to detect anthelmintic residues in bovine liver over an extended concentration range. Talanta 83:14–24.

Kolakowski BM, Mester Z (2007). Review of applications of high-field asymmetric waveform ion mobility spectrometry (FAIMS) and differential mobility spectrometry (DMS). Analyst 132:842–864.

Le Bizec B, Pinel G, Antignac J-P (2009). Options for veterinary drug analysis using mass spectrometry. J Chromatogr A 1216:8016–8034.

Lehotay SJ, Mastovska K, Amirav A, Fialkov AB, Alon T, Martos PA, de Kok A, Fernández-Alba AR (2008). Identification and confirmation of chemical residues in food by chromatography-mass spectrometry and other techniques. Trends Analyt Chem 27(11):1070–1090.

Lehotay SJ, Lightfield AR, Geis-Asteggiante L, Schneider MJ, Dutko T, Ng C, Bluhmd L, Mastovska K (2012). Development and validation of a streamlined method designed to detect residues of 62 veterinary drugs in bovine kidney using ultrahigh performance liquid chromatography-tandem mass spectrometry. Drug Test Anal 4(Suppl 1):75–90.

Li H, Kijak PJ, Turnipseed SB, Cui W (2006). Analysis of veterinary drug residues in shrimp: A multi-class method by liquid chromatography–quadrupole ion trap mass spectrometry. J Chromatogr B 836:22–38.

Little JL, Cleven CD, Brown SD (2011). Identification of "known unknowns" utilizing accurate mass data and chemical abstracts service databases. J Am Soc Mass Spectrom 22:348–359.

Little JL, Williams AJ, Pshenichnov A, Tkachenko V (2012). Identification of "known unknowns" utilizing accurate mass data and ChemSpider. J Am Soc Mass Spectrom 23:179–185.

Lombardo-Agüí M, García-Campaña AM, Gámiz-Gracia L, Cruces-Blanco C (2012). Determination of quinolones of veterinary use in bee products by ultra-high performance liquid chromatography–tandem mass spectrometry using a QuEChERS extraction procedure. Talanta 93:193– 199.

Lopez MI, Feldlaufer MF, Williams AD, Chu P-S (2007). Determination and confirmation of nitrofuran residues in honey using LC-MS/MS. J Agric Food Chem 55:1103–1108.

Lopez MI, Pettis JS, Smith IB, Chu P-S (2008). Multiclass determination and confirmation of antibiotic residues in honey using LC-MS/MS. J Agric Food Chem 56:1553–1559.

Malone EM, Dowling G, Elliott CT, Kennedy DG, Regan L (2009). Development of a rapid, multi-class method for the confirmatory analysis of anti-inflammatory drugs in bovine milk using liquid chromatography tandem mass spectrometry. J Chromatogr A 1216:8132–8140.

Malone EM, Elliott CT, Kennedy DG, Regan L (2010a). Rapid confirmatory method for the determination of sixteen synthetic growth promoters and bisphenol a in bovine milk using dispersive solid-phase extraction and liquid chromatography-tandem mass spectrometry. J Chromatogr B 878(15–16):1077–1084.

Malone EM, Elliott CT, Kennedy DG, Regan L (2010b). Screening and quantitative confirmatory method for the analysis of glucocorticoids in bovine milk using liquid chromatography-tandem mass spectrometry. J AOAC Int 93(5):1656–1665.

Marazuela MD, Bogialli SA (2009). Review of novel strategies of sample preparation for the determination of antibacterial residues in foodstuffs using liquid chromatography-based analytical methods. Anal Chim Acta 645:5–17.

Marchesini GR, Hooijerink H, Haasnoot W, Buijs J, Campbell K, Elliott CT, Nielen MWF (2009). Resonance biosensing combined with bioaffinity-assisted nano HILIC liquid chromatography/time-of-flight mass spectrometry identification of paralytic shellfish poisons. Trends Analyt Chem 28(6):792–803.

Martos A, Lehotay SJ, Shurmer B (2008). Ultratrace analysis of nine macrolides, including tulathromycin a (Draxxin), in edible animal tissues with minicolumn liquid chromatography tandem mass spectrometry. J Agric Food Chem 56:8844–8850.

McDonald M, Mannion C, Rafter P (2009). A confirmatory method for the simultaneous extraction, separation, identification and quantification of Tetracycline, Sulphonamide, Trimethoprim and Dapsone residues in muscle by ultra-high-performance liquid chromatography–tandem mass spectrometry according to Commission Decision 2002/657/EC. J Chromatogr A 1216:8110–8116.

Mcmullen SE, Schenck FJ, Vega VA (2009). Rapid method for the determination and confirmation of fluoroquinolone residues in catfish using liquid chromatography/fluorescence detection and liquid chromatography-tandem mass spectrometry. J AOAC Int 92(4):1233–1240.

Michelangelo A, Lehotay SJ, Štajnbaher D, Schenck FJ (2003). Fast and easy multiresidue method employing acetonitrile extraction/partitioning and "dispersive solid-phase extraction" for the determination of pesticide residues in produce. J AOAC Int 86(2):412–431.

Moreno-Bondi MC, Marazuela MD, Herranz S, Rodriguez E (2009). An overview of sample preparation procedures for LC-MS multiclass antibiotic determination in environmental and food samples. Anal Bioanal Chem 395:921–946.

Mottier P, Khong S-P, Gremaud E, Richoz J, Delatour T, Goldmann T, Guy PA (2005). Quantitative determination of four nitrofuran metabolites in meat by isotope dilution liquid chromatography–electrospray ionisation–tandem mass spectrometry. J Chromatogr A 1067:85–91.

Nielen MWF, van Engelen MC, Zuiderent R, Ramaker R (2007). Screening and confirmation criteria for hormone residue analysis using liquid chromatography accurate mass time-of-flight, Fourier transform ion cyclotron resonance and orbitrap mass spectrometry techniques. Anal Chim Acta 586:122–129.

Olejnik M, Szprengier-Juszkiewicz T, Jedziniak P (2009) Multi-residue confirmatory method for the determination of twelve coccidiostats in chicken liver using liquid chromatography tandem mass spectrometry. J Chromatogr A 1216:8141–8148.

Oxford Dictionaries (2013). Definition of *confirm* in English. Oxford: Oxford University Press. Available at Oxforddictionaries.com. Accessed June 23, 2013.

Quon D, Carson MC, Nochetto C, Heller DN, Butterworth F (2006). Peer validation of a method to confirm chloramphenicol in honey by liquid chromatography-tandem mass spectrometry. J AOAC Int 89(2):586–593.

Rezende CP, Almeida MP, Brito RB, Nonaka CK, Leite MO (2012). Optimisation and validation of a quantitative and confirmatory LC-MS method for multi-residue analyses of β-lactam and tetracycline antibiotics in bovine muscle. Food Addit Contam 29(4):541–549.

Samanidou V, Nisyriou S (2008). Multi-residue methods for confirmatory determination of antibiotics in milk. J Sep Sci 31:2068–2090.

SANCO (2012). Method validation and quality control procedures for pesticide residues analysis in food and feed (Document N° SANCO/12495/2011). Available at http://www.crl-pesticides.eu/library/docs/fv/SANCO12495-2011.pdf. Accessed July 2, 2013.

Schneider MJ, Lehotay SJ, Lightfield AR (2012). Evaluation of a multi-class, multi-residue liquid chromatography – tandem mass spectrometry method for analysis of 120 veterinary drugs in bovine kidney. Drug Test Anal 4(Suppl 1):91–102.

Schürmann A, Dvorak V, Crüzer C, Butcher P, Kaufmann A (2009). False-positive liquid chromatography/tandem mass spectrometric confirmation of sebuthylazine residues using the identification points system according to EU directive 2002/657/EC due to a biogenic insecticide in tarragon. Rapid Commun Mass Spectrom 23:1196–1200.

Shao B, Jia X, Wu Y, Hu J, Tu X, Zhang J (2007). Multi-class confirmatory method for analyzing trace levels of tetracycline and quinolone antibiotics in pig tissues by ultra-performance liquid chromatography coupled with tandem mass spectrometry. Rapid Commun Mass Spectrom 21:3487–3496.

Smith S, Gieseker C, Reimschuessel R, Decker C-S, Carson MC (2009). Simultaneous screening and confirmation of multiple classes of drug residues in fish by liquid chromatography-ion trap mass spectrometry. J Chromatogr A 1216:8224–8232.

SOFT/AAFS (2006). SOFT/AAFS forensic toxicology laboratory guidelines. Available at http://www.soft-tox.org/files/Guidelines_2006_Final.pdf. Accessed July 2, 2013.

Stoev G, Xuan Y, Peycheva M, Scigelova M (2012). Quantitative assessment of the contribution of high resolution mass spectrometric analysis to the reliability of compound confirmation. Talanta 98:19–27.

Stolker AAM, Rutgers P, Oosterink E, Lasaroms JJP, Peters RJB, van Rhijn JA, Nielen MWF (2008). Comprehensive screening and quantification of veterinary drugs in milk using UPLC–TOF MS. Anal Bioanal Chem 391:2309–2322.

Toaldo IM, Gamba GZ, Picinin LA, Rubensam G, Hoff R, Bordignon-Luiz M (2012). Multiclass analysis of antibacterial residues in milk using RP-liquid chromatography with photo diode array and fluorescence detection and tandem mass spectrometer confirmation. Talanta 99:616–624.

Trullols E, Ruisánchez I, Xavier Rius F (2004). Validation of qualitative analytical methods. Trends Analyt Chem 23(2):137–145.

Turnipseed SB, Roybal JE, Pfenning AP, Kijak PJ (2003). Use of ion-trap liquid chromatography–mass spectrometry to screen and confirm drug residues in aquacultured products. Anal Chim Acta 483:373–386.

Turnipseed SB, Roybal JE, Andersen WC, Kuck LR (2005). Analysis of avermectin and moxidectin residues in milk by liquid chromatography–tandem mass spectrometry using an atmospheric pressure chemical ionization/atmospheric pressure photoionization source. Anal Chim Acta 529:159–165.

Turnipseed SB, Andersen WC, Karbiwnyk CM, Madson MR, Miller KE (2008). Multi-class, multi-residue liquid chromatography/tandem mass spectrometry screening and confirmation methods for drug residues in milk. Rapid Commun Mass Spectrom 22:1467–1480.

Turnipseed SB, Clark SB, Karbiwnyk CM, Andersen WC, Miller KE, Madson MR (2009). Analysis of aminoglycoside residues in bovine milk by liquid chromatography electrospray ion trap mass spectrometry after derivatization with phenylisocyanate. J Chromatogr B 877:1487–1493.

Turnipseed SB, Storey JM, Clark SB, Miller KE (2011). Analysis of veterinary drugs and metabolites in milk using quadrupole time-of-flight liquid chromatography-mass spectrometry. J Agric Food Chem 59:7569–7581.

USDA (2013). USDA, Agricultural Marketing Service, Science & Technology: Pesticide Data Program. Available at http://www.ams.usda.gov/AMSv1.0/getfile?dDocName=stelprdc5083988. Accessed July 2, 2013.

Verdon E, Couedor P, Sanders P (2007). Multi-residue monitoring for the simultaneous determination of five nitrofurans (furazolidone, furaltadone, nitrofurazone, nitrofurantoine, nifursol) in poultry muscle tissue through the detection of their five major metabolites (AOZ, AMOZ, SEM, AHD, DNSAH) by liquid chromatography coupled to electrospray tandem mass spectrometry—in-house validation in line with Commission Decision 657/2002/EC. Anal Chim Acta 586:336–347.

Villar-Pulido M, Gilbert-López B, García-Reyes JF, Martos NR, Molina-Díaz A (2011). Multiclass detection and quantitation of antibiotics and veterinary drugs in shrimps by fast liquid chromatography time-of-flight mass spectrometry. Talanta 85:1419–1427.

WADA (2010). Identification criteria for qualitative assays incorporating column chromatography and mass spectrometry. WADA Technical Document TD2010IDCR. Available at http://www.wada-ama.org/en/World-Anti-Doping-Program/Sports-and-Anti-Doping-Organizations/International-Standards/Laboratories/Technical-Documents/. Accessed July 2, 2013.

Wang J (2004a). Confirmatory determination of six penicillins in honey by liquid chromatography/electrospray ionization–tandem mass spectrometry. J AOAC Int 87(1):45–55.

Wang J (2004b). Determination of five macrolide antibiotic residues in honey by LC-ESI-MS and LC-ESI-MS/MS. J Agric Food Chem 52:171–181.

Wang J, Leung D (2007). Analyses of macrolide antibiotic residues in eggs, raw milk, and honey using both ultra-performance liquid chromatography/quadrupole time-of-flight mass spectrometry and high-performance liquid chromatography/tandem mass spectrometry. Rapid Commun Mass Spectrom 21:3213–3222.

Wang J, Leung D (2009). Determination of spiramycin and neospiramycin antibiotic residues in raw milk using LC/ESI-MS/MS and solid-phase extraction. J Sep Sci 32:681–688.

Wang J, Leung D (2012). The challenges of developing a generic extraction procedure to analyze multi-class veterinary drug residues in milk and honey using ultra-high pressure liquid chromatography quadrupole time-of-flight mass spectrometry. Drug Test Anal 4(Suppl 1):103–111.

Wang J, Leung D, Butterworth F (2005). Determination of five macrolide antibiotic residues in eggs using liquid chromatography/electrospray ionization tandem mass spectrometry. J Agric Food Chem 53:1857–1865.

Wang J, Leung D, Lenz SP (2006). Determination of five macrolide antibiotic residues in raw milk using liquid chromatography-electrospray ionization tandem mass spectrometry. J Agric Food Chem 54:2873–2880.

Wehling P, LaBudde RA, Brunelle SL, Nelson MT (2011). Probability of detection (POD) as a statistical model for the validation of qualitative methods. J AOAC Int 94(1):335–347.

Whelan M, Kinsella B, Furey A, Moloney M, Cantwell H, Lehotay SJ, Danaher M (2010). Determination of anthelmintic drug residues in milk using ultra high performance liquid chromatography–tandem mass spectrometry with rapid polarity switching. J Chromatogr A 1217:4612–4622.

Wu X, Zhang G, Wu Y, Hou X, Yuan Z (2007). Simultaneous determination of malachite green, gentian violet and their leuco-metabolites in aquatic products by high-performance liquid chromatography–linear ion trap mass spectrometry. J Chromatogr A 1172:121–126.

14

THE FOOD ANIMAL RESIDUE AVOIDANCE DATABANK: AN EXAMPLE OF RISK MANAGEMENT OF VETERINARY DRUG RESIDUES

Thomas W. Vickroy[1], Ronald E. Baynes[2], Lisa Tell[3] and Jim E. Riviere[4]

[1]*Department of Physiological Sciences, College of Veterinary Medicine, University of Florida, Gainesville, FL, USA*
[2]*Department of Population Health and Pathobiology, College of Veterinary Medicine, North Carolina State University, Raleigh, NC, USA*
[3]*Department of Medicine and Epidemiology, University of California, Davis, CA, USA*
[4]*Department of Anatomy and Physiology, College of Veterinary Medicine, Kansas State University, Manhattan, KS, USA*

It is an undeniable fact that food safety is a highly important societal issue that impacts people in nearly every country and region of the modern world. In the United States, consumer concerns over food safety have had a significant impact on the production, marketing, and sale of nearly all types of food products, including those that are derived from production animals. Such concerns among American consumers have provided a substantial impetus for strengthening many aspects of governmental oversight and, undoubtedly, have contributed to a recent shift by regulatory agencies to adopt a more proactive posture when developing strategies to enhance overall public food

Strategies for Reducing Drug and Chemical Residues in Food Animals: International Approaches to Residue Avoidance, Management, and Testing, First Edition.
Edited by Ronald E. Baynes and Jim E. Riviere.

safety. To that end, passage of the Food Safety Modernization Act (FSMA) of 2012 along with a myriad of regulations that have been enacted or proposed by the U.S. Food and Drug Administration (FDA) has sent a clear message that the future focus of food safety will place substantially greater responsibility on producers, both domestic and international, as a means to reduce overall risks and to ensure the safety and economic vigor of our nation's food system.

Most domestic food products of animal origin, notably meat, milk, and eggs, are generally considered to be safe by the majority of American consumers. Nevertheless, reasonable and justified concerns persist regarding the potential for contamination of food products by harmful microorganisms or unsafe chemical residues. In response to persistent consumer concerns and heightened regulatory oversight, producers have moved to expand internal quality assurance programs as a means to enhance the safety of food products. To this end, the Food Animal Residue Avoidance and Depletion (FARAD) program has become widely recognized and heavily relied upon by food animal veterinarians, farmers, producers, and state regulators as an invaluable resource for keeping abreast of ever-changing government regulations. In addition, FARAD has developed into a unique resource for expert-mediated advice to mitigate risk in situations involving accidental chemical contamination of food-producing animals or circumstances that entail legal extralabel use of drugs by food animal veterinarians.

14.1 ORIGINS OF FARAD

In the three decades since FARAD was created, the program has developed into a unique and highly valued national risk prevention and mitigation program. FARAD began in 1982 as an outgrowth of the Residue Avoidance Program (RAP) and was supported early on by the U.S. Department of Agriculture (USDA) Food Safety Inspection Service (FSIS). Since its inception, a central goal of FARAD has been to fill a critical vulnerability in the U.S. food animal industry by providing expert advice to veterinarians on matters related to food animal exposures to a wide array of chemicals, including drugs, pesticides, agrochemicals, and environmental and industrial agents. In order to do this, the FARAD program has amassed the single most comprehensive source of scientific data related to the depletion of drugs and chemicals from tissues of major and minor food animal species. Data within these expansive FARAD databases have been collected from a wide array of sources that includes regulatory information about U.S.-approved food animal drugs, peer-reviewed scientific reports, unpublished data from drug manufacturers, as well other proprietary resources. In 1986, FARAD published the first ever compendium of drug products approved for use in the

United States for food production animals; several years later, in 1992, FARAD launched a website (www.farad.org—see Section 14.3) that is still operational today. In its present-day configuration, FARAD is overseen by a small group of highly trained veterinary pharmacologists, pharmacokineticists, toxicologists, and food animal specialists who are located at colleges/schools of veterinary medicine at the University of California–Davis (UCD), the University of Florida (UF), Kansas State University (KSU), and North Carolina State University (NCSU). Owing to the multicenter structure of FARAD in four separate geographical regions of the United States and the sharing of responsibilities among these regions, FARAD is well suited to address food animal issues throughout the country and to provide uninterrupted service in the event of natural or man-made disasters that might otherwise limit communication with a single centralized location (Fig. 14.1).

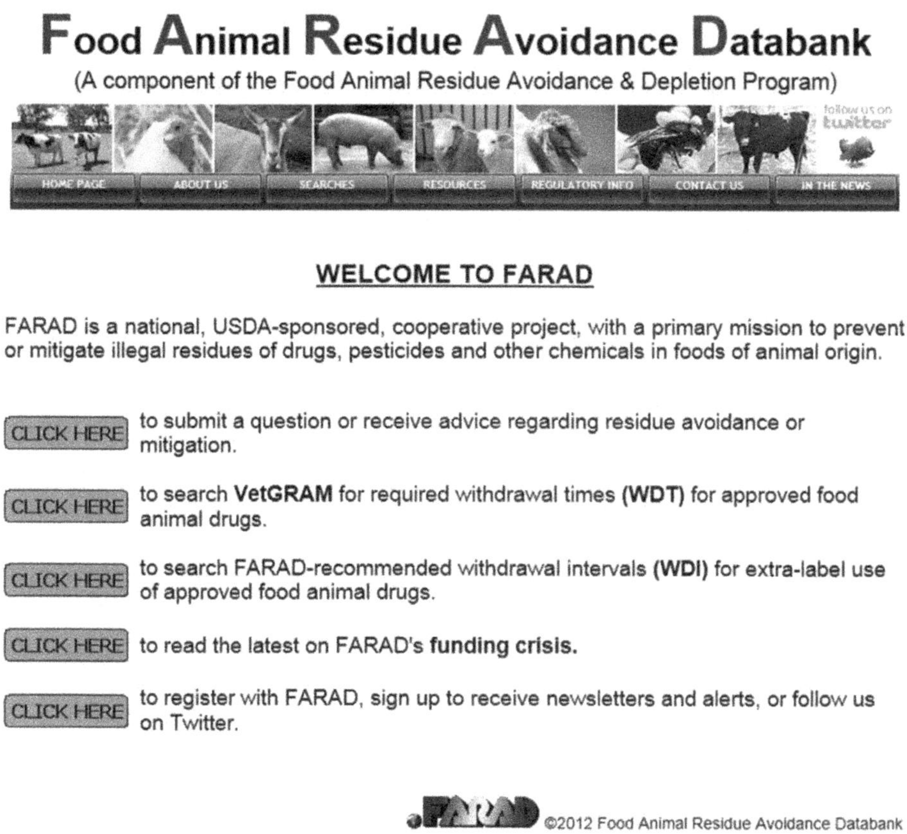

FIGURE 14.1 Home page of the FARAD website. The interactive FARAD website (www.farad.org) provides access to the latest regulations for approved food animal drugs as well as many user-defined search options and tools.

14.2 THE ROLE OF FARAD

In 1994, the U.S. Congress passed a landmark legislation known as the Animal Medicinal Drug Use Clarification Act (AMDUCA). Under AMDUCA, veterinarians were granted legal authority to use drugs in an extralabel manner in food-producing animal species so long as certain requirements and safeguards were met. Foremost among these requirements were four specific stipulations that veterinarians must follow, including:

1. Making a careful diagnosis and evaluation of conditions for which the drug is to be used
2. Establishing a substantially extended withdrawal period prior to marketing of mink, meat, eggs, or other edible products supported by appropriate scientific information, if applicable
3. Instituting procedures to assure that the identity of the treated animal or animals is carefully maintained
4. Taking appropriate measures to assure that assigned timeframes for withdrawal are met and no illegal drug residues occur in any food-producing animal subjected to extralabel treatment

Although AMDUCA created the legal framework for limited extralabel drug treatments in food-producing animals, the legislation did not identify any specific means by which veterinarians should or could use in order to establish a "...substantially extended withdrawal period that is based on scientific evidence." In practical terms, this AMDUCA-based requirement helped establish FARAD as the *de facto* scientific resource for veterinarians to use in order to establish safe extended withdrawal periods for food-producing animals following extralabel drug use (ELDU) or unintended exposures to a wide array of chemicals. With its comprehensive collection of pharmacokinetic databases combined with a substantial breadth of highly relevant scientific expertise, FARAD was and still is the sole program that is capable of combining advanced pharmacokinetic modeling studies with population-based and physiologically based approaches for deriving drug and chemical withdrawal intervals for food animal species. Under the Agricultural Research, Extension, and Education Reform Act (AREEA) of 1998, FARAD was authorized for funding by the U.S. Congress, although the program has been funded perpetually at far less than half of the authorized level. In recent years, funding through the USDA's National Institute of Food and Agriculture (NIFA) has been awarded with few interruptions on an annual basis, although the tenuous funding situation has threatened programmatic stability, limited development in several key areas, and weakened

FARAD's ability to deliver optimal service to producers, veterinarians, and extension specialists.

14.3 ACCESS TO REGULATORY DRUG INFORMATION VIA THE FARAD WEBSITE

In the United States, the FDA has authority to establish safety limits for residues of drugs or related (marker) compounds in foods derived from animal sources that are intended for human consumption. The maximum level of a drug residue(s) or tolerance drug is based upon a multitude of factors, including the concentration of drug residues in edible products from treated animals, the estimated acceptable daily intake (ADI) of total drug residues, and the relationship between the marker analyte for a drug and total tissue residues. For all new food animal drugs, the FDA requires that sponsors submit a new animal drug application (NADA) that addresses both the safety and efficacy of the drug for its intended use(s) in target animals as well as the drug's safety for human consumption. In order to satisfy the latter requirement, a sponsor must provide complete studies that address specific issues related to toxicity, residue chemistry, and, in some cases, microbial impact studies. Key step in determining a drug's relative safety are to establish a *No Observable Effect Level* (NOEL) in test animals and to apply various safety factors to translate that NOEL into a safe level for human consumption. If and when the drug is approved for use in one or more food-producing animal species, the established tolerances for the drug (marker) residues in meat, milk, or eggs are published and updated in the Code of Federal Regulations (CFR Title 21, Volume 6, Part 556 Subpart B). In contrast to the United States, regulatory agencies in other countries frequently use slightly different or additional factors to establish maximum residue levels (MRLs), including different edible tissue consumption factors and broader consideration of animal husbandry standards such as good agricultural practices.

Based upon the established tolerance for a food animal drug, a mandatory waiting period or withdrawal time (WDT) is established. Following treatment of a food-producing animal species with an approved drug in accordance with label instructions, it is illegal to use or market food products (meat, milk, eggs, etc.) from treated animals at any time prior to passage of the WDT. As mentioned earlier, the WDT applies exclusively to approved (labeled) uses of an approved drug product, including limitations related to dose, route of administration, duration of treatment, approved species, and, in certain cases, specified indications for drug use. Owing to inherent differences in the rates of drug accumulation and elimination from various tissues across different species, a single drug product may have a wide range of WDT values for different food products that are derived from different species. As an example of the complexities of regulatory WDTs, consider the Type A medicated article *Chloratet*™ (NADA 048-480), a product that was approved long ago for use in several species. The active ingredient in this product is chlortetracycline, which is formulated at different concentrations (50, 90, or 100 g/lb) in products that are approved for use in cattle (beef and nonlactating dairy), calves (excluding veal calves), chickens (excluding layers), turkeys (excluding layers), and all use classes of swine. When fed to mature cattle (over 400 lb bwt) at levels up to 70 mg/head/day, the WDT for meat products is zero. However, if *Chloratet*™ is used to control bacterial pneumonia associated with shipping fever complex or to treat active infection by *Anaplasma marginale*, cattle are treated at higher dose rates (350 mg/head/day for cattle under 700 lb bwt; 0.5 mg/lb bwt/day for cattle over 700 lb bwt), which are associated with a mandatory minimum WDT of 48 h for any meat products. An extended meat WDT of 10 days is required when the product is used to treat bacterial enteritis associated with *E. coli* or bacterial pneumonia caused by *Pasteurella multocida* in cattle (10 mg/lb bwt for up to 5 days). As shown by this single example, veterinarians and farmers can be faced with a myriad of approved uses and multiple WDTs for any given drug product.

Several years after the FARAD website was launched, a novel intuitive search interface called the Veterinarian's Guide to Residue Avoidance Management (VetGRAM) was developed and made available through the FARAD website (http://www.farad.org/vetgram/search.asp). At that time, VetGRAM was the only online resource that could be used to conduct user-defined searches for regulatory information about FDA-approved food animal drugs. A sample screenshot from the front page of VetGRAM is shown in Figure 14.2. In its current configuration, VetGRAM allows users to limit their searches based on multiple search parameters, including a product's trade name or active ingredient(s), the animal species use class, the route of administration, the drug classification, or the NADA number. Information within the resulting search engine report table (SERT) can be sorted and organized according to species, active ingredient, route of administration, or other user-selectable parameters. Hyperlinks within a user-generated SERT provide immediate access to additional detailed information about any listed product, including available drug formulations, approved species, approved indications for use, label dosing instructions, warnings or restrictions, and approved regulatory tolerances for the drug or marker residues in different food products. Since regulatory WDTs are predicated upon specific conditions of drug use (e.g., dose, duration, route, etc.), it is critical that all relevant information be included with SERT outputs if it is pertinent to a drug WDT value. In 2013, FARAD launched its first application for mobile devices (mobile app) with the release of the VetGRAM mobile app for Android mobile devices

FIGURE 14.2 Home page of VetGRAM. The VetGRAM is an intuitive online search interface located on the FARAD website (http://www.farad.org/vetgram/search.asp). VetGRAM allows users to conduct user-defined searches of all U.S. drug approvals for food-producing animals.

(see Fig. 14.3). This free app is currently available through the Google store (https://play.google.com/store/apps/details?id=com.farad.vetgram), and a similar app for Apple mobile devices is currently under development.

14.4 EXPERT-MEDIATED CONSULTATIONS BY FARAD

As mentioned earlier, the limited availability of approved drugs for treating diseases in food-producing animal species forces veterinarians to consider using drugs in an extralabel manner. However, in order to be in compliance with AMDUCA, veterinarians must establish an appropriate withdrawal period that is scientifically based for any ELDU involving food-producing animals. Highly trained FARAD personnel are available to provide science-based answers to such inquiries by using real-time access to proprietary databases that contain more than 43,000 entries extracted from over 8,000 sources.

FIGURE 14.3 Mobile phone VetGRAM application for the Android operating system. In the Spring of 2013, FARAD launched a free mobile app for use on touchscreen mobile devices that use the Android operating system, including smartphones and tablet computers. The new product is a native app with an updatable database that provides users with full access to key information about all FDA-approved drugs for use in food-producing animal species.

Veterinarians, extension specialists, regulators, or producers may submit questions to FARAD via a nationwide toll-free hotline (1-888-USFARAD or 1-888-873-2723) or through a Web-based online submission portal (https://cafarad.ucdavis.edu/FARMWeb/) that is pictured in Figure 14.4. Although the hotline remains in existence to this day, the vast majority of questions are submitted through the Web portal. On an annual basis, FARAD handles approximately 1500 inquiries that are estimated to impact more than 6 million individual animals. A breakdown by species of annual inquiries to FARAD is shown in Figure 14.5, and a breakdown by individual drug or drug class during a single year is summarized in Table 14.1. Inspection of the data reveals that antibiotic agents represent the largest single group of agents that comprise inquiries to FARAD (49%) and that a substantial number of inquiries involve cases in which animals have been treated with or exposed to more than one drug (Table 14.1).

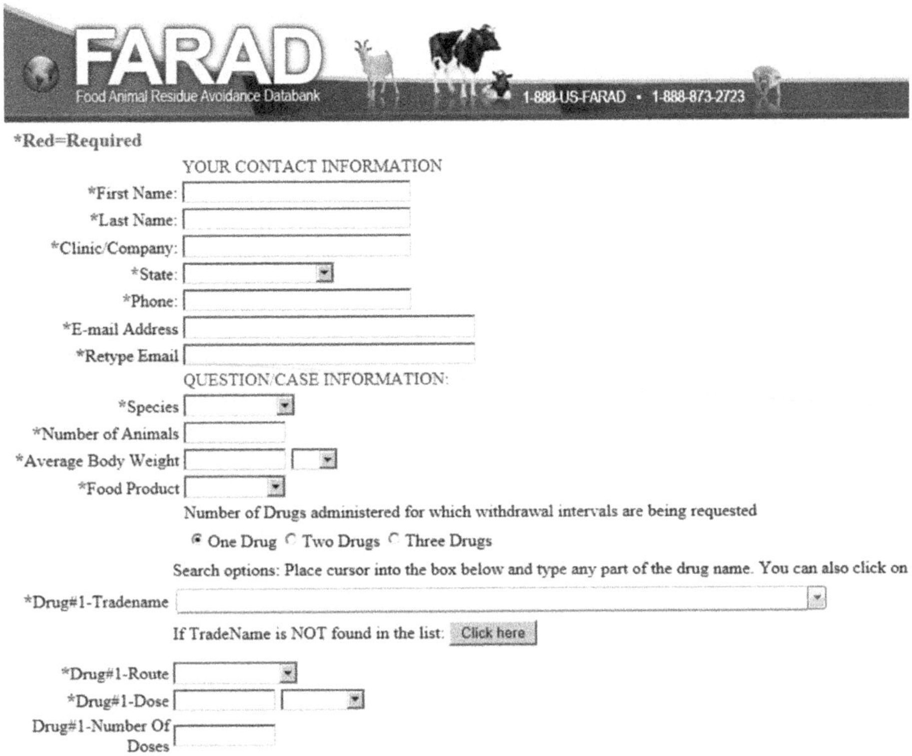

FIGURE 14.4 Screen capture from Web portal for submission of residue-related questions to FARAD. The U.S. FARAD online request system is operated at UCD and provides an easy conduit for veterinarians to submit questions about accidental chemical exposures or ELDU in food-producing animals. This service as well as the toll-free hotline is a free service for licensed U.S. veterinarians.

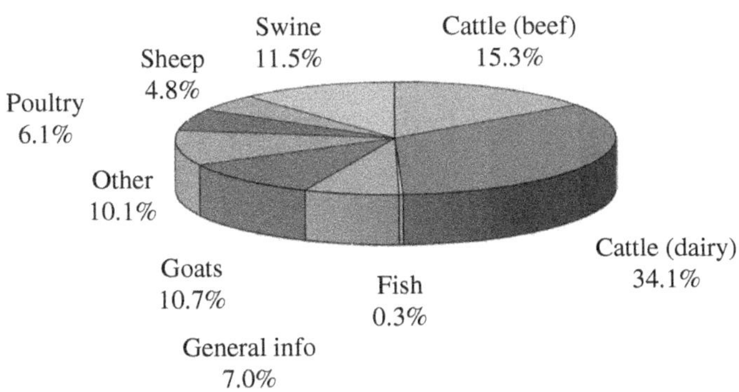

FIGURE 14.5 Submission statistics by species for questions submitted to FARAD. The categories cover a 5-year period (2008–2012) and are reported as a percentage of all submissions rather than numbers of animals involved.

TABLE 14.1 Residue-related inquiries by agent or drug class to FARAD during 2012

Parameter	Total count	Only agent	Total percent
Total inquiries in 2012	1591	—	—
Contacts via Web submission	1295	—	81
Contacts via phone or email	296	—	19
Analgesics—alone or in combination with other agent(s)	—	—	—
NSAIDs	189	135	—
Flunixin	72	38	—
Meloxicam	64	47	—
Phenylbutazone	21	19	—
Alpha-2 agonists (xylazine, detomidine, (dex)medetomidine, tolazoline, atipamezole)	41	21	—
Opiates (butorphanol, morphine, buprenorphine, tramadol, nalbuphine)	21	15	—
Local anesthetics (lidocaine, bupivacaine, mepivacaine, tetracaine)	14	9	—
Antibiotics—alone or in combination with other agent(s)	773	691	49
Penicillins (penicillin G = 126 total)	203	156	—
Tetracyclines (chlortetracycline, doxycycline, oxytetracycline, tetracycline)	121	89	—
Cephalosporins	109	81	—
Sulfonamides	96	77	—
Macrolides (tulathromycin, tylosin, tilmicosin, azithromycin, gamithromycin, erythromycin)	85	68	—
Florfenicol	66	49	—
Miscellaneous	—	—	—
Corticosteroids—alone or in combination with other agent(s)	135	114	8
Dexamethasone	101	89	—
Parasiticides—alone or in combination with other agent(s)	245	187	15
Avermectins (ivermectin, moxidectin, selamectin)	85	51	—
Benzimidazoles (albendazole/fenbendazole)	73	54	—

Analysis of calls for this calendar year is categorized according to broad pharmacological classes. The average response time for calls was less than 1 day (0.86 ± 0.02 days) across all three regional access centers

In addition to the typical annual traffic of inquiries to FARAD, the program has been involved with a handful of high-profile cases with significant economic and social impact. In each of these cases, FARAD's expertise and unique databases have proven instrumental in using a sound scientific rationale to resolve chemical food safety crises. FARAD continues to serve as source of information for stakeholders in livestock and pharmaceutical industry as they strive to increase U.S. meat exports to Asia and Europe. Because FARAD serves as an academic and independent nongovernmental organization, its computations and simulations contribute significant transparency on issues that can influence international trade. FARAD's value to U.S. animal agriculture and trade has been difficult to quantify simply because its primary role is to provide information and guidance to producers, veterinarians, and other stakeholders that has prevented significant economic losses due to condemnation of animal products and, in addition, has helped protect the public from exposure to hazardous drugs and chemical contaminants. Examples of high-profile cases where FARAD has served as a *silent collaborator* or *invisible shield* in mitigating chemical contamination of U.S. meat and milk are documented elsewhere in this text in the chapter on environmental contaminants.

14.5 FARAD PUBLICATIONS AND PRESENTATIONS

FARAD scientists publish peer-reviewed articles in scientific journals, including a regular series of FARAD Digests in the Journal of the American Veterinary Medical Association. Many of the published FARAD Digests have provided detailed analyses and extended withdrawal interval (WDI) recommendations for a limited number of drugs that are used off-label very frequently in some food-producing animal species. Since these WDI recommendations have been peer-reviewed and made available in the scientific literature, FARAD has compiled all standard WDI recommendations into a searchable database that was made available on our website in 2010. The WDI Lookup Tool (see Fig. 14.6) receives a significant number of weekly visits and is likely to have reduced the number of inquiries about ELDU of these selected drugs that are commonly submitted by veterinarians. In addition to peer-reviewed publications, FARAD scientists have written books that provide insights into the unique aspects of FARAD's kinetic databases as well as its novel computational approaches and methods that could be used to estimate safe WDIs under several practical scenarios (Baynes et al., 1999; Buur et al., 2006a, 2008; Martin-Jimenez et al., 2002). FARAD members make regular presentations at local, national, and international scientific meetings and are available to meet with interested groups.

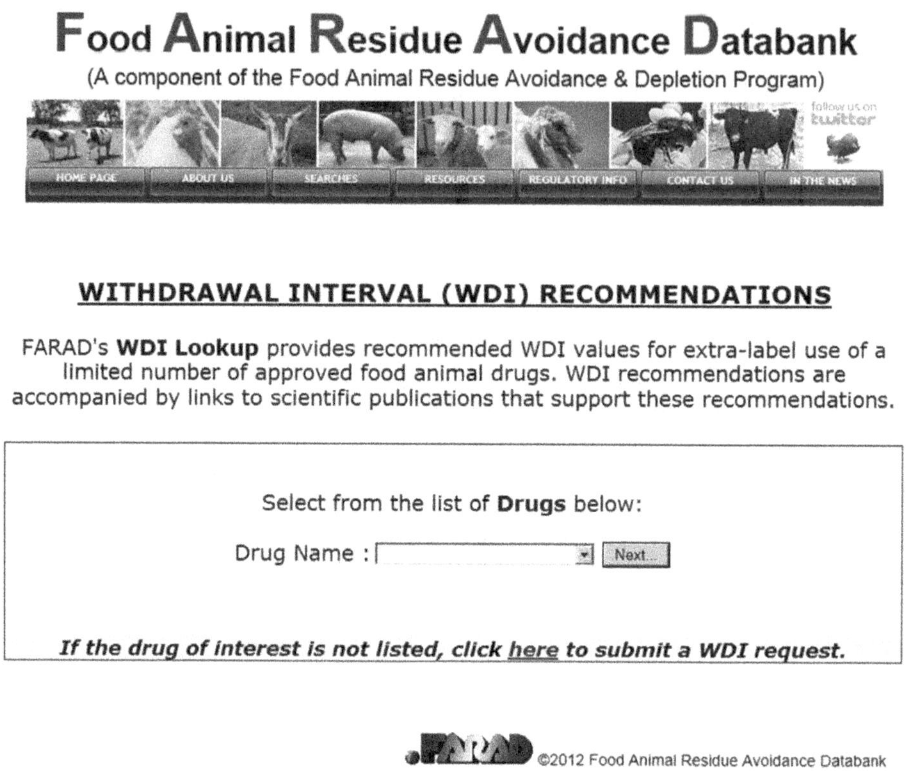

FIGURE 14.6 WDI Lookup Tool on the FARAD website. This online searchable database of FARAD-recommended safe WDIs is limited to a select group of animal drugs that are approved but commonly used in an extralabel manner. The WDI tool was launched in late 2010 and currently includes recommendations for 31 drugs in major and minor food animal species. The recommendations are based on analyses of peer-reviewed published data and help fulfill the AMDUCA mandate to veterinarians regarding ELDU.

14.6 GLOBAL FARAD

In light of the internationalization of food products, FARAD has established an international program called Global FARAD (gFARAD), with the general goal of fostering global food safety standards. The gFARAD program was originally recognized in 2001 by the Food and Agriculture Organization (FAO) of the United Nations. This action launched gFARAD programs, funded by each host country, in France, China, Taiwan, and Canada, the latter of which is in existence today. Despite the obvious need for such a global database, a truly integrated gFARAD has not been pursued in a systematic matter owing to a lack of funding by public or private entities. The latest efforts have focused on working with the CAB International and several countries to establish virtual platform that will be a clearinghouse for drug and chemical contaminant data relevant to chemical residues in food animal products. This effort is ongoing.

REFERENCES

Baynes RE, Martin-Jimenez T, Craigmill AL, Riviere JE (1999). Estimating provisional acceptable residues for extra-label drug use in livestock. Regul Toxicol Pharmacol 29(3):287–299.

Buur JL, Baynes RE, Smith G, Riviere JE (2006a). Use of probabilistic modeling within a physiologic based pharmacokinetic model to predict sulfamethazine residue withdrawal times in edible tissues in swine. Antimicrob Agents Chemother 50(7):2344–2352.

Buur JL, Baynes RE, Riviere JE (2008). Estimating meat withdrawal times in pigs exposed to melamine contaminated feed using a physiologically-based pharmacokinetic model. Regul Toxicol Pharmacol 51:324–331.

Martin-Jimenez T, Baynes RE, Craigmill A, Riviere JE (2002). Extrapolated withdrawal-interval estimator (EWE) algorithm: A quantitative approach to establishing extra-label withdrawal times. Regul Toxicol Pharmacol 36(1):131–137.

FURTHER READING

Baynes RE, Smith G, Mason SE, Narrett E, Barlow BM, Riviere JE (2008). Pharmacokinetics of melamine in pigs following an intravenous bolus. Food Chem Toxicol 46:1196–1200.

Buur JL, Baynes RE, Smith G, Riviere JE (2006b). Pharmacokinetics of flunixin meglumine in swine after intravenous dosing. J Vet Pharmacol Therap 29:437–440.

Haskell SR, Gehring R, Payne MA, Craigmill AL, Webb AI, Baynes RE, Riviere JE (2003). Update on FARAD food animal drug withholding recommendations. J Am Vet Med Assoc 223(9):1277–1278.

Mason SE, Baynes RE, Buur JL, Riviere JE, Almond GW (2008). Sulfamethazine water medication pharmacokinetics and contamination in a commercial pig production unit. J Food Protect 71(3):584–589.

Mason SE, Baynes RE, Almond GW, Riviere JE (2012). Variables important to the population pharmacokinetic modeling of tetracycline water medication in swine. Am J Vet Res 73(10):1641–1649.

Needham ML, Webb AL, Baynes RE, Riviere JE, Craigmill AL, Tell LA (2007). FARAD Digest: Current update on drugs for game bird species. J Am Vet Med Assoc 231:1506–1508.

Payne MA, Craigmill AL, Riviere JE, Baynes RE, Webb AI, Sundlof SF (1999). The Food Animal Residue Avoidance Databank (FARAD). Past, present and future. Vet Clin North Am Food Anim Pract 15(1):75–88.

Smith GW, Davis JL, Tell LA, Webb AI, Riviere JE (2008). FARAD Digest: Extra-label use of nonsteroidal anti-inflammatory drugs in cattle. J Am Vet Med Assoc 232:697–701.

Smith GW, Davis JL, Baynes RE, Yeatts JL, Barlow BM, Riviere JE (2009). Elimination kinetics of tilmicosin following intramammary administration in lactating dairy cattle. J Am Vet Med Assoc 234:245–248.

15

RISK MANAGEMENT OF CHEMICAL CONTAMINANTS IN LIVESTOCK

Ronald E. Baynes[1] and Jim E. Riviere[2]

[1] *Department of Population Health and Pathobiology, College of Veterinary Medicine, North Carolina State University, Raleigh, NC, USA*
[2] *Department of Anatomy and Physiology, College of Veterinary Medicine, Kansas State University, Manhattan, KS, USA*

15.1 INTRODUCTION

Chemical contamination of livestock products other than by veterinary drug contamination is a great concern. These contaminants include biotoxins (e.g., botulin, mycotoxin), heavy metals (e.g., Pb, Hg, Cd), polyaromatic hydrocarbons (e.g., PCBs, PBBs), insecticides (e.g., organophosphates, organochlorines), herbicides (e.g., 2,4-D, paraquat), and petroleum products (e.g., gasoline, diesel). While these cases are rare, they often involve large exposed animal populations and are often associated with extensive media and public attention resulting in reduced public confidence and thus sales and consumption of livestock commodities. The PBB accidental contamination of many dairies in Michigan in 1974 is recognized as one of the first major chemical contaminations of livestock and livestock products in modern history. There were significant catastrophic economic losses to farmers and potential adverse health effects in humans. Because it is well documented, it will not be discussed in this chapter. The more recent contamination cases in

Strategies for Reducing Drug and Chemical Residues in Food Animals: International Approaches to Residue Avoidance, Management, and Testing, First Edition.
Edited by Ronald E. Baynes and Jim E. Riviere.

the last 30 years have been as a result of accidental or intentional exposure of livestock to these contaminants. This chapter will focus on exposure and mitigation of (i) pesticide exposures in the 1980s in the United States, (ii) dioxin exposure in the EU in 1999, (iii) melamine (ME) exposure in United States in 2007, (iv) radioactive fallout in Japan in 2009, and (v) exposure to fracking by-products in the United States in 2012. These incidents involved mitigation by the USDA-supported Food Animal Residue Avoidance and Depletion (FARAD), the records of which have been used to describe these events. The pharmacokinetic files that supported many of these interventions have been published in compendium form (Craigmill et al., 2006).

15.2 HEPTACHLOR

Two major incidents of mass contamination of livestock by the chlorinated hydrocarbon insecticide, heptachlor, occurred some two decades ago. The first incident, in 1982, was confined to Hawaii and resulted from feeding dairy cattle pineapple green-chop by-products contaminated with the insecticide. The EPA earlier had granted an emergency exemption to pineapple growers for the use of heptachlor on their crop. After harvesting the pineapples, the remainder of the plant was processed for use as cattle feed. Widespread contamination of the Hawaiian milk supply followed, and losses were estimated at several million dollars (Craigmill AL, personal communication, 1988) if the milk was condemned due to a lack of a scientific body that could recommend a withdrawal interval.

In early 1986, residues of heptachlor, along with several other chlorinated hydrocarbon pesticides, were detected in livestock from five Midwestern states. The source of contamination was traced to a feed mill in Van Buren, Arkansas, that had used by-products from an industrial ethanol plant to manufacture animal feeds. The ethanol plant routinely purchased surplus seed grain that had been treated with insecticides and fungicides. Following the production of ethanol, the spent distiller's grains were used in the manufacture of animal feeds. When heptachlor residues first were detected in milk samples, more than 50 dairies in southwest Missouri were quarantined. Several beef and swine herds also were quarantined. Losses undoubtedly would have totaled several million dollars if FARAD had not been available to provide scientific advice regarding a withdrawal interval for these animals.

FARAD examined the heptachlor data from the Hawaii and Missouri cases and was able to make estimations of withholding intervals based on biopsy and postmortem sampling. These analyses by FARAD allowed the farmers to keep the animals alive for the estimated withdrawal interval and allowed the chemical to be cleared from the animal. Ultimately, the farmers did not incur severe economic losses as the animals were not condemned because of exposure to heptachlor. FARAD also served to prevent contaminated meat and milk from entering the human food chain.

15.3 DIOXIN

In spring of 1999, dioxin was detected in animal feed supplied to Belgian, French, and Dutch farms. High levels of dioxin were detected in meat and egg products in these European countries. Government and media reports resulted in reduced confidence in the quality of egg and meat products and subsequent banning of related Belgian agriculture products.

FARAD scientists were consulted by officials within the U.K. Ministry of Agriculture and were able to utilize its database to derive a safe withholding time of U.K. livestock that were exposed to dioxin in feed. Dioxins are actually a family of chemical congeners (e.g., TCDD, TCDF). They are extremely potent toxicants with some laboratory animal LD_{50}s as low as 1 μg/kg, therefore resulting in very low tolerances. Twenty percent of ingested dioxins are excreted in the milk. In the European countries affected, this impacted production of high-priced confectionary consumables such as chocolate. Most data on such compounds available at the time were on PCBs with less on actual dioxins and DBFs. However, for these types of compounds with low safety tolerances, available data was analyzed using traditional pharmacokinetic models that indicated very long half-lives measured in months and not days. This resulted in economically nonviable withdrawal times.

15.4 MELAMINE

Good agriculture practices encourage the prevention of drug or chemical contaminant residues at or above the safe or tolerance levels. While many of the residue violations can be attributed to poor farmer compliance with labeled withholding times, etc., there are many other variables such as chemical and biological interactions that influence the clearance of these substances from the body of the animal before slaughter for meat and/or harvest of eggs and milk. ME is a good case study of a contaminant chemical, and associated biological interactions that can complicate estimates of when the animal has cleared the contaminant.

In the spring of 2007, there was a major pet food recall in the United States following complaints that pet foods contaminated with ME and several of its analogs were probably responsible for renal failure as a result of crystal formation that resulted in illness and/or deaths in dogs and cats. ME, a by-product of plastic manufacturing, was intentionally added to protein supplements because it is a nitrogen-rich molecule that protein quality control tests register as real protein. These supplements are then added to pet foods to fool protein quality control standards. It is the first widely occurring economic adulterant. Crystal formation in the kidney was believed to be associated with pet food contaminated with mixtures of ME and its analogs and not just ME alone (Burns, 2007a, b), a situation that occurs when industrial ME products are employed. A similar episode had occurred in Asia in 2004 (Brown et al., 2007).

Pet food scraps can typically comprise of up to 5–10% of swine feed; however, FDA investigations during the recent ME contamination scare determined that some swine herds were fed 50–100% pet food scraps (US FDA, 2007a). The concentration of ME and related contaminants in these pet food scraps ranged from 9.4 to 1952 ppm ME, while the highest for cyanuric acid (CA), ammelide, and ammeline were 2180, 10.8, and 43.3 ppm, respectively. The final swine feed samples ranged from 30 to 120 ppm with triazine analogs also present at lower concentrations. This would have translated into a food consumption of 4% body weight of a 200 lb (91 kg) pig, which can result in exposure to (i) 5 mg/kg BW of ME using the high-end concentration (120 ppm) of confirmed swine feed samples or (ii) 78 mg/kg BW of ME using the highest concentration in pet food scraps. At the time of FDA investigation, the "limit of detection" (LOD) for the analytical method was 50 ppb (0.05 ppm) for ME only. Tissues from swine believed to have been exposed to contaminated pet food scraps rarely contained ME levels above 50 ppb. There were few details regarding the exact exposure and duration of exposure, which would allow for reliable correlation between ME exposure and tissue levels.

These levels of feed contamination were estimated by USDA to unlikely place human health at risk (USDA, 2007), yet the public is still concerned about consuming meat from swine and poultry exposed to ME and/or its analogs.

In early fall of 2008, China reported four infant deaths and related nephrotoxicity in more than 53,000 other infants after they drank baby formula tainted with ME. There is now general consensus among the scientific community that in both human and pet exposures, renal disease was associated with exposure to mixtures of two triazines, ME and CA. This has been supported by data from recent experimental mixtures studies in pigs and fish (Reimschuessel et al., 2008). *Thus, exposure to mixtures of melamine and cyanuric acid (ME + CA) is a public health concern and NOT just sole exposure to ME.* This fact delayed identifying ME as the toxicological agent since most research studies were conducted with analytical-grade pure ME, and not the "dirty" industrial product used as an adulterant.

However, there is limited information on how to guide livestock farmers on how long to withhold their animals from slaughter, milk from dairy cows, or their eggs after production animals have been exposed to related mixtures that can also alter their own clearance from the body of the animal exposed. This applies to many drugs in addition to ME. The recent ME cases represented animal and human exposures to mixtures of ME and ME-related analogs (e.g., CA) and not exposure to ME alone. It is now confirmed that animal and human nephrolithiasis and impaired renal function were associated with *mixture* exposure and not ME alone. It is plausible to assume that livestock exposure to these ME+CA mixtures could significantly alter the clearance of these triazine contaminants and other drugs cleared by the kidney and ultimately the predicted time for when meat, milk, and eggs from exposed animals are safe for human consumption.

Our laboratory was among the first to determine the pharmacokinetics of ME in pigs (Fig. 15.1) as well as develop a physiologically based pharmacokinetic (PBPK) model to estimate the meat withdrawal time should pigs be exposed to ME (Baynes et al., 2008; Buur et al., 2008).

The PBPK model was used to predict a tissue withdrawal time of 21 h for a 5.12 mg/kg single oral bolus, although we have no *in vivo* pig studies to validate these estimates. Further studies are also needed to test our model assumptions such as species differences in tissue/plasma partitioning, rates of intestinal absorption, renal clearance mechanisms such as differences in GFR, dose linearity, and tissue dosimetry.

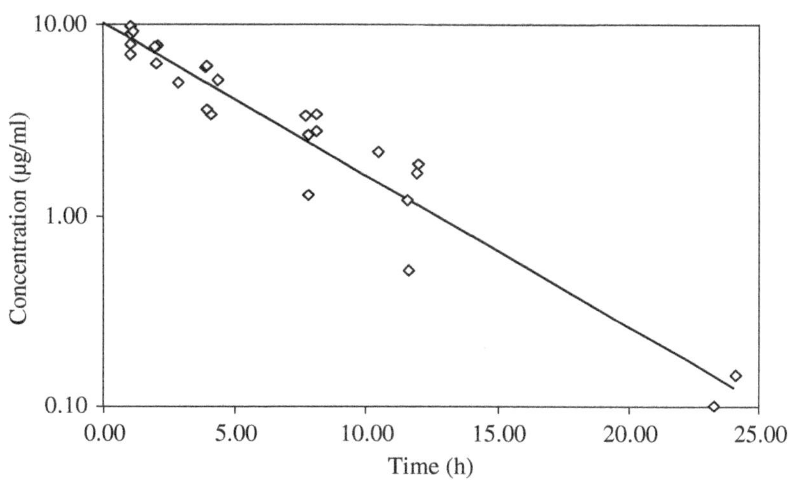

FIGURE 15.1 Plasma concentration–time profiles of ME in swine given 6 mg/kg ME IV.

15.5 RADIOACTIVE CONTAMINATION AND MANAGEMENT CONSIDERATION

The recent natural disaster and unfolding events at the Fukushima Daiichi nuclear facility in Japan raise concerns over the potential for contamination of production livestock and food products by radioactive fallout. Following a similar incident in 1986 at the Chernobyl nuclear power facility in Ukraine, FARAD played a role in determining withdrawal times after radionuclide exposure. Based upon available studies that examined animals exposed to Chernobyl fallout, it is clear that a number of issues must be considered for animals in proximity of such a disaster. Since data were scarce from Japan at the time of the incident, specific recommendations were impossible to make as the dose, duration, and route of exposure must be known. A number of potential considerations are listed below. Acute and chronic radiation poisoning of animals is beyond the scope of this book. In addition, contamination of livestock and grazing areas by tsunami-contaminated debris (chemicals, petroleum products, etc.) introduces the real risk for chemical contamination of animals and feed, a situation also seen in hurricane-flooded areas. These threats are not considered here.

Radionuclides presently implicated in the Japanese incident include iodine-131, cesium-137, and strontium-90. The specific element involved is one of the primary determinants of animal tissue(s) that are targeted as well as the route and rate of biological clearance and elimination from the body. For example, iodine accumulates in the thyroid, cesium uniformly distributes throughout the body similar to that of potassium but can concentrate in muscle tissues, while strontium mirrors the biodistribution and clearance of calcium. The radioisotope determines half-life for radioactive decay, which is very long for many of the radioisotopes of concern here (e.g., half-lives for cesium-135 and cesium-137 are 2.3 million years and 30 years, respectively). A point of confusion in mitigating this incident was the confusion between biological elimination half-life, a pharmacokinetic parameter, and radioactive decay half-life, a metric of nuclear physics.

To estimate exposure, radioactivity must be quantified using appropriate units, including curies and becquerels. Units of radiation exposure use different metrics, including roentgens, sieverts, REM, or coulombs. The level of radiation exposure correlates with health effects at specific levels and durations of exposure. These are not units of radioactivity that are required to calculate absorption and clearance of radioisotopes in an exposed animal.

When considering disposition in animals, the radioactive decay "physical" half-life has no relation to pharmacokinetic elimination or "biological" half-life. As an example, cesium-137 has a "physical" half-life of 30 years but has a "biological" half-life on the order of weeks to months. The actual dose and duration of exposure, often very difficult to obtain in field studies, would significantly impact these values. There are three distinct scenarios for radioactive fallout exposure to food-producing animals:

1. Direct contamination by exposure to radioactive fallout (skin, inhalation, food)
2. Exposure via consumption of contaminated feed or forage
3. Exposure from contaminated drinking water

All three have different considerations and remediation strategies since the routes of exposure and exposed doses are very different. Management generally involves the latter two scenarios.

The first consideration is to remove livestock and feed from potential fallout by going indoors. One must secure contaminant-free feedstuff. Bales of hay must be covered. If bales of hay are contaminated, outer layers may be removed and uncontaminated hay obtained from the center. These are emergency management issues and are most effective if instituted as soon as possible.

Determination of the absorption and subsequent fate of radioactive fallout is both a function of the specific radionuclide and the radioactive dose. Direct exposure to high-level radiation that results in clinical signs in animals should lead to immediate carcass contamination with appropriate disposal, taking into consideration protection against human radiation exposure. Such tissues should not enter the human food chain.

Therapy of exposed or ill animals could be considered only on a humane basis. For example, use of Ca–DTPA (diethylenetriaminepentaacetate), a systemic decorporation and chelating agent, is designed to increase excretion of already absorbed radionuclide. It is very expensive and not practical for food animals. Other approaches to protect exposed animals from biological effects are calcium salts and alginate after strontium exposure and potassium iodide for radioiodine exposure. There is evidence of their marginal effectiveness and some potential adverse safety issues. They are not viable methods to increase withdrawal times for animal products destined for human consumption.

The most common postexposure scenario is one of managing nonexposed grazing animals on long-term low-level contaminated pastures or when exposed to low-level contaminated feed or water. Purification strategies are available for decontaminating water using ion-exchange and filtering approaches. The primary management step postexposure is feeding animals uncontaminated feed and water. An effective remediation strategy after cesium exposure in Chernobyl was to feed contaminated pigs clean fodder 2 months prior to slaughter, allowing depletion of absorbed compound to occur. This approach is presently not acceptable.

Therapeutic compounds are used to prevent animals from further radioisotope absorption after exposure to such contaminated feedstocks. Compounds such as colloidal Prussian blue (ammonium ferric cyanoferrate (AFCF)) and clay minerals (e.g., bentonite) trap the radionuclide before absorption can occur, thereby significantly reducing radiation levels in milk and meat. Prussian blue irreversibly binds cesium in the gastrointestinal tract and is not absorbed after oral administration with 99% being excreted in feces in pigs. Also note that feces from treated animals would be expected to have increased radiation due to excretion of the bound radionuclide. In humans, AFCF reduces cesium's biological half-life by almost 50%. AFCF administered at 2–3 g/day has been shown to be economically effective in treating cesium contamination of feedstuff contaminated by Chernobyl in Europe by reducing cesium-137 levels in milk and meat up to 80–90%. At low cesium contamination levels (e.g., 10 Bq/day), milk levels were below detection. Such a regimen was approved in West Germany and Austria in the late 1980s. Prussian blue has been shown to be relatively safe to animals and considered safe and effective for use in humans. Reinforcing feed with excessive clay minerals is less effective than using Prussian blue and may alter mineral and trace element homeostasis.

Milk is a significant source of human exposure in postcrisis contamination areas. Radionuclides of strontium and cesium are the primary concerns for milk contamination. Strontium's biological half-life in milk is from 10 to 40 h, while cesium has a reported half-life of up to 9 days. Transfer coefficients (the equilibrium ratio between radionuclide activity concentration in milk or meat and the daily intake of radionuclide) are used to predict contamination of animal products following the release of radionuclides into the environment. This is complex, for as in the case of strontium, dietary calcium alters this ratio. Soil type has a major effect on these values, with transfer from soil to plants favored in peat versus clay soils. After exposure, radioisotope uptake into plant roots determines how long fields, and thus feedstuffs, stay contaminated. This must be monitored on the ground. Finally, there have been some reported practical approaches to removing radionuclides from milk.

15.6 BY-PRODUCTS OF FRACKING

The final source of accidental contamination to be discussed, one that could become more prevalent in the future, is livestock feed and water contamination near gas hydraulic fracturing operations. The so-called "fracking" operations involve drilling over mile-long L-shaped well shafts and pumping millions of gallons of a water–sand–chemical additive (<1/2%) slurry to fracture deep beds of shale to release trapped natural gas. Ten to twenty percent of the water flows back up the well shaft and is placed in holding ponds and more recently sealed tanks for reuse (Spellman, 2013). This is the point where leakages have occurred that could potentially contaminate livestock water sources. There have been anecdotal reports of livestock exposure from surface holding pond leakage near gas fracturing sites resulting in pet and livestock illness (Bamberger and Oswald, 2012). FARAD was involved in a 2011 potential livestock contamination incident in Pennsylvania where a few dozen cattle were exposed to fluid that was assayed to contain barium, iron, manganese, and strontium. Organics were not assessed.

Assessing the potential risk of tissue residues resulting from exposure to fracking fluids is extremely complex, because composition is unknown and exposure is difficult to quantitate. Fracking fluid is composed of a number of different and sometimes proprietary additives (formulations are trade secrets). These include up to 400 different chemicals and include potassium chloride, acids, various organic and inorganic gels, biocides, clay stabilizers, corrosion inhibitors, foamers and defoamers used at different stages, friction reducers, scale controllers, and surfactants (Colborn et al., 2011; Spellman, 2013). There have been no data on what was found in field exposures. In addition, backwash fluid also contains raw petroleum hydrocarbons, minerals, heavy metals, and radioactive nuclides (e.g., radium, strontium) that are released from the fracking sites. It has also been estimated that the temperature at the underground site of fracture may be hundreds of degrees, functioning as a chemical retort for all artificially added fracking additives to react with the natural mineral and petroleum compounds. These types of complex mixture exposures are the most difficult to handle, for even if analytical data were available, the number and potential interactions of contaminants are staggering.

REFERENCES

Bamberger M, Oswald RE (2012). Impacts of gas drilling on human and animal health. New Solut 22:51–77.

Baynes RE, Smith G, Mason SE, Barrett E, Barlow BM, Riviere JE (2008). Pharmacokinetics of melamine in pigs following intravenous administration. Food Chem Toxicol 46(3):1196–2000.

Brown CA, Jeong KS, Poppenga RH, Puschner B, Miller DM, Ellis AE, Kang KI, Sum S, Cistola AM, Brown SA (2007). Outbreaks of renal failure associated with melamine and cyanuric acid in dogs and cats in 2004 and 2007. J Vet Diagn Invest 19:525–531.

Burns K (2007a). Recall shines spotlight on pet foods. J Am Vet Med Assoc 230(9):1285–1286.

Burns K (2007b). Events leading to the major recall of pet foods. J Am Vet Med Assoc 230(11):1600–1620.

Buur J, Baynes RE, Riviere JE (2008). Estimating meat withdrawal times in pigs exposed to melamine contaminated feed using a physiologically based pharmacokinetic model. Regul Pharmacol Toxicol 51:324–331.

Colborn T, Kwiatkowski C, Schultz K, Bachran M (2011). Natural gas operations from a public health perspective. Hum Ecol Risk Assess 17(5):1039–1056.

Craigmill AL, Riviere JE, Webb AI (2006). Tabulation of FARAD Comparative and Veterinary Pharmacokinetic Data. Ames: Blackwell Publishing.

Reimschuessel R, Gieseker CM, Miller RA, Ward J, Boehmer J, Rummel N, Heller DN, Nochetto C, de Alwis GK, Bataller N, Andersen WC, Turnipseed SB, Karbiwnyk CM, Satzger RD, Crowe JB, Wilber NR, Reinhard MK, Roberts JF, Witkowski MR (2008). Evaluation of the renal effects of experimental feeding of melamine and cyanuric acid to fish and pigs. Am J Vet Res 69:1217–1228.

Spellman FR (2013). Environmental Impacts of Hydraulic Fracturing. Boca Raton: CRC Press.

USDA (2007). Disposition of hogs and chickens from farms identified as having received pet food scraps contaminated with melamine and melamine-related compounds and offered for slaughter. Federal Register, 2007; Vol. 72, No. 103, Food Safety Inspection Service, Docket No. FSIS 2007-0018.

United States Food and Drug Administration (US FDA) (2007). Interim melamine and analogues safety/risk assessment. Available at www.fda.gov/OHRMS/DOCKETS/98fr/07n-0208-ra00001.pdf. Accessed August 12, 2010.

INDEX

Note: Page numbers in *italics* refer to Figures; those in **bold** to Tables.

Strategies for Reducing Drug and Chemical Residues in Food Animals: International Approaches to Residue Avoidance, Management, and Testing, First Edition.
Edited by Ronald E. Baynes and Jim E. Riviere.